Mapping Medieval Geographi

Mapping Medieval Geographies explores the ways in which geographical knowledge, ideas and traditions were formed in Europe during the Middle Ages. Leading scholars reveal the connections between Islamic, Christian, biblical and classical geographical traditions from Antiquity to the later Middle Ages and Renaissance. The book is divided into two parts: Part I focuses on the notion of geographical tradition and charts the evolution of celestial and earthly geography in terms of its intellectual, visual and textual representations; whilst Part II explores geographical imaginations; that is to say, those 'imagined geographies' that came into being as a result of everyday spatial and spiritual experience. Bringing together approaches from art, literary studies, intellectual history and historical geography, this pioneering volume will be essential reading for scholars concerned with visual and textual modes of geographical representation and transmission, as well as the spaces and places of knowledge creation and consumption.

KEITH D. LILLEY is Reader in Historical Geography at Queen's University, Belfast.

Mapping Medieval Geographies

Geographical Encounters in the Latin West and Beyond, 300–1600

Edited by

Keith D. Lilley

CAMBRIDGE
UNIVERSITY PRESS

CAMBRIDGE
UNIVERSITY PRESS

University Printing House, Cambridge CB2 8BS, United Kingdom

Cambridge University Press is part of the University of Cambridge.

It furthers the University's mission by disseminating knowledge in the pursuit of education, learning and research at the highest international levels of excellence.

www.cambridge.org
Information on this title: www.cambridge.org/9781316620274

© Cambridge University Press 2013

First published 2013
First paperback edition 2016

A catalogue record for this publication is available from the British Library

Library of Congress Cataloguing in Publication data
Mapping medieval geographies : geographical encounters in the Latin west and beyond / edited by Keith D. Lilley.
     pages   cm
Includes bibliographical references.
ISBN 978-1-107-03691-8 (Hardback)
1. Geography, Medieval.   2. Geography in literature.   3. Cartography–Early works to 1800.   I. Lilley, Keith D.
G89.M38 2013
910.9′02–dc23   2013011081

ISBN 978-1-107-03691-8 Hardback
ISBN 978-1-316-62027-4 Paperback

Dedicated to the memory of
Denis E. Cosgrove

Contents

Illustrations

Acknowledgements

The essays within this book arose from a conversation I had with Denis Cosgrove during 2006–7. We recognized that across different humanities and arts disciplines in the Anglophone world there is developing scholarly momentum on the topic of 'medieval geographies', in which geographers such as ourselves are largely on the periphery. Partly to learn more from these 'other' scholars of geography, and partly to stimulate cross-disciplinary debate on medieval geographies, we applied for and received generous financial support from the Ahmanson Foundation to convene a gathering of academics working in this field. So a conference was organized and held at the Center for Medieval and Renaissance Studies (CMRS) at the University of California, Los Angeles (UCLA) in May 2009. I am indebted to the Ahmanson Foundation and to the Historical Geography Research Group (of the RGS-IBG) for providing funds to enable us to convene this gathering, and to the CMRS for hosting it, particularly Karen Burgess and Brett Landenberger, whose organizational efficiency and professionalism were appreciated by all who attended. I am very grateful to Denis for the enthusiasm he showed for the project right from the start, and deeply saddened that he did not live to see the results of the conversation we initiated. The volume is dedicated to Denis's memory. My thanks are due also to the contributors here whose essays have provided me with such stimulation and insight into mapping medieval geographies, as well as to Liz Friend-Smith at the Press for her editorial thoughts and guidance, and the anonymous readers who commented on an earlier draft. Without their inputs, their efforts and scholarship, this volume would not have been possible.

KDL
1 October 2012

Contributors

DANIEL BIRKHOLZ is Associate Professor of English at the University of Texas at Austin (USA).

VERONICA DELLA DORA is Professor of Human Geography at Royal Holloway, University of London (UK).

MARCIA KUPFER works as an independent scholar based in Washington, D.C. (USA).

KATHY LAVEZZO is Associate Professor of English at the University of Iowa (USA).

KEITH D. LILLEY is Reader in Historical Geography at Queen's University, Belfast (UK).

NATALIA LOZOVSKY is a Visiting Scholar of the Office for History of Science and Technology at the University of California, Berkeley (USA).

ANDY MERRILLS is Lecturer in Ancient History at the University of Leicester (UK).

KAREN C. PINTO is Assistant Professor of History at Gettysburg College, Pennsylvania (USA).

AMANDA POWER is Senior Lecturer in Medieval History at the University of Sheffield (UK).

MEG ROLAND is Associate Professor and Department Chair in English at Marylhurst University, Oregon (USA).

CAMILLE SERCHUK is Professor of Art at Southern Connecticut State University (USA).

JESSE SIMON is a graduate student in the Oxford Centre for Late Antiquity at the University of Oxford (UK).

MARGARET SMALL is Lecturer in Europe and the Wider World in the School of History and Cultures at the University of Birmingham (UK).

SARA V. TORRES is a graduate student in the Department of English at the University of California, Los Angeles (USA).

Introduction
Mapping medieval geographies

Keith D. Lilley

> The discipline imprinted in a sequentially unfolding narrative predisposes the reader to think historically, making it difficult to see the text as a map, a geography of simultaneous relations and meanings that are tied together by a spatial rather than a temporal logic.[1]

In the early years of the fourteenth century an English minor lord by the name of Henry de Bray set down an account of those lands in Northamptonshire that belonged to his family. Descriptions of where these lands lay, what they were worth, as well as copies of charters relating to when they were acquired by the de Bray family, survive in two British Library manuscripts.[2] One manuscript, Cotton Nero C XII, differs slightly from the other, however, in that its second folio contains a description of the world followed by a list of all the counties of England.[3] Like contemporary *mappaemundi*, de Bray's 'descriptio mundi' follows a tripartite formula, dividing the world into three continents, relating their origins to the sons of Noah, and listing within each the provinces they comprise. Likewise, with the named English counties de Bray's approach is to group them geographically. Following a sequence that reveals a regional pattern and underlying spatial logic, he starts with the counties situated in southern England, then moves north, listing those of the Midlands and northern England, before turning southwards to his own region of England, ending with counties of the eastern lowlands. Orientating himself and his readers firstly to a 'world map' and then to a national geography, only then does de Bray set out to record his own particular localized geographies of Harlestone Manor with its fields and habitations.

Through its geographical contents and its spatial ordering de Bray's estate book reveals sensitivity both to space and scale, mapping his desire

---

[1] Edward Soja, *Postmodern Geographies: Reassertion of Space in Critical Social Theory* (London: Verso, 1989), 1.
[2] British Library (BL) MSS. Cotton Nero C xii; Lansdown 761.
[3] Dorothy Willis (ed. and tr.), *The Estate Book of Henry de Bray of Harlestone, Co. Northants (c. 1289–1340)*, Camden Third Series 27 (London: Camden Society, 1916).

1

to place Harlestone and its lands and people within wider worlds, some familiar, others more distant. The folios in de Bray's Cotton Nero manuscript thus map myriad medieval geographies, both those personal to him, by being lived and experienced first-hand, and those that came from elsewhere, reflecting knowledge of certain geographical sources – now unknown – that he thought as worthy as the records of lands, charters and accounts. The essays contained in this volume echo de Bray's geographical encounters, ranging from those concerned with geographies of the wider world, through to more incidental geographies that similarly map out particular worlds within the world.

Mapping Medieval Geographies brings together the work of a group of scholars from different disciplines whose common interest lies in how the world was understood in the Middle Ages. Their contributions here provide a measure of the concern and current engagement of medievalists in ideas of space and place, maps and mappings, and geography and cartography, reflecting an ever increasing preoccupation with the spatial within contemporary humanities discourse.[4] The volume here, like de Bray's estate book, comprises two particular facets of 'medieval geography': providing, firstly, a sense of *geographical traditions* across the Middle Ages, in which essays in Part I map out intellectual and scholarly engagements with 'geography' as a subject, a field of enquiry, tracing chronologically the widespread transmission and circulation of geographical sources in the Latin West from Late Antiquity to the Renaissance; and then, secondly, focusing on *geographical imaginations*, essays in Part II deal with Arabic, Judaic and Latin and Orthodox Christian encounters with the geographical through subjectively experienced and embodied spatial practices, using images and texts to map out these 'imagined geographies'. There are connections to be made between 'traditions' and 'imaginations', for these two dimensions of medieval geographies existed not as discrete entities but as integrated views of the world, each influencing the other, as de Bray's account also reveals.

The complex and myriad nature of 'medieval geographies' is apparent in collections of essays published over the past decade or so, for example in Talbert and Unger's *Cartography in Antiquity in the Middle Ages*, which examines the relationship between medieval maps and texts, in Allen's *Eastward Bound*, an exploration of travel accounts linking eastern and western cultural traditions of the Middle Ages, and in those volumes that have considered imagined and material spaces and places, such as

---

[4] On the 'spatial turn' see Barney Warf and Santa Arias (eds.), *The Spatial Turn. Interdisciplinary Perspectives* (New York: Routledge, 2008).

Hanawalt and Kobialka's *Medieval Practices of Space*, and Tomasche and Gilles's *Text and Territory: Geographical Imagination in the European Middle Ages.*[5] Where *Mapping Medieval Geographies* differs from these previous studies is in its attempt to approach the subject by embedding the volume's content within contemporary geographical discourse. Indeed, 'geographical traditions' and 'geographical imaginations' are commonplace phrases in modern-day (Anglophone) geography and critical thought, thanks especially to the titles of two influential books by two key geographers, *The Geographical Tradition* by David Livingstone and *Geographical Imaginations* by Derek Gregory.[6] During the past two decades these two themes have been well explored by human geographers, by those concerned with geography's history and the trajectory of western geography as a 'contested enterprise', as well as by those interested in visual and textual representations of places, lands and landscapes, and what these reveal of past perceptions of the world.[7] These are themes addressed explicitly by the essays that make up the two parts of this book, and form the rationale for its structure and organization.

By taking both a temporal and a spatial approach to 'mapping' medieval geographies this volume aims to address the issue, identified by Edward Soja in his opening quotation, of how to combine the historical and the geographical without privileging one over the other. Although geographers often work historically, few have so far attempted to do what this volume does. Looking both at traditions of geography and at geographical imaginations in the Middle Ages, *Mapping Medieval Geographies* contributes not only towards our understanding of the medieval world but also offers insights that will benefit more generally those interested in the spatial and the geographical. These are concerns not just for geographers

---

[5] Richard J. A. Talbert and Richard W. Unger (eds.), *Cartography in Antiquity and the Middle Ages: Fresh Perspectives, New Methods* (Leiden: Brill, 2008); Sylvia Tomasch and Sealy Gilles (eds.), *Text and Territory: Geographical Imagination in the European Middle Ages* (Philadelphia: University of Pennsylvania Press, 1997); Rosamund Allen (ed.), *Eastward Bound: Travel and Travellers 1050–1550* (Manchester University Press, 2004); Barbara A. Hanawalt and Michal Kobialka (eds.), *Medieval Practices of Space* (Minneapolis: University of Minnesota Press, 2000).

[6] David Livingstone, *The Geographical Tradition: Episodes in the History of a Contested Enterprise* (Oxford: Blackwell, 1992); Derek Gregory, *Geographical Imaginations* (Cambridge, Mass. and Oxford: Basil Blackwell, 1994).

[7] For example, see Morag Bell, Robin Butlin and Mike Heffernan (eds.), *Geography and Imperialism, 1820–1940* (Manchester University Press, 1995); Felix Driver, *Geography Militant: Cultures of Exploration and Empire* (Oxford: Blackwell, 2001); Denis Cosgrove, *Geography and Vision: Seeing, Imagining and Representing the World* (London: I. B. Tauris, 2008); Miles Ogborn, *Global Lives. Britain and the World, 1550–1800* (Cambridge University Press, 2008).

but for all those historians who have embraced the humanities' 'spatial turn', including medievalists from different specialisms and fields, a number of whom are contributing authors here. Collectively, their essays represent an interdisciplinary mapping of medieval geographies, covering a long temporal span as well as a broad geographical and multicultural compass, from the Byzantine and Arabic worlds of the Mediterranean through to Christian Europe. But in their individual *mappings* of medieval geographies, both literal and figurative, the volume's authors also seek self-consciously to reflect different, multi-disciplinary perspectives. Thus the volume itself is edited by a geographer, while its various contributors as historians of art, literature, science, geography and cartography, not only take as their focus different objects and objectives of historical (and geographical) enquiry but also employ scholarly approaches and apparatus that are constitutive of their own particular fields and specialisms.

Dividing between 'geographical traditions' and 'geographical imaginations' allows us to map from the volume's essays two distinctive emerging agendas among medievalists interested in the geographical and spatial, with, on the one hand, those concerned primarily with narrating histories and historiographies of medieval geography and cartography, and on the other, those concerned with understanding spatial relationships through exploring histories of spaces and places. Separating out *geographical traditions* from *geographical imaginations* allows us to explore those links and connections that ran between them. To help demonstrate this, what follows is in part a historiographical contextualization of the volume's essays, through the two themes they address, as well as an attempt to bridge these themes to show how the contributions and their contributors speak to one another. Taken together they provide a sense of the myriad geographies that permeated all aspects of cultural life in the Middle Ages, within the Latin West and beyond.

Geographical traditions

What sources of geographical thought and knowledge circulated in and around Europe through the Middle Ages, and how was geography understood at the time, whether in the form of 'textual geographies', such as those descriptions of the world compiled by scholars in particular centres of learning, or through 'visual geographies', those maps and depictions of the terrestrial and celestial worlds that modern scholars for so long judged to be poor examples of cartography? The essays in this first half of the volume address these key questions, and by doing so challenge myths that have dogged the study of medieval geographies and cartography for over a century.[8] Far from simply repeating geographical content derived from classical sources, or producing 'inaccurate' maps, those in the Middle Ages writing geography – regardless of whether they saw their work as geographical – created new geographies, reflecting the norms and values of their own age, while in various ways the cartographic forms that visualized the world were 'truthful' for the particular purposes the maps served.[9] So it is possible not only to expose the view that geography saw little study in medieval Europe but also to challenge the oft-repeated orthodoxy that the Middle Ages represents a geographic and cartographic lacuna in a western history of intellectual and scientific endeavour.

Ptolemy casts a long shadow over not just medieval geography but the Middle Ages generally, for there are still those who write of the modern 'discovery' of Ptolemy in the West, and bind this into a story about a new 'cartographic consciousness' through which the Old World 'discovered' the Americas.[10] Yet as some historians are now beginning to demonstrate, Ptolemy was neither 'lost' to the West in the Middle Ages nor was the Renaissance characterized by a switch from 'medieval' to 'modern' modes of cartographic and geographic representation; rather the revisionist history of medieval geography is decidedly more complex.[11] Knowing this is in part important in countering continued simplistic claims about the naïvety of geographical knowledge in the Middle Ages, and traditional western-orientated histories of science and discovery. It signals too that further detailed study of geography and cartography in the Middle Ages is required, revisiting medieval visual and textual geographical sources, including Ptolemaic ones. Here the work of Patrick Gautier Dalché and Margriet Hoogvliet, in particular, has undermined

---

[8] For example, see Raymond Beazley, *The Dawn of Modern Geography*, vol. 2: *A History of Exploration and Geographical Science from the Close of the Ninth to the Middle of the Thirteenth Century (c. ad 900–1260)* (London: Henry Frowde, 1901), 549–642. Cf. Natalia Lozovsky, 'Telling a new story of pre-modern geography: challenges and rewards', *Dialogues in Human Geography* 1 (2011), 178–82.

[9] See Keith D. Lilley, '*Quid sit mundus*? Making space for medieval geographies', *Dialogues in Human Geography* 1 (2011), 191–7.

[10] For the repeated orthodoxy, see Norman J. W. Thrower, *Maps and Civilization: Cartography in Culture and Society*, 3rd edn. (Chicago University Press, 2007), 58; John Pickles, *A History of Spaces: Cartographic Reason, Mapping and the Geo-coded World* (London: Routledge, 2004), 96–9, who refers to a new 'cartographic consciousness' emerging in Renaissance Europe. See also Donald K. Smith, *The Cartographic Imagination in Early Modern England: Re-writing the World in Marlowe, Spenser, Raleigh and Marvell* (Aldershot: Ashgate, 2008), 41–3.

[11] For a summary of such revisionist views, see Keith D. Lilley, 'Geography's medieval history: a neglected enterprise?', *Dialogues in Human Geography* 1 (2011), 147–62.

the old order, revealing the ways in which Ptolemy was understood in the Middle Ages, in different spatial contexts, as well as demonstrating the assimilation of different geographical sources and traditions.[12] As an example of this, Hoogvliet uses a printed edition of Johann Reger's *Geographia* of 1486, pointing out how it included earlier works, such as Jean Germain's *La mappemonde spirituelle* of c. 1450, Vincent of Beauvais's *Speculum Naturale* of the thirteenth century, and Isidore of Seville's *Etymologiae* and *De natura rerum*, both c. 600.[13] Through exploring these 'geographical traditions', essays in the first half of this volume similarly examine this spectrum of textual and visual geographical sources in the Middle Ages, as well as locales of their production and consumption.

The work of Ptolemy is considered by Jesse Simon (Chapter 1), not as a source of 'mathematical geography', however, but as a source of the more descriptive 'chorography' that formed one of three constituent components in his *Geographia*, dealing with 'the form and character of localised spaces and places'.[14] Simon's essay examines the relationships between cartography and chorography in late antique contexts and in so doing explores the roots of traditions of later, medieval descriptions of regions and places, whether depicted in graphic or textual form.[15] This raises the issue of continuity in language and nomenclature, of whether the apparent absence of use of a particular word – of Greek origin in the case of 'geography' or 'chorography' – should be taken also to mean that likewise they failed to exist as ideas or concepts, for one of the reasons why 'geography' is assumed (wrongly) to have disappeared in the Latin West between Antiquity and the Renaissance is because of the apparent paucity of written sources that use this particular word. Caution is required here, however, since 'geography' was not quite as invisible in medieval texts as some have suggested. For example, Gautier Dalché points to a Latin translation of an Arabic astronomical text undertaken by Hermann of Carinthia, who, writing c. 1140 in southern France and Spain, inserted a 'textual *mappamundi*' and placed it 'under the patronage of those whom he calls *geographi*, that is to say those who draw up

---

[12] Patrick Gautier Dalché, 'The reception of Ptolemy's *Geography* (end of the fourteenth to beginning of the sixteenth century)', in *The History of Cartography*, vol. 3: *Cartography in the European Renaissance, Part 1*, David Woodward (ed.) (University of Chicago Press, 2007), 285–364; Patrick Gautier Dalché, *La Géographie de Ptolémée en Occident (IVe–XVIe siècle)* (Turnhout: Brepols, 2009); Margriet Hoogvliet, 'The medieval texts of the 1486 Ptolemy Edition by Johann Reger of Ulm', *Imago Mundi* 54 (2002), 7–18.

[13] Hoogvliet, '1486 Ptolemy Edition'.

[14] Denis Cosgrove, *Apollo's Eye: A Cartographic Genealogy of the Earth in the Western Imagination* (Baltimore: Johns Hopkins University Press, 2001), 103.

[15] Jesse Simon, ch. 1, this vol.

maps'.[16] Similarly, Natalia Lozovsky has shown that with ninth- and tenth-century manuscripts and glosses on late antique geographical and cosmographical sources, such as Martianus Capella's *Marrriage of Philology and Mercury*, 'Carolingian commentators usually explain the words *geographicus* and *geographia* by translating them from the Greek', and finds specific use of *geographia* in two manuscripts which explain, '*GEOGRAPHIA id est terrae scriptio*', both of them originating from important monastic centres in northern France, Auxerre and Cluny.[17]

Descriptions of the world, then, defined what 'geography' was to Latin authors and authorities in the West, and such examples confirm that the word 'geography' was not unknown to medieval scholars. However, geographical sources, whether visual depictions or textual descriptions of the world, also existed independently from specific usage in naming them as 'geography', and such sources circulated widely throughout the Middle Ages. The dominance of particular sources of geographical knowledge is a subject explored in Chapter 2 in Andrew Merrills's examination of Isidore of Seville's *Etymologies*, 'arguably the most influential book, after the Bible, in the learned world of the Latin West for nearly a thousand years'.[18] The *Etymologies* assimilated geographical knowledge from earlier encyclopaedic works, and provided an accessible accumulation of these for medieval scholars, describing for them the wider cosmos and its parts, as well as the earth, its constituent continents and their provinces. Through giving information on lands and cities, and through the text's spatial ordering, the *Etymologies* presented a model for other *terrae scriptio* that came later. This is a point discussed by Merrills who, in building on his earlier study of early-medieval geographical authorities, examines the composition of the *Etymologies* and its sources, reminding us of the influences exerted by Isidore throughout the Middle Ages (as Reger's *Geographia* of 1486 demonstrates).[19] Merrills argues that in his geography in the *Etymologies* Isidore was 'presenting the world on its own terms'.[20]

---

[16] Patrick Gautier Dalché, 'Maps in words: The descriptive logic of medieval geography, from the eighth to the twelfth century', in *The Hereford World Map. Medieval World Maps and Their Context*, Paul D. A. Harvey (ed.) (London: The British Library, 2006), 223–42, at p. 230.

[17] Natalia Lozovsky, *'The Earth is Our Book'. Geographical Knowledge in the Latin West ca. 400–1000* (Ann Arbor: University of Michigan Press, 2000), 9, n. 5.

[18] Stephen A. Barney, W. J. Lewis, J. A. Beach and Oliver Berghof (eds. and trs.), *The Etymologies of Isidore of Seville* (Cambridge University Press, 2007), 3.

[19] Andrew H. Merrills, *History and Geography in Late Antiquity* (Cambridge University Press, 2005).

[20] Andy Merrills, ch. 2, this vol.

Some indication of how such authorities on geography were subsequently appropriated, interpreted and understood in monastic contexts is taken up by Natalia Lozovsky (Chapter 3) in her close study of codices and manuscripts belonging to the library of the monastery of St Gall, now in Switzerland. Here Lozovsky draws upon her earlier acclaimed study, *The Earth is Our Book*, but looks in more detail specifically at St Gall manuscripts with an aim to understand how the monastery's incumbents *read* geography.[21] This is a difficult task, but one made possible thanks to annotations and glosses that appear on a codex of Orosius' *History against the Pagans* (St Gall Stiftsbibliothek MS 621), in particular additions made, dating to the eleventh century, 'identified as that of Ekkehard IV (*c.* 980–1060)'.[22] The significance of these, for us, lies in Ekkehard's desire to add to the textual description of the Holy Land a visual depiction – a map – placed in the margins as an aid to comprehending this geographical description, while other textual amendments to Orosius in effect suggest an updating by Ekkehard to reflect contemporary geographical knowledge. Far from accepting uncritically his sources therefore, Ekkehard's additions and amendments show how geographies were being remade by scholars in centres of learning, not just in monasteries but elsewhere too, in Europe's nascent universities, for example.

Canonical sources of early medieval geography such as those examined by Merrills and Lozovsky also characterized later medieval scholarship too, reflected, for example, in the ways in which geographical knowledge was assimilated into textual and visual geographies of the thirteenth and fourteenth centuries. Two essays address this enduring aspect of geographical traditions, one by Amanda Power (Chapter 4) on the cosmography of Roger Bacon, and one by Marcia Kupfer (Chapter 5) on the now lost but no less celebrated Ebstorf *mappamundi*. Both essays serve to reinforce the growing realization among historians of medieval geography and cartography that to divorce maps from texts is a mistake, and that instead 'maps in texts' and 'texts in maps' were mutually constitutive as descriptions and depictions of the world.[23] Thus with Roger Bacon, a theologian and natural philosopher in Oxford writing in the later thirteenth century, Power looks at 'the constitutive elements of the way in which Bacon imagined the cosmos, the world, and the

---

[21] Lozovsky, *'The Earth is Our Book'*.   [22] Natalia Lozovsky, ch. 3, this vol.

[23] See Dalché, 'Maps in words'; Patrick Gautier Dalché, *La* Descriptio mappe mundi *de Hughes de Saint-Victor: Texte inedit avec introduction et commentaire* (Paris: Etudes Augustiniennes, 1988); Peter Barber, 'The Evesham world map: a late medieval English view of god and the world', *Imago Mundi* 47 (1995), 13–33; Evelyn Edson, *Mapping Time and Space: How Medieval Mapmakers Viewed Their World* (London: British Library, 1997).

Church's role within it', which means looking at those written sources that influenced his thinking as well as the insights Bacon himself showed through his acute observations.[24] Looking at Bacon's writings this way yields again the complexities of defining the geographical in contemporary sources, yet for Bacon this clearly meant looking at the world graphically as well as textually, for as Power notes, in his *Opus maius* Bacon recognized when it came to trying to describe the world there were limits in 'verbal description alone' and that 'a map must be used to make them clear to our senses'.[25] To do so he seems to draw on Ptolemy's *Almagest*, translated into Latin in the twelfth century.

While Bacon's cosmography reveals an Oxford scholar viewing the world in Aristotelian terms, the Ebstorf *mappamundi* points instead to the continued importance of neo-Platonist thought in Christian theology and cosmography. Kupfer assesses this in her close reading of this world map of *c.* 1300, seeking to situate it both in its particular local, conventual setting (in Lower Saxony), as well as in its broader cosmological and doctrinal setting. Tracing through the image the unwritten influences that acted on its cartographic and visual content, but with an eye too on how the map was understood by those (women) who viewed it, Kupfer argues that the map's dual embodiment of the world and of Christ forges an indexical relationship between God above and humanity below, where 'The staring effigy reminds the novice to comport herself as if always under divine surveillance in accordance with the founding document of western monasticism, the Benedictine Rule'.[26] In creating geographical knowledge in the thirteenth and fourteenth centuries, once again, then, there is a mixing of the old and the new.

The two chapters on Roger Bacon and the Ebstorf *mappamundi* thus continue with themes developed in previous chapters, on the assimilation and reproduction of geographical knowledge, and on the influences exerted by the locales of production and consumption. These two dimensions of geographical traditions are explored further by the final chapters that make up the first half of the volume, where the focus moves chronologically forward in time to the fifteenth and sixteenth centuries, to that period of European history typically seen as a transition from a pre- to post-Ptolemaic world, a so-called 'age of discovery'.[27] Chapters 6 and 7 by Meg Roland and Margaret Small immediately set a challenge to any such reductive views of geography and cartography in the later Middle Ages, and in this sense both provide support of those other historians of geography who have sought to undermine the modern

---

[24] Amanda Power, ch. 4, this vol. [25] Ibid.
[26] Marcia Kupfer, ch. 5, this vol. [27] See above, n. 10.

hubris that a sharp divide exists between the 'medieval' and 'modern'. In his monumental book *The Tropics of Empire*, Wey Gómez argues for such a reappraisal of cosmographical and geographical learning in the Columbian era and describes the cultural collision between the Old and New Worlds creating new geographical accounts alongside the old.[28] Here, Roland and Small explore similar connections, between print cultures and circuits of geography's dissemination, through their examination of the production and consumption of geographical knowledge in Tudor England and Renaissance Italy respectively.

Taking in not just maps but their relationships to texts, Roland and Small help to show again the persistence of earlier geographical knowledge within new arenas of learning and scholarship, as well as the continued duplicity of geography as a definable field of study, in some ways distinctive in content yet at the same time crossing into cognate areas such as cosmography and astronomy, just as it had in previous centuries. As Roland demonstrates, this is a period 'in which geographical thought and print culture productively co-developed, refashioning literary genres, geographic writing, and, eventually, cartography', as in England, in the fifteenth and sixteenth centuries, where English geographical thought circulated through print editions of Thomas Malory's Arthurian *Le Morte Darthur* (first printed in 1485), William Caxton's translation of *Mirrour of the World* (first printed in 1481), and editions of *The Kalender of Shepherds* and *The Compost of Ptholomeus* (each examined by Roland in her chapter).[29] In Italy at this same time, Ptolemaic geography was being assimilated too into new textual and visual geographies, but as Small's chapter on Giovanni Battista Ramusio's *Navigazioni e viaggi* shows, again the picture that emerges is far from a wholesale 'revolution', either in geography or in cartography. Between 1550 and 1559 in his compilation of *Navigazioni e viaggi*, Ramusio showed more interest in textual accounts of geography than with cartography, despite living in a world characterized now by many as one which saw 'the birth and growth of mapmaking',[30] and Small thus concludes that: 'he built up a geography that had more connection with Isidore, and his encyclopaedic coverage and careful systematic ordering, than with the post-Ptolemaic cosmographies of his own era'.[31]

What this long-established tradition of geography and cartography in medieval Europe reveals is an enduring spatial sensibility. The wider

---

[28] Nicholás Wey Gómez, *The Tropics of Empire: Why Columbus Sailed South to the Indies* (Cambridge, Mass.: MIT Press, 2008).
[29] Meg Roland, ch. 6, this vol.   [30] Pickles, *History of Spaces*, 99.
[31] Margaret Small, ch. 7, this vol.

world is thus seen through geography: *terrae scriptio*, writing the Earth. Through tracing genealogies of geographical sources and authorities during the Middle Ages, the chapters in the first half of the volume make clear how geography mattered in the intellectual efforts of medieval scholars: firstly, as evidenced by the geographical content to be found in their sources, mediated both visually and textually, and secondly in terms of where, geographically, this mediation occurred. Then as now, knowing the world is geographically contingent. This view has strong resonance too with geography's later history and historiography, and western science more generally, for as historical geographers and historians of science are now revealing, the production and consumption of knowledge – whether geographic or not – is inherently a geographical process.[32] Moreover, by exploring these geographical traditions over the *longue durée*, the Middle Ages included, certain characteristics and traits emerge in what constituted 'medieval geography', traits such as encyclopaedism, and chorography, cosmography and cartography.[33] Not only does this help us to understand medieval European geographical thought and knowledge, and reveal complex and important connections between visual and textual geographies during the Middle Ages, it challenges, too, the idea that a transition occurred from a pre- to post-Ptolemaic world, highlighting similarities between geography across historical periods, thus questioning whether 'the origins of modern geography' really do lie 'in the century after Columbus', as some still suggest.[34]

Geographical imaginations

What geographies permeated the lives of individuals and groups during the Middle Ages, and how was their 'spatial sensibility' reflected and reinforced by the texts and images that describe these lived and represented worlds? Such questions lead us to explore the geographies of texts and images, and the geographical imaginations they construct and convey. Mapping medieval geographies encompasses more than just particular intellectual traditions, for as is evident from de Bray's account

---

[32] For example, David Livingstone, *Putting Science in its Place: Geographies of Scientific Knowledge* (Chicago University Press, 2003); Charles Withers, 'Eighteenth-century geography: texts, practices, sites', *Progress in Human Geography* 30 (2006), 711–29; Charles Withers, *Placing the Enlightenment: Thinking Geographically about the Age of Reason* (University of Chicago Press, 2007).

[33] For examples, see Keith D. Lilley, 'Medieval geography', in *International Encyclopaedia of Human Geography*, vol. 7, Nigel Thrift and Rob Kitchin (eds.) (Oxford: Elsevier, 2009), 21–31.

[34] Mike Heffernan, 'Histories of geography', in *Key Concepts in Geography*, Sarah Holloway, Stephen Rice and Gill Valentine (eds.) (London: Sage, 2003), 3–22, at 4.

of his estate, and indeed as some past historians of medieval geography, such as John Kirkland Wright and George Kimble, have pointed out, an implicit geography can always be gleaned from close readings of medieval sources, whether labelled 'geography' or not.[35] Hence the essays in the volume's second half focus on 'thinking geographically', which means dealing with those texts and images that are implicitly geographical in their nature, rather than just those authoritative sources characterizing the geographical tradition considered earlier. Exploring these geographical imaginations might be understood as a journey through medieval subjectivity, touching the everyday spatial lives of those living beyond cloister and cathedral as well as those within. The second part of the volume therefore considers the imagined geographies contained in various medieval maps and texts, and uses these as a means to explore the geographical consciousness of those who themselves inhabited and represented multiple medieval worlds.

How have medievalists sought to explore medieval geographical imaginations? For many it seems 'mapping' provides a *modus operandi*, but mapping in a figurative sense, and not necessarily in the literal sense of 'making a map', an important distinction that Cosgrove reflects upon when he writes:

The measure of mapping is not restricted to the mathematical; it may equally be spiritual, political or moral. By the same token, the mapping's record is not confined to the archival; it includes the remembered, the imagined, the contemplated. The world figured through mapping may thus be material or immaterial, actual or desired, whole or part, in various ways, experienced, remembered or projected.[36]

The power of 'mapping' as a rhetorical device for interpreting imagined worlds is used widely by those working in humanities traditions, and is evident, too, in recent work – especially by art and literary historians – on medieval cultures. Thus Daniel Lord Smail's study of the activities of notaries in recording property in late-medieval Marseille shows how 'a cartographic lexicon' enables us to explore the 'mental geography' (as he puts it) of the city and its people.[37] His 'non-graphic' view of maps, like Cosgrove's, is also paralleled by Nicholas Howe in his book,

---

[35] John K. Wright, *The Geographical Lore at the Time of the Crusades: A Study in the History of Medieval Science and Tradition in Western Europe* (New York: American Geographical Society, 1925; reprinted, London: Constable, 1965); George H. T. Kimble, *Geography in the Middle Ages* (London: Methuen, 1938).

[36] Denis Cosgrove, 'Introduction: mapping meaning', in *Mappings*, Denis Cosgrove (ed.) (London: Reaktion Books, 1999), 1–23, at 2.

[37] Daniel Lord Smail, *Imaginary Cartographies: Possession and Identity in Late Medieval Marseille* (Ithaca: Cornell University Press, 1999), 1–41.

Writing the Map of Anglo-Saxon England, in which he observes that 'the fact that almost no visual maps survive from the Anglo-Saxons [in England] does not mean that they had no geographic imagination or, yet more strongly, felt no imperative to map the world'.[38] In this sense texts are 'maps' – and maps are 'texts' – two interpretive strategies that medievalists have used to open up medieval worlds, particularly in thinking through the connections between place and identity, as is evident, for example, in Kathy Lavezzo's book, *Angels on the Edge of the World*, on the relationship between geography, literature and English identity, and also Naomi Reed Kline's study of the Hereford *mappamundi* in her *Maps of Medieval Thought*, which explores how both history and geography are woven into the map.[39] Particular spatial practices, embodied geographies such as travel and pilgrimage, worship and property-holding, are thus embedded in representing places and defining cultural identities. Here we see historians and geographers working with maps in a metaphoric as well as a literal sense, and in so doing yielding insights into geographical imaginations.[40] This is the paradigm that ties together essays that comprise the second half of *Mapping Medieval Geographies*, where we find that maps reveal much more than first appears, and where mapping through texts reflects again a 'spatial sensibility' permeating medieval lives and cultures.

Such a link between geography, mapping and cultural identity is considered in the first two chapters, starting with Chapter 8 and Camille Serchuk's examination of a geographical depiction – a map to some – showing France and its constituent *pays* and cities. Dating to around the mid-fifteenth century, and created at a time when France was in a state of flux as a country (let alone a 'nation'), this map (Bibliothèque Nationale MS. fr 4991 f. 5v) formed part of a broader historical narrative, 'a manuscript of the French genealogy and chronicle called *A tous nobles*'.[41] As Serchuk explains, through tying together history and territory, the map and text of *A tous nobles* performs a dual role of explaining the origins of the land of France as well as the destiny of its people, thus promoting a

---

[38] Nicholas Howe, *Writing the Map of Anglo-Saxon England: Essays in Cultural Geography* (New Haven: Yale University Press, 2008), 5.

[39] Kathy Lavezzo, *Angels on the Edge of the World: Geography, Literature, and English Community, 1000–1534* (Ithaca: Cornell University Press, 2006); Naomi Reed Kline, *Maps of Medieval Thought: The Hereford Paradigm* (Woodbridge: Boydell Press, 2001).

[40] See for example, Daniel K. Connolly, *The Maps of Matthew Paris: Medieval Journeys through Space, Time and Liturgy* (Woodbridge: Boydell Press, 2009); Katherine Breen, 'Returning home from Jerusalem: Matthew Paris's first map of Britain in its manuscript context', *Representations* 89 (2005), 50–93; Keith D. Lilley, *City and Cosmos: The Medieval World in Urban Form* (London: Reaktion Books, 2009).

[41] Camille Serchuk, ch. 8, this vol.

'unity that would permit France to endure'.[42] While such 'national' maps are, apparently, relatively unusual in the medieval West, historians of cartography have long recognized that all such maps represent claims made on territory.[43] This is a theme further explored by Karen Pinto's essay (Chapter 9), not through maps *from* the West however, but rather maps *of* the West, specifically medieval Islamic maps depicting the Mediterranean world. The Surat al-Maghrib map Pinto describes represents a cartographic genre of the eleventh century onwards in which maps of the Mediterranean share similar stylistic conventions, for example in their use of geometrical shapes. From this, Pinto develops the case for the maps' gendering of those lands situated around the western Mediterranean. To fulfil a Muslim desire to unify the region culturally, geographically and politically, the Surat al-Maghrib maps both describe and prescribe an 'imagined geography' of the West, anthropomorphizing Islamic claims over neighbouring (and contested) Christian territories.[44] In both chapters, then, the map is seen to be a common means of defining cultural identities by spatial means, as well as charting the thinking of those who created them.

The place of cartography in defining lands and peoples is a topic that has resonance not just in the Middle Ages but in later historical periods too. Within medieval contexts historians have begun to explore the close connections between maps, sovereignty and rule, for example with the maps of Britain by Matthew Paris of the mid-thirteenth century, and the 'Gough Map' of the later fourteenth century, but what often is missed is how maps themselves were formulated, partly because mapmakers in the Middle Ages say so little about their work.[45] To address this, some detailed analysis of particular maps has proved fruitful, showing the complexity of medieval cartographic production, whether *mappaemundi*,

---

[42] Ibid.

[43] J. Brian Harley, 'Silences and secrecy: the hidden agenda of cartography in early modern Europe', *Imago Mundi* 40 (1988), 57–76; J. Brian Harley, *The New Nature of Maps: Essays in the History of Cartography*, John H. Andrews (ed.) (Baltimore: Johns Hopkins University Press, 2002); Peter Barber, 'England I: pageantry, defense, and government: maps at court to 1550', in *Monarchs, Ministers, and Maps: The Emergence of Cartography as a Tool of Government in Early Modern Europe*, David Buisseret (ed.) (University of Chicago Press, 1992), 26–56; Daniel Birkholz, *The King's Two Maps: Cartography and Culture in Thirteenth-Century England* (New York: Routledge, 2004).

[44] Karen Pinto, ch. 9, this vol.

[45] See Connolly, *The Maps of Matthew Paris*; Birkholz, *King's Two Maps*; Keith D. Lilley and Christopher D. Lloyd (with Bruce M. S. Campbell), 'Mapping the realm: a new look at the Gough Map of Britain (*c.*1360)', *Imago Mundi* 61 (2009), 1–28; Elizabeth Solopova, 'The scribes of the Gough Map', and 'Linguistic evidence', in *Linguistic Geographies: The Gough Map of Great Britain*, e-resource, www.goughmap.org (accessed 1 July 2011).

'national maps', or local topographic maps and plans. Through focusing on one particular centre of cartographic production (and consumption), Daniel Birkholz's essay (Chapter 10) continues with this desire to enter the mapmaker's world(s), but he does so by taking a different approach. By choosing an important centre of scientific learning in the Middle Ages – Hereford, a city situated on the disputed borders of Wales and England – and through using a combination of biography and cartography, Birkholz's focus is not the renowned Hereford *mappamundi per se* but instead a study of how Hereford itself played a role in connecting maps and mapmakers. He thus charts different 'mapping lives' across three centuries in Hereford, linking Gerald of Wales, Roger de Breynton and Walter Mybbes, and so within one particular locale reveals how cartographic and geographic encounters shaped the life-worlds of each. There are parallels between Birkholz's biographical approach and that taken by Kathy Lavezzo, in Chapter 11, though looking this time at Norwich – another English provincial city – where the lives she maps are those of Jewish inhabitants as perceived through the 'geographical imagination' of a Christian hagiographer.

Connections between place and identity, particularly liminality and marginality in the medieval world, have attracted attention among medievalists interested in questions of race, gender and ethnicity. Lavezzo herself, in *Angels on the Edge of the World*, sought to show how textual and cartographic images of medieval England had long reconciled the marginal geographical location of the English with their (self-)perceived place as a nation at the centre of their culturally constructed world.[46] The answer, she explains, lies not in a straightforward mapping of oppositional binaries of 'inside and outside' onto 'centre and edge', but in what she calls 'geographic alterity'; that is, the mediated and contingent nature of place(s) in the geographical imagination. In her essay on anti-Semitic narratives in later-medieval Norwich, Lavezzo further explores this theme using a mid-twelfth-century English hagiography, Thomas of Monmouth's *Life and Miracles of St William of Norwich*, as a 'map' of contemporary attitudes to Jews in an English city.[47] This mapping deals again with biography – an account of a Christian boy in the city allegedly killed by a Jewish family – and just as Birkholz's 'analysis of the "life-map" of Roger de Breynton leads to the destabilization of "lay/clerical, financial/spiritual, civic/ecclesiastical, local/universal"' binaries, so Lavezzo's 'close analysis of Thomas of Monmouth's imagined geography of Norwich, England and the Christian West offers a means of problematizing the

---

[46] Lavezzo, *Angels on the Edge of the World.*
[47] On 'geographic alterity', see ibid. 20.

opposition of Jew and Christian'.[48] In Norwich, then, the placing of Jew and Christian seen in Thomas of Monmouth's account does not simply map cultural difference onto the city's urban spaces by marginalizing the Jews spatially and socially, but rather Monmouth uses the local landscape to legitimize Christian authority in the context of Judaeo-Christian history. In doing so the experienced and embodied geographies Monmouth describes reinforce a sense of difference in which 'place' still defines and shapes cultural identities in Norwich.

Questions of place and identity, of ordering the world geographically, are further explored by the volume's two final essays, where again liminality is mapped through image and text. Religious experiences in place, and the subjective geographies they generate, are examined by Veronica della Dora (Chapter 12) and by Sara Torres (Chapter 13) by focusing on pilgrimage to places on the margins, to the ends of the world, both in eastern (Greek) and western (Latin) Christian traditions. Their essays, like Lavezzo's, concern the power of place, and the associations in the mind between place and meaning. For della Dora, these places are those that are elevated physically, particularly peaks and mountain tops, while for Torres the margins she considers are not arranged vertically but horizontally, yet both are similarly spatially detached by virtue of their peripheral locations. While connections between pilgrimage and geography, for example through pilgrims' guides and maps, is reasonably well-studied by medievalists, the focus here is on the experience of place *through* pilgrimage, and imagined geographies of both familiar and unfamiliar lands, thus engaging pilgrims' geographical imaginations.[49]

In the Greek Orthodox tradition that della Dora describes, the hierarchically, spatially ordered world of Christian cosmology, linking heaven and earth, means that high places are of particular significance in defining divine relationships between God, the saints and humanity. Combining historical accounts and depictions of landscapes in Orthodox iconography along with examining evidence of the material culture of monasticism in the Christian East, reveals a landscape imagined both as a Garden of Eden and as a 'ladder to heaven', a connection forged on earth, in a middle place, a 'frontier' between material and spiritual life. As a liminal site of contemplation and vision, della Dora concludes, 'Byzantine holy mountains were symbolic spaces that allowed the pious

---

[48] Daniel Birkholz, ch. 10, this vol.; Kathy Lavezzo, ch. 11, this vol.

[49] E.g. Breen, 'Returning home from Jerusalem'; Catherine Delano Smith, 'The intelligent pilgrim: maps and medieval pilgrimage to the Holy Land', in *Eastward Bound*, Allen (ed.), 107–30.

reader to map out his or her own ascetic path'.[50] For the Catalan Christian pilgrimage account of the late fourteenth century that forms the principal focus for Torres, journeying to a liminal place removed from the everyday is also a spiritual destiny. Here though the far north-west of Ireland, on Europe's periphery, is the point at which earthly and divine realms meet, a geographical *and* holy frontier in the mind of Ramon de Perellós. His *Pilgrimage to St Patrick's Purgatory* describes his experiences and in doing so, Torres explains, 'The mental mappings and imagined geographies created by Perellós's itinerary guide its readers through geographical spaces marked by historical record and textual precedent'.[51] Again, the peripheral location of St Patrick's Purgatory echoes other mystical places 'on the edge' of the medieval world, includ-ing the Garden of Eden as Torres notes, but as with the essays of Lavezzo and Birkholz, this suggests Perellós's world is not simply constructed through a series of spatial and cultural binaries, rather 'His descriptions bear out what Kathryn L. Lynch describes as "a richer and more dialogic, rather than sharply oppositional, relationship between the cultures we think of as "West" and "East" throughout much of the Middle Ages"'.[52] The two essays together allow us to reflect on these meanings associated with 'centre' and 'edge', and on the geographical and spatial ordering of the world in both eastern and western Christian traditions.

Where places are, what they mean, and to whom, as revealed in the textual and visual accounts examined by the essays here, collectively show medieval authors and their subjects 'thinking geographically' about the worlds they inhabit and encounter. It is not just that they are simply describing peoples and lands in what may be labelled today as 'geog-raphy', but rather that they make sense of the material world through a mutable 'geographical imagination'. For medieval authors such as Mon-mouth and Perellós the relative placing of objects and people is a means for constructing and reinforcing a sense of cultural difference *or* unity (depending on the particular agenda), something evident too in the maps of France and the Mediterranean discussed earlier by Serchuk and Pinto. Taking journeys through medieval subjectivities, then, and encountering the spatial lives of clerics and pilgrims, nobles and townsfolk, we see in the accounts a reflection of the spatial ordering of these worlds, an underlying geographically grounded 'spatial sensibility' perceived through placed-relationships observable in the sentient world. Through these visual and textual representations of the medieval world – or worlds

---

[50] Veronica della Dora, ch. 12, this vol.   [51] Sara V. Torres, ch. 13, this vol.
[52] Ibid., citing Kathryn L. Lynch (ed.), *Chaucer's Cultural Geography* (New York: Routledge, 2002), 5.

within the world – are made yet others: thus 'mapping begets mappings', as Cosgrove observes.[53] The accounts of those various authors and authorities considered here, as well as the essays themselves, are an accumulation of such 'mappings', each engendering past and present 'geographical imaginations'. Through these personal encounters with the geographical certain 'imagined geographies' emerge, each a lens through which to see not just medieval worlds but the 'geographical imaginations' of those describing and depicting them.

★★★

Today's geographers who examine 'geography' as it was studied and represented in the past emphasize that not only did visual and textual geographies construct and frame contemporary views of the world, but also that these reflected the particular interests and concerns of those who sought to map it and describe it.[54] For those contemporaries whose accounts form the focus of the essays here, there was similarly no neat line separating 'geographical traditions' from 'geographical imaginations', for both 'geographical thinking' and 'thinking geographically' occur as mutually constitutive ways through which the world is understood and perceived. This is evident throughout the period in the examples covered by this volume, from the writing of Isidore of Seville in the early seventh century through to that of Giovanni Battista Ramusio in the mid-sixteenth. It is evident too across medieval cultures, whether Arabic, Judaic or Christian – Byzantine or Latin – both in maps of the period, whether Ekkehard's in St Gallen, or the Surat al-Maghrib maps of the Mediterranean, and in 'mappings' too, of Christian and Jew in Norwich, or pilgrims journeying to the holy mountain. Any division between 'geographical traditions' and 'geographical imaginations' is thus a heuristic device, more a reflection of where emphasis is placed in interpreting contemporary sources.

What emerges through reading across the two parts of the volume is geography's dual identity and slippery character as a subject: a geographical discourse which is in part about 'geography *per se*', those geographical texts and images circulating and enduring across time and space through geographical traditions upheld in scholarly institutions across medieval

---

[53] Cosgrove, 'Introduction', 13.

[54] E.g. Livingstone, *Geographical Tradition*; Driver, *Geography Militant*; Ogborn, *Global Lives*; Cosgrove, *Apollo's Eye*; Robert Mayhew, *Enlightenment Geography: The Political Languages of British Geography, 1650–1850* (New York: Palgrave, 2000); Charles Withers and Robert Mayhew, 'Rethinking disciplinary history: geography in British universities, c.1580–1887', *Transactions of the Institute of British Geographers*, NS 27 (2002), 11–29; Lesley B. Cormack, *Charting an Empire: Geography at the English Universities, 1580–1620* (Chicago University Press, 1997).

Europe; and in part about 'geographical knowledges', not always formal or intellectual, but rather everyday, culturally embedded placed experiences and spatial practices of those inhabiting the medieval world.[55] Such traits have long defined 'writing the Earth' in western geography. Through both, in contemplation, movement and encounter, geographies are produced and consumed, as Amanda Power makes evident in her chapter on Roger Bacon's 'cosmographical imagination', for example, where we see the geographies that Bacon constructs in part are a product of his intellectual engagement with various sources and traditions, through his scientific work, but also a reflection of his geographical encounters, both in Paris and Oxford, all combining to shape his thinking and writing, conditioning his 'geographical imagination'. Therefore the geographical and cartographic works of the Middle Ages are mapped geographic lives, and geographic lives map geographies and cartographies. In his chapter on Hereford, Daniel Birkholz reveals this too for Gerald of Wales, whose particular 'life-world' was as much enshrined in his cartographic and topographic works as were the ideas and thought of those geographical sources that he, like many others, drew upon in their teaching and studies.

There are, then, geographies embedded in the making of geography. The influence of locale in creating and consuming geographical knowledge, and the spatial practices of those involved in this process, reflect and reinforce cultural norms of their time. The complexity and myriad nature of these medieval geographies are revealed by the essays offered here. At the same time their content and analysis reveal complementary ways of mapping medieval geographies, for what follows is as much a methodological perspective on working with medieval geographies as a statement of current agendas and concerns among those scholars involved in studying them. For those concerned with the spatial and the geographical in general, and for medievalists in particular, these are agendas with great potential. Those traditions of geographical thought mapped out in them point to continuities through time, both in the use of particular sources and forms of geographical representation as well as in how geographical knowledge is treated, rendering evident its inherent spatial sensibilities. The multiple worlds of medieval men and women – such as those revealed in Henry de Bray's account of his lands in Cotton

[55] On geography's 'slippery nature', see Mike Heffernan, 'Histories of geography', in *Key Concepts in Geography*, Holloway et al (eds.), 19; on 'geographical knowledge' and 'geography *per se*', see Lilley, 'Geography's medieval history'; cf. Robert Mayhew, 'The effacement of early modern geography (c.1600–1850): a historiographical essay', *Progress in Human Geography* 25 (2001), 383–401.

Nero C XII and many other countless manuscripts – are thus opened up through 'thinking geographically' ourselves when studying both textual and visual sources, while at the same time, too, the 'geographical thinking' of our own day – whether derived from everyday practices or learned treatises – is here culturally embedded in mapping out these medieval geographies.[56]

[56] Particular thanks are due to Veronica della Dora, Marcia Kupfer, Natalia Lozovsky and Karen Pinto for their comments on earlier drafts of this introductory essay, but in so many ways all of the volume's contributors have helped to shape its content and genesis.

Part I

Geographical traditions

1 Chorography reconsidered
An alternative approach to the Ptolemaic definition

Jesse Simon

Medieval geography begins and ends with Ptolemy, or so modern scholarship would have us believe. According to the narrative which has become accepted in the last century, it is the loss of Ptolemaic knowledge at the end of Classical Antiquity which casts geography into a wilderness period lasting some thousand years; only when Ptolemy is 'rediscovered' in the fifteenth century is the western world provided with a practical and theoretical basis for geographical and cartographical expression.[1] Because medieval Europe appears to lack a cartographic tradition which can be explicitly linked to Ptolemaic teachings – that is to say, it did not produce recognizable maps based upon astronomical observations, coordinates and projections – the long period between Late Antiquity and the Renaissance is often characterized as a period of intellectual stagnation, during which the development of geography essentially ground to a halt. Such a position, however, is only tenable if one takes Ptolemy at his word and believes that every approach to geography not rooted in mathematical observation must be inherently inferior.

Ptolemy is openly critical of what he perceives to be 'lesser' forms of geographical expression and many of his biases have survived intact into the present day. Perhaps the most notable example comes in the very opening chapter of his *Geographia*, where he attempts to explain why geography – that is, the precise mathematical apprehension and delineation of the known world – is superior to the practice known as chorography; the latter, he explains, is a decidedly inferior pursuit which focuses on a smaller area, requires no particular scientific skills and is

---

[1] For a conventional account of Ptolemy's reintroduction to the Latin world, see O. A. W. Dilke, *Greek and Roman Maps* (London: Thames and Hudson, 1985), 160–6. For a reappraisal of the notion of Ptolemy's complete disappearance in the West, see P. Gautier Dalché, *La Géographie de Ptolémée en Occident (IVe–XVIe siècle)*, Terrarum Orbis 9 (Turnhout: Brepols, 2009); P. Gautier Dalché, 'The reception of Ptolemy's Geography (end of the fourteenth to beginning of the sixteenth century)', in *The History of Cartography*, vol. 3, D. Woodward (ed.) (University of Chicago Press, 2007), 285–364.

best left to those who are good at drawing or painting.[2] While Ptolemy was not the only classical author to mention chorography, he was certainly the only one who bothered to define it. For this reason alone, his reading has come to characterize most modern interpretations of the word. During the medieval period, chorography seemingly disappears from the geographic vocabulary for several centuries, emerging again in the Renaissance with the rediscovery of Ptolemy in the West.[3] However, while the word may have taken on a distinct meaning for cartographers in the early modern period – and while it remains in marginal use even today – one must be cautious about defining the chorography of Ptolemy's time according to our modern understanding.[4]

Modern scholarship has tended to view chorography as an imprecise midpoint between 'topography' – the specific description of a location – and 'geography', the scope of which encompasses the whole world. For the most part, Ptolemy's definition has gone unchallenged. Indeed, in a recent translation of Ptolemy, we find chorography rendered as 'regional cartography', which, to be sure, makes sense in the context of Ptolemy's argument but, as a definition, is somewhat limiting.[5] If we attempt to understand chorography through the Ptolemaic lens, then we are always going to find a subdiscipline unworthy of being considered alongside the scientific pursuit of geography.

Few attempts have been made to look beyond Ptolemy's definition or to evaluate how chorography may have contributed to geographical understanding at the end of antiquity.[6] The goal of the present study,

---

[2] Ptolemy, *Geog.* I.1. *Claudii Ptolemaei Geographia*, K. Nobbe (ed.) (Hildesheim: Olms, 1966). See also A. Stückelberger and G. Graßhoff (eds.), *Ptolemaios Handbuch der Geographie (Griechisch–Deutsch)* (Basle: Schwabe Verlag, 2006).

[3] The sixteenth-century geographer Egnatio Danti seems to have based his definition of chorography directly on that of Ptolemy; see T. Frangenberg, 'Chorographies of Florence: the use of city views and city plans in the sixteenth century', *Imago Mundi* 46 (1994), 41–64, at 55–6.

[4] On the use of chorography in sixteenth- and seventeenth-century Britain, see R. Helgerson, 'The land speaks: cartography, chorography, and subversion in Renaissance England', *Representations* 16 (1986), 50–85; S. Mendyk, 'Early British chorography', *The Sixteenth Century Journal* 17 (1986), 459–81. For its currency in modern-day geography, see R. J. Johnston, D. Gregory, G. Pratt and M. Watts (eds.), *The Dictionary of Human Geography*, 4th edn. (Oxford: Blackwell, 2000), 79–80.

[5] J. L. Berggren and A. Jones, *Ptolemy's Geography: an annotated translation of the theoretical chapters* (Princeton University Press, 2000), 57ff.

[6] There have been a few recent studies on chorography: C. Nicolet, 'De Vérone au Champ de Mars: *Chorographia* et Carte d'Agrippa', *Mélanges de l'école française de Rome* 100 (1988), 127–38, considers non-Ptolemaic evidence for the role of the chorographer in the Roman world; K. Olwig, 'Has "geography" always been modern? *choros*, (non) representation, performance, and the landscape', *Environment and Planning* 40 (2008), 1843–61, treats chorography as a precursor to the modern idea of 'landscape'; two articles

therefore, is to reconsider the place of chorography within the classical tradition and to suggest how this largely ignored legacy may offer us a new context by which to assess the emergence of medieval geography. By making a close examination of available sources from classical and Late Antiquity, we may discover that chorography did not, in fact, represent a loss of scientific knowledge, but was, rather, the rejection of one mode of representation for an equally viable alternative which better suited the perceptual realities of the early-medieval period. Specifically, the present study hopes to argue that chorography, while not a mathematical discipline, was neither limited to a regional scope nor unskilled in its execution; however, whereas Ptolemy's approach involved the projection of coordinates onto a surface, chorography was an attempt to catalogue and represent the world based on perception and experience. This fundamental difference in perceptual approach may be one of the characteristics which most strongly define the break between classical and medieval geographical traditions. By understanding the origins of chorography, we may find ourselves with a context for better understanding medieval geographical thought and the development of its distinctive characteristics, as will be examined in subsequent chapters of this volume.[7] By approaching chorography as a geographical approach unto itself – rather than merely an unskilled subdiscipline – we shall demonstrate that the origins of medieval geography were not the result of stagnation following the loss of Ptolemy but, rather, an active rejection of Ptolemaic teaching and the continuation of a parallel intellectual tradition with roots in antiquity. Indeed, by the end of antiquity, chorography had superseded the pure geography of Ptolemy as the dominant mode of geographical expression and would, thus, have provided a foundation for both the apprehension and representation of the larger world in the medieval period.

Ptolemy and the geographical tradition

Ptolemy opens his *Geographia* by differentiating between geography and chorography. For him geography is a mathematical science that involves the determination – through measurement and sightings – of the proportions and distances which define the accurate shape of the

by J. F. Moffitt – 'Medieval *mappaemundi* and Ptolemy's *Chorographia*', *Gesta* 32 (1993), 59–68, and 'The Palestrina Mosaic with a "Nile Scene": Philostratus and Ekphrasis; Ptolemy and Chorographia', *Zeitschrift für Kunstgeschichte* 60 (1997), 227–47 – examine the relationship between chorography and medieval geography largely in light of Ptolemy's definition.

[7] Cf. Merrills, ch. 2, this vol.; Lozovsky, ch. 3, this vol.

world. The science, however, is inseparable from its cartographic output; the goal of the geographer should always be 'the imitation, through drawing, of the whole of the inhabited world (the *oikoumene*) as a single continuous entity'.[8] For Ptolemy, the word geography (γεωγραφία) implied the creation of a map and, thus, the bulk of his treatise contains the names and coordinates of the places which make up the known world, prefaced by a set of instructions on how the geographer may translate those coordinates into an accurate projection. By his own admission, he wrote his work with the intention of correcting measurements given in previous geographical works, primarily that of Marinos of Tyre who would have been writing maybe half a century before Ptolemy.[9] Indeed, one of Ptolemy's criticisms of Marinos is that the latest edition of his work isn't accompanied with any kind of graphical representation.[10]

Ptolemy's desire to envisage the world in cartographic terms represents the continuation of a tradition dating back at least six centuries. By the fourth century BCE, there was already an established model for understanding the shape and the content of the world, in which the world was imagined as a globe divided into five zones: the northern and southernmost zones were uninhabitable due to extreme cold, while the central zone remained inaccessible due to extremes of heat. This left two temperate zones suitable for human habitation, of which the northern zone contained the *oikoumene*.[11] The *oikoumene* itself was imagined as an island, approximately twice as long as it was high, and surrounded on all sides by ocean. In the centre of the island was the Mediterranean, which emptied into the ocean at the Pillars of Hercules. The land, which formed a ring around the central sea, was divided into three continents: from the northern pillar to the Tanaïs (Don) was Europe; from the Tanaïs to the Nile was Asia; and from the Nile to the southern pillar was Libya. While the arrangement and content of these lands would be the subject of much debate, the basic tripartite structure provided a

---

[8] ἡ γεωγραφία μίμησίς ἐστι διὰ γραφῆς τοῦ κατειλημμένου τῆς γῆς μέρους ὅλου μετὰ τῶν ὡς ἐπίπαν αὐτῷ συνημμένων. Ptolemy, *Geog.* I.i.1. Also: Προκειμένου δ' ἐν τῷ παρόντι καταγράψαι τὴν καθ' ἡμᾶς οἰκουμένην σύμμετρον ὡς ἕνι μάλιστα τῇ κατ' ἀλήθειαν. Ptolemy, *Geog.* I.ii.2.

[9] Ptolemy, *Geog.* I.iv.1. Although Marinos' exact dates are unknown, Ptolemy describes his work as 'recent' (I.vi.1) and indications within the text allow us to suggest that Marinos was writing between 107 and 115 CE; see H. Cancik and H. Schneider (eds.), *Brill's New Pauly* (Leiden: Brill, 2002–).

[10] Ptolemy, *Geog.* I.xvii.1.

[11] See Aristotle, *Meteor.* II.5. Strabo (*The Geography of Strabo* in 8 vols. ed. and trans. H. L. Jones (Cambridge, Mass.: Harvard University Press, 1917–32)). II.ii–iv provides a summary of the different theories put forth by his predecessors, including Eratosthenes, Posidonius, Hipparchus and Polybius.

durable foundation for geographical understanding which lasted from Classical Antiquity through the end of the medieval period.[12]

The goal of Greek mathematical geographers was to refine this model of the world by means of actual measurement.[13] In the third century BCE, Eratosthenes of Cyrene – who may be best remembered for his remarkably accurate calculation of the world's circumference – wrote a three-volume work on geography, the final volume of which dealt with the creation of a world map.[14] The map was almost certainly not the first of its kind – Eratosthenes himself described it as a revision[15] – however it may have been the first to be projected based upon a system of meridians. If the numerous criticisms levelled by Hipparchus and Strabo are anything to go by, the map of Eratosthenes may not have been especially accurate as a geographical representation;[16] it seems, however, to have provided a solid foundation for subsequent geographical undertakings. As more of the *oikoumene* became known, the job of the geographer was to integrate the knowledge of the present with the measurements of the past, in order to create an ever more accurate graphical representation.

Alongside refinements in distances and measurements, there were also developments in the techniques used to project a segment from a sphere onto a flat surface. According to Strabo, it was permissible to draw the world onto a flat surface using an orthogonal system of parallels and meridians.[17] This practice was apparently still current a century later when Marinos of Tyre was writing his geography: Ptolemy criticizes Marinos for his reliance on an orthogonal projection and goes on to propose his own system which would take into account the curvature of the earth.[18] Ptolemy's text represents a culmination of mathematical knowledge and, at the time of writing, the most current

---

[12] The tripartite structure of the *oikoumene* provides a starting point for nearly every classical text on geography; examples may be found in Strabo, Pliny and Pomponius Mela (first century CE), Dionysius Periegetes (second century) and Julius Honorius (fourth/fifth century). Examples from the medieval period may be found in Dicuil (ninth century), the Geography of Ros Ailithir (tenth century) and Eustathius of Thessalonica (twelfth century).

[13] For an introduction to the early mathematical geographers, see: J. O. Thomson, *History of Ancient Geography* (Cambridge University Press, 1948), 152–68; C. Nicolet, *Space, Geography and Politics in the Early Roman Empire* (Ann Arbor: University of Michigan Press, 1991), 57–74 and Dilke, *Greek and Roman Maps*, 21–38.

[14] The work of Eratosthenes does not survive, but the existence of his geographical work is attested in Strabo who presents (and often refutes) direct quotations and measurements. See Strabo II.i *passim*.

[15] Strabo II.i.2.

[16] Like Eratosthenes, the geographical works of Hipparchus (second century BCE) have not survived and are known only through quotations in Strabo.

[17] Strabo II.v.10.   [18] Ptolemy, *Geog.* I.xxii.

and informed attempt to provide the tools necessary for an accurate graphical representation of the known world.

Although Ptolemy was insistent that geography should be accompanied by a pictorical representation of the *oikoumene*, there are numerous texts from antiquity which sought to describe – rather than depict – the contents of the known world. Eratosthenes was famously sceptical of leaving geography in the hands of poets: he claimed that they were less interested in instruction than in entertainment and that they were prone to fabrication.[19] However, for Hipparchus and Strabo, there was no question that Homer was the father of geography, and that the poet's descriptions provided a geographical framework for the Mediterranean world.[20] Indeed, there is evidence to suggest that geographical verse was not uncommon in the classical world: Alexander of Ephesus was said to have written a descriptive poem in three parts – one for each of the continents – which in turn inspired the poet Varro Atacinus to write a descriptive geography of his own.[21] Cicero, as well, contemplated writing a geographical work, and even mentions that he had been reading Eratosthenes in preparation;[22] he appears, however, to have given up not long after having started, claiming that he was ill-suited to the task.[23] The only major geographical poem from antiquity to survive in its entirety is the *Orbis descriptio* of Dionysius Periegetes, probably composed in the time of Hadrian.[24]

Literary geography was hardly restricted to poets. The beginning of the first century CE would have seen the publication of Strabo's exhaustive *Geographia*, which, for all its engagement with the mathematical tradition, is equally concerned with topography, ethnography and history. More succinct prose encapsulations of the world may be found in the work of Pomponius Mela[25] – which favours straightforward regional description over geographical theory – and in the four geographical books

---

[19] See, for instance, Strabo I.ii.3, 7, 12 and 17.
[20] See, in particular, Strabo I.i.2, although the presence of Homer looms large throughout the entire first book.
[21] Alexander's poem is attested in Strabo XIV.i.25 and in a letter from Cicero, but the poem itself has not survived. Cicero is uncomplimentary towards the poet, but admits that the poem is not without its uses: *neglegentis hominis et non boni poetae sed tamen non inutilis*. (*Ad Atticum*, ed. D. R. Shackleton Bailey (Cambridge University Press, 1965–70), II.xxii.7). Varro's poem has survived only in fragments, which are collected in A. S. Hollis (ed. and trans.), *Fragments of Roman Poetry c.60 BC–AD 20* (Oxford University Press, 2007).
[22] Cicero, *Ad Atticum* II.iv.3 and II.vii.1.
[23] *et hercule sunt res difficiles ad explicandum et* ὁμοειδε ς *nec tam possunt* ἀνθηρουραφε σθαι *quam videbantur*. Cicero, *Ad Atticum* II.vi.1.
[24] *Geographi Graeci Minores*, vol. 2, ed. C. Müller (Paris: A. Firmin Didot, 1861), 104–76.
[25] The most recent edition is *Chorographie*, ed. A. Silberman (Paris: Les Belles Lettres, 1988), although see also *Pomponii Melae De Chorographia libri tres*, ed. P. Parroni (Rome: Edizioni di storia e letteratura, 1984).

at the beginning of Pliny's *Natural History*. Of course, these different literary geographies would have served very different purposes: Strabo's exhaustive work, like those of Eratosthenes and Polybius before him, would have been written for specialists – historians, perhaps, who required accurate information about distant places.[26] The shorter, more schematic works like those of Pomponius Mela and Dionysius Periegetes would have probably been used to provide a basic geographical education.[27] The geographical books of Pliny, on the other hand, are neither usefully brief nor mathematically rigorous; although they contain some distances and a few historical digressions, they are primarily a summary of the world's more important topographic contents.

An emergence of literary geography in the Roman world during the first century BCE and the first century CE should hardly come as a surprise. Under Julius Caesar and, later, Augustus, the limits of the Roman world were being continually expanded and, for perhaps the first time in all of human history, the *oikoumene* was on the cusp of being known in its entirety, not merely from poetry, but from exploration and observation. Strabo, who criticizes his predecessors for having a hazy knowledge of the more distant places, conveys in his own work a sense of at last being able to write about the whole world with some authority.[28] There seems little doubt that the expansion of Roman territory would have opened up geographic possibilities and that, in a sense, geography was inseparable from conquest. We know, for instance, that sightings and measurements were taken in the military field, and we are informed that, in the time of Augustus, Marcus Agrippa was in possession of a vast array of geographical information compiled during his tours of the nascent empire.[29]

---

[26] Geography was important to the historian of antiquity: both Polybius and Ephorus are said to have written geographical works to complement their histories (Strabo VIII.1.1); Strabo himself wrote an extensive work of history which has not survived (Strabo I.1.22–3).

[27] The poem of Dionysius found use as an educational tool almost immediately: it was twice translated into Latin during antiquity, once by Rufus Festus Avienus in the fourth century and again by Priscian in the sixth (see *Geographi Graeci Minores*, ed. K. Müller (Paris: Firmin-Didot, 1855–61), ii, xxix–xxxii), and it remained known in the original Greek as late as the twelfth century. Less is known about the early use of Mela's text, although it seems probable that it was devised for educational use: see the differing views of A. Silberman, 'Le premier ouvrage latin de géographie: la Chorographie de Pomponius Méla et ses sources grecques', *Klio* 71 (1989), 571–81; R. Batty, 'Mela's Phoenician Geography', *Journal of Roman Studies* 90 (2000), 70–94.

[28] See, for instance, Strabo II.i.41, where he claims that Eratosthenes was largely ignorant of distant regions such as Iberia and Celtica. Eratosthenes offers a similar criticism of Homer.

[29] A reference to measurement-taking may be found in Caesar, *Belli Gallici* V.xiii.4 and the presence of military surveyors is well attested in the epigraphic record. For an overview, see R. K. Sherk, 'Roman geographical exploration and military maps', in *Aufstieg und Niedergang der römischen Welt* II.1 (Berlin: De Gruyter, 1974), 534–62.

In Strabo, in Mela and in Pliny, there is a strong sense that the increased knowledge available to them was due, in no small part, to the increased frontiers of the Roman world.

While we may wish to suggest that the tradition of mathematical geography was pursued more avidly by Greeks, and that literary geography flourished more in the hands of Latin authors, such a statement is inevitably misleading. In reality, it is impossible to separate these two strands of geographical enquiry. Eratosthenes accompanied his more purely mathematical works with works of descriptive geography; and Strabo, while considering himself an heir to the mathematically inclined works of Eratosthenes and Hipparchus, produced the greatest literary geography to survive from the classical world. If anything Ptolemy's *Geographia* may seem at odds with the tradition in its complete rejection of topographic and ethnographic description. This absence may, however, have been intentional, a reaction against what Ptolemy perceived as an unwelcome bifurcation in the science: the literary geographies to come out of the Roman world were largely uninterested in the theoretical and scientific underpinnings of the discipline and, thus, in writing a text which removed the description and focused purely on the mathematical, Ptolemy may have been attempting to reassert what he perceived as the true basis of geography. However, Ptolemy's complaint may not have been directly aimed at the emphasis of descriptive over mathematical, but rather, at the use of purely descriptive material as a source for cartographic representation. For Ptolemy, the shape and content of the world was immutable and objective; the job of the geographer was to reproduce this shape as accurately as possible. However, as we shall see, a parallel cartographic tradition based on subjective experience rather than detached observation may have started not only to undermine the mathematical exactitude for which Ptolemy and his predecessors had worked, but also to change the perceived shape of the world itself.

Chorography: the secret evolution

Ptolemy defines the practice of geography as the imitation of the world by means of drawing, but it is clear that the graphical representation must be derived from a foundation of mathematical observation. In order to create a clear sense of the discipline, Ptolemy defines geography by comparing it with chorography, another form of cartographical expression which is characterized as being generally unscientific in nature. Chorography, he tells us, deals with the particular contents of a local landscape – he lists towns, harbours and tributaries – and is generally restricted to a smaller region; as a practice it requires no particular scientific knowledge and thus it need only be undertaken by a man capable of drawing or painting.[30] Both geography and chorography involve an element of cartographical representation; what differentiates the two disciplines, according to Ptolemy, would seem to be the level of detail, the scope of the artefact and the skill of the craftsman.

On the surface, Ptolemy's assessment would seem to agree with the etymology of the word. The constituent parts of chorography – χῶρος or χώρα, plus the verb γραφέω, to write or draw – suggest a kinship with the related descriptive practices of topography, geography and cosmography. The first part of the word, however, presents us with something of a grey area: χώρα often refers to land or country in an indefinite, non-political sense, while χῶρος is neither as specific as a local place (τόπος) nor as all-encompassing as γῆ or κόσμος. It has been recently suggested that χῶρος may be understood in a similar way to the modern concept of 'landscape', that is, an area of land that can be apprehended and experienced by an observer.[31] The defining characteristic of χῶρος, however, seems to be its limits. The word appears most often in classical sources to describe a space which is finite and bounded, land which need not necessarily be perceived in a single glance, but whose extent is nonetheless known. Etymology, however, will only take us so far, and to get a sense of the word, we must look at how it was employed.

Chorography – in its various Greek and Latin forms – appears sporadically in the written record from the second century BCE through to around the sixth century, after which there are only a handful of appearances until the incorporation of the word into the early modern geographical vocabulary. The earliest attested usage – from a long inscription found at Magnesia on the Maeander – may be reliably dated to 139 BCE.[32] Here the word χωρογραφιῶν is used to mean the plans of an area, in which that area may be taken in at a single glance, a usage which would seem to conform to the Ptolemaic definition. A number of inscriptions from the first century CE attest the presence of chorographers as an occupation within the military: one from Pselcis in Egypt (dated to 35 CE) lists a soldier who worked as a chorographer,[33] and a funerary inscription from Verona tells us of a soldier who was a chorographer and engraver

[30] ὅθεν ἐκείνη μὲν δεῖ τοπογραφίας, καὶ οὐδὲ εἰς ἂν χωρογραφήσειεν, εἰ μὴ γραφικὸς ἀνήρ. Ptolemy, *Geog.* I.i.5.
[31] Olwig, 'Has "geography" always been modern?', 1843–61.
[32] W. Dittenberger, *Sylloge Inscriptionum Graecarum*, 3rd edn. (Leipzig: S. Herzelium, 1917), 685.71. The inscription is dated to 139 BCE.
[33] R. Cagnat, *Inscriptiones Graecae ad Res Romanas pertinentes* (Paris: E. Leroux, 1911), I.1365.

(*chorographiarum item caelatori*).[34] Another funerary inscription from Carthage lists the occupation of one soldier as *chorograpus* (sic),[35] and in fragmentary tax records from the second century BCE, we find further evidence for the presence of chorographers.[36]

The epigraphic evidence for chorography suggests that the chorographer was responsible for both description and representation of the land. We must imagine military chorographers collected topographic details (most probably in writing) that would help to shape perceptions of newly explored lands. In some cases – the soldier from the Verona inscription being one example – the chorographer was responsible for creating a pictorial representation from information gathered in the field. The role of some chorographers, however, may have been limited to textual assessment: Diogenes Laertius, for instance, informs us of a chorographer named Archelaus who described the lands crossed by Alexander.[37]

Many instances of the word in Strabo suggest that chorography was a textual rather than pictorial undertaking. Strabo quotes Polybius as saying: 'but I [...] shall show the facts as they now are, as regards both the position of places and the distances between them; for this is the most appropriate function of chorography'.[38] Interestingly, Polybius' understanding of the word, and his insistence on position and distance, do not sound too far removed from Ptolemy's assessment of geography. However, when Strabo uses chorography as part of his contemporary vocabulary, it is often used to denote a textual rather than a pictorial document: he is critical of a description in Apollodorus, saying that 'it is ignorance and not knowledge of chorography to call such a four sided figure triangular. Yet he published a chorography in comic metre called *A Description of the Earth*'.[39] At one point Strabo also refers to his own work as a chorography,[40] further emphasizing the sense of descriptive text. In books five and six, Strabo also draws upon a specific chorographer (referred to simply as ὁ χωρογράφος) as an authority for distances

---

[34] Originally published in *Epigraphica* 7 (1945), 35–38. For a discussion of the inscription, see Nicolet, 'De Vérone au Champ de Mars', whose proposed reading I have followed here.

[35] Corpus Inscriptionum Latinarum VIII, 12914.

[36] J. G. Tait and C. Préaux, *Greek Ostraca in the Bodleian Library at Oxford*, vol. 2 (London: Egypt Exploration Society, 1955), nos. 1725, 1738 and 1759.

[37] Diogenes Laertius II.17. (*Diogenis Laertii Vitae philosophorum*, ed. H. S. Long (Oxford: Clarendon Press, 1964)). This Archelaus is mentioned only in passing, lest we confuse him with the fifth-century BCE philosopher.

[38] ἡμεῖς δέ, φησί, τὰ νῦν ὄντα δηλώσομεν καὶ περὶ θέσεως τόπων καὶ διαστημάτων; τοῦτο γάρ ἐστιν οἰκειότατον χωρογραφίᾳ. Strabo X.iii.5. The translation quoted is that of H. L. Jones.

[39] Strabo XIV.v.22.   [40] Strabo VIII.iii.17.

in Italy and Sicily;[41] in one instance, a measurement is taken directly from the chorographer's document (τῇ χωρογραφίᾳ).[42] It has traditionally been assumed that Strabo's chorographer is none other than Agrippa, and that the document is Agrippa's famed world map, built in the Porticus Vipsania in Rome. This is probably not the case: it has been argued that Strabo had compiled most of his text by 7 BCE and thus would not have had access to Agrippa's map, which was probably completed in 2 BCE.[43] Even if Strabo was using information from Agrippa, the source appears to be textual rather than pictorial.

There are several passages in Strabo, however, where chorography appears to refer not merely to a pictorial document, but to a representation of the entire *oikoumene*. In the second book, where he describes the projection of the inhabited world onto a flat surface, he tells us how the sea, most of all, but also rivers and mountains give shape to the *oikoumene*:

It is through such natural features that we gain a clear conception of the continents, nations, favourable positions of cities and all the other diversified details with which our chorographical panel (ὁ χωρογραφικὸς πίναξ) is filled.[44]

The word πίναξ, meaning a board or flat surface, carries with it the implication of a pictorial artefact and is known to have been used in a distinctly cartographical sense, as in Herodotus, where πίναξ refers to an image of the world engraved in bronze.[45] Elsewhere, Strabo uses it in his discussion of Eratosthenes, making reference to a 'panel of the *oikoumene*' (τὸν τῆς οἰκουμένης πίνακα) or, more confusingly, to the ancient geographical panel (τὸν ἀρχαῖον γεωγραφικὸν πίνακα) which Eratosthenes hopes to revise.[46] Was there, in fact, a difference between the geographical and the chorographical image in Strabo's mind, or was he using the qualifiers interchangeably? If anything, the evidence from Strabo suggests that chorography was a term which could be applied to a variety of geographical pursuits. It could refer to a textual document which sought to enumerate the contents of a particular area, but it could also refer to a pictorial representation of those contents. Moreover – and here is where

---

[41] Strabo V.ii.8, VI.i.11, VI.ii.11 and VI.iii.10.   [42] Strabo VI.ii.1.

[43] See E. Pais, *Ancient Italy* (University of Chicago Press, 1908), 384ff. and R. Syme, *Anatolica: Studies in Strabo* (Oxford: Clarendon Press, 1995), 356–67. An argument for a post-Augustan dating may be found in D. Dueck, 'The date and method of composition of Strabo's "Geography"', *Hermes* 127 (1999), 467–78.

[44] διὰ γὰρ τῶν τοιούτων ἤπειροί τε καὶ ἔθνη καὶ πόλεων θέσεις εὐφυεῖς ἐνενοήθησαν καὶ τἆλλα ποικίλματα, ὅσων μεστός ἐστιν ὁ χωρογραφικὸς πίναξ. Strabo II.v.17. The translation is adapted from Jones, who misleadingly renders the final words as 'geographical map'.

[45] Herodotus V.49; πίνακες of a cartographic nature also appear in Diogenes Laertius Vit. Phil. V.51.

[46] Strabo II.i.1 and II.i.2.

our idea of chorography begins to diverge from Ptolemy's more rigid definition – chorography could display the contents of a single area, but it could also encompass the entire known world.

Further evidence for the existence of world chorography may be found in a passage in Vitruvius – who, like Strabo, would have been writing in the latter half of the first century BCE. When describing a theory that rivers travel underground, he backs up the point by saying: 'Evidence of this may be found in the sources of rivers as they are depicted and described in chorographies of the world'.[47] There can be little doubt that the chorographies he is referring to not only encompass the whole of the *oikoumene*, but are also pictorial in nature. Indeed, there is a vaguely discernible tradition of what we might term 'vernacular maps' in the Roman world. These pictorial representations probably would have taken the form of wall painting and would have contained some degree of cartographical information. Livy, for instance, describes a tablet dedicated by Tiberius Sempronius Gracchus in 174 BCE, on which was inscribed 'the shape of Sardinia, along with paintings of battles which took place there'.[48] Although the Latin word *forma* can be used to mean 'map', Livy may not have been describing a strictly cartographic document; the attempt to situate historical events within the representative framework of a local landscape does, however, suggest that it may have had some cartographic ambitions. Elsewhere, Varro's manual of agriculture features a scene in which the author arrives at a temple to find his relatives looking at a 'painting of Italy' on the wall.[49] The subsequent conversation – which makes direct reference to Eratosthenes – suggests that the painting may have been conceived with cartographic representation in mind. Although the examples provided by Livy and Varro may depict a regional space – Sardinia and Italy respectively – the passage in Vitruvius allows for the possibility that such representations could, in fact, be expanded to take in a much larger area.

This proposed tradition of world chorography in the time of Vitruvius and Strabo may well have reached its culmination in the so-called 'map of Agrippa'.[50] The map itself has not survived, but we know that it was

---

[47] Haec autem sic fieri testimonio possunt esse capita fluminum, quae orbe terrarum chorographiis picta itemque scripta ... Vitruvius, *De Arch.* VIII.ii.6 (Vitruvius, *De architectura*, F. Granger (ed.) (Cambridge, Mass.: Harvard University Press, 1931)).

[48] *Sardiniae insulae forma erat, atque in ea simulacra pugnarum picta.* Livy XLI.xxviii.10.

[49] *spectantes in pariete pictam Italiam.* Varro, *Rer. Rust.* I.ii.1.

[50] On which, see: J. J. Tierney, 'The map of Agrippa', *Proceedings of the Royal Irish Academy* 63 (1963), 151–66; Nicolet, *Space, Geography and Politics*, 95–122; O. A. W. Dilke, 'Itineraries and geographical maps in the early and late Roman Empires', in *The History of Cartography*, vol. 1: *Cartography in Prehistoric, Ancient, and Medieval Europe and the Mediterranean*, J. B. Harley and D. Woodward (eds.) (University of Chicago Press, 1987), 207–9.

located in the Porticus Vipsania, that it was built by Agrippa's sister and Augustus sometime between 7 and 2 BCE, and that it was based on information collected by Agrippa. Pliny used it as a source for his geographical chapters, but stops to mentions it only when he discovers a particular figure to be inaccurate.

Indeed, Agrippa was, in many ways, a scrupulous man and devoted great attention to his work. Thus, when he was about to display [an image of] the whole world for Rome to see, who would believe that both he and the deified Augustus made a mistake? For it was Augustus who completed the portico which had been started by Agrippa's sister from the specification and commentaries of Agrippa himself.[51]

Dio, who does not specifically mention the map, tells us that the portico was located in the Campus Agrippae – remains are thought to have been found on the east side of the Via del Corso – and that it was still incomplete by 7 BCE.[52] Although attempts have been made to reconstruct the distances and figures presented in the map, drawing on evidence from Strabo, Pliny and geographical fragments from Late Antiquity, the form of the map remains completely unknown.[53] We do not know, for instance, if the map was carved – like the later *Forma urbis romae* – or painted. We also don't know if the map presented an Eratosthenean *oikoumene* about twice as wide as it was tall, or if, like the later Peutinger map, it stretched out to fill an entire wall of a portico.[54] For that matter, we cannot even be sure if the map contained the figures which Pliny found so useful, or if these were recorded in a separate textual document.[55] We may, however, suggest that the map was an attempt to display, on a large scale, the contents of the world with an authority that was not previously available: the authority of experience.

If χῶρος is a piece of land whose extent is known, and if chorography is the elaboration – either textual or pictorial – of that land, then logically,

---

[51] *Agrippam quidem in tanta viri diligentia praeterque in hoc opere cura, cum orbem terrarum urbi spectandum propositurus esset, errasse quis credat et cum eo divum Augustum? is namque conplexam eam porticum ex destinatione et commentariis M Agrippae a sorore eius inchoatam peregit.* Pliny, *Nat. Hist.* III.i.17.

[52] Dio LV.viii.4.

[53] The most extensive reconstruction is A. Klotz, 'Die geographischen commentarii des Agrippa und ihre Überreste', *Klio* 24 (1931), 386–466.

[54] Cf. Tierney, 'Map of Agrippa', 152; Dilke, 'Itineraries and geographical maps', 208; Nicolet, *Space, Geography and Politics*, 108; P. Trousset, 'La "carte d'Agrippa": nouvelle proposition de lecture', *Dialogues d'histoire ancienne* 19 (1993), 137–57.

[55] It has even been suggested that the 'map of Agrippa' was actually a monumental inscription; that is, a purely textual document with no cartographic accompaniment; see K. Brodersen, *Terra Cognita: Studien zur römischen Raumerfassung* (Hildesheim: G. Olms, 1995, 2nd edn. 2003), 275ff.

a chorography of the world would be possible only once the limits of the world itself had been properly assessed. By representing the world not according to mathematical speculation, but according to information collected first-hand, Agrippa's map – and those world chorographies that may have preceded it – may be understood as an affirmation that the limits of the world were no longer hypothetical, for their existence had been proven by experience. However, the image of the world based solely on chorographic sources may have also represented a point of divergence in the geographic tradition. Instead of populating an immutable shape with known contents, the contents may have started to dictate the shape of the land. Ptolemy warns that the display of gathered information (ἐφοδευόμενα, or information which has been collected through exploration) will result in an unbalanced map, with some areas crowded with description and others virtually empty. 'To get around this', he tells us, 'a great many are often compelled, on account of the panels themselves, to distort the measurements and the shapes of the lands.'[56] Ptolemy specifically cites geographers who have drawn Europe larger than Asia or Libya, simply because there was more information to be accommodated. If Ptolemy's criticisms are true, it would seem that the desire for chorographical information had superseded the desire for geographical accuracy.

Traditions and mythologies: chorography in Late Antiquity

Although the seeds of world chorography may have been planted in early imperial times, it is in Late Antiquity that we can best sense the presence of a chorographic tradition. Apart from the instance in Vitruvius, the majority of our reliable instances of the word in Latin can be dated to the fourth to the sixth century, although evidence from manuscript traditions allows us to suggest that the word may have still been employed and understood as late as the tenth century. During the late antique period, chorographies of a textual nature were used as sources for commentaries. Lactantius Placidus, probably writing in the late fourth century, makes reference to one in his commentary on the *Thebaid* of Statius.[57] Servius' commentary on Virgil, likewise, contains numerous references to chorography, although these point mostly towards textual, rather than

---

[56] Ὅπερ οἱ πλεῖστοι περιῖστάμενοι πολλαχῇ διαστρέφειν ἠναγκάσθησαν τά τε μέτρα καὶ τὰ σχήματα τῶν χωρῶν ὑπὸ τῶν πινάκων αὐτῶν. Ptolemy, *Geog.* VIII.1.2.

[57] Lactantius Placidus, *In Statii Thebaida commentum*, R. D. Sweeney (ed.) (Stuttgart: Teubner, 1997), II.34.

pictorial sources. In one instance he verifies his point with evidence he has read in the works of the chorographers;[58] elsewhere he draws upon a chorography to back up a historical, rather than a topographical point.[59] There is also, however, a passage in which he talks about 'chorographers and geometers who say that the world is spherical', which suggests that chorography, as a discipline, would still have been concerned with visualizing the world on a large scale.[60] Servius also mentions Pomponius Mela by name, and it seems probable that Mela's text would have been used as a source.[61] Known for a long time as either the *Cosmographia* or *De orbis situ*, the ninth-century manuscript which contains the earliest version of Mela's text actually bears the title *De chorographia*.[62]

While we may only speculate about what name the text may have had in Late Antiquity – Mela never employs the word chorography himself – there are other instances of earlier texts becoming known as chorographies. The sixth-century grammarian Priscian refers to the poem of Varro Atacinus as *De chorographia* and elsewhere attributes that title to a work by Cicero.[63] Another chorographical text from Late Antiquity, which is attested but has not survived, is the 'Chorography of the World' (Χωρογραφία οἰκουμενική) written by Pappos of Alexandria, a philosopher who lived in the time of Theodosius I.[64] While it is impossible to determine what this work may have contained, the title alone is enough to contribute to our idea of late antique chorography as a practice with global rather than regional ambitions.

Interestingly, in the manuscript traditions of both Mela and Priscian, the word chorography appears in manuscripts up to around the tenth or eleventh centuries, after which it becomes corrupted in a number of intriguing ways. In manuscripts of Mela, it is replaced with *cosmographia* – a word ill-suited to a description of lands – and in Priscian it becomes *chronographia* and even *hortographia*. Such transcriptions suggest that, at some point, the word must have fallen completely out of common use within the Latin West. This may not, however, have been the case in the

---

[58] Servius, *Ad Aen.* (*Servii Gramatici qui feruntur in Vergilii carmina commentarii*, ed. G. Thilo and H. Hagen (Leipzig: Teubner, 1881–1902), III.104.

[59] Servius, *Ad Aen.* IV.42.

[60] *secundum chorographos et geometras, qui dicunt terram σφαιροειδῆ esse.* Servius, *Ad Aen.* VI.532.

[61] Servius, *Ad Aen.* IX.30.

[62] Vat. Lat. 4929. For a discussion of the manuscript tradition, see P. Parroni *Pomponii Melae*, pp. 55–81.

[63] Priscian (*Grammatici Latini*, vol. 2, ed. H. Keil (Leipzig: Teubner, 1855–80)), III.25 (Varro) and VI.82 (Cicero).

[64] The work is mentioned in the tenth-century *Suidae Lexicon*, 4 (Leipzig: Teubner, 1935), 26.

Byzantine East, where, as late as the twelfth century Eustathius of
Thessalonica was able to provide a Ptolemaic definition of chorography
in his commentary on the geographical poem of Dionysius Periegetes.[65]
While the word itself may not appear with great frequency, there is
a number of surviving geographical texts from Late Antiquity which
display a chorographical approach. The *Cosmographia Iulii Caesaris*[66] –
a fragment which can be dated to no earlier than the fifth century – even
goes so far as to situate its approach to geographical collection within the
context of early imperial Rome.[67] The text opens with the line: 'In the
consulship of Julius Caesar and Mark Antony, the whole world was
travelled by four chosen men of great wisdom'.[68] The four men are said
to have worked for between twenty and thirty years each and at the end
were able to report that the world contained 28 seas, 74 islands, 35
mountains, 70 provinces, 264 towns, 52 rivers and 129 peoples.[69]

The idea that an emperor would send out four men (one for each
cardinal point) rather than three (one for each continent) immediately
strikes one as anachronistic: while the cardinal points would have been
important within land surveying and theoretical geography, they may not
have had as much relevance to the prevailing concept of the world as an
island of three continents. Furthermore, if the wise men had been commissioned by Caesar, he would not have lived to see the results, as the
consulship in question would have been 44 BCE, the year of his death. If
these wise men did, in fact, exist, it is possible that their notes would have
made their way into Agrippa's files, or could have been turned into one of
the pre-Agrippan world chorographies described by Vitruvius. It is also
possible that the wise men were commissioned by Augustus, and not
Caesar: in the thirteenth-century Hereford *mappamundi*, there is a depiction of a similar event where Augustus is the emperor and the wise men
are three in number.[70] In fact, what seems most probable is that the event

---

[65] *Geographi Graeci Minores*, Müller (ed.), ii, 212.
[66] *Geographi Latini Minores*, A. Riese (ed.) (Heilbronn: Henningeros fratres, 1878), 21–3.
   On the dating, see xxi.
[67] For an introduction to this text, see C. Nicolet and P. Gautier Dalché, 'Les "quatre
   sages" de Jules César et la "mesure du monde" selon Julius Honorius: réalité antique et
   tradition médiévale', *Journal des Savants* 4 (1986), 157–218.
[68] *Iulio Caesare et Marco Antonio consulibus omnis orbis peragratus est per sapientissimos et electos
   viros quattuor. Geographi Latini Minores*, Riese (ed.), 21.
[69] The length of time spent by each of the men is provided in the text, but the figures are
   most probably incorrect. See Nicolet, *Space, Geography and Politics*, 96–7 and Dilke,
   Greek and Roman Maps, 40.
[70] It has been argued that the presence of Augustus may have come from a misreading
   of Luke 2.1; see T. P. Wiseman, 'Julius Caesar and the Hereford World Map', *History
   Today* 37 (1987), 53–7; A. Hiatt, 'The cartographic imagination of Thomas Elmham',
   Speculum 75 (2000), 859–86.

and the wise men are fabrications from the late antique period, a mythology perhaps based on several dimly remembered traditions of imperial land assessment. However, the truth of the event is perhaps less important to us than the fact that a fifth-century author would situate the origin of geographical enquiry in an imperially sanctioned project of collection and recording. It suggests that, by Late Antiquity, the desire to conceive of the world may have become separated from the notion of mathematical speculation. Instead, the very idea of knowing the world was bound to the contents found within.

Another geographical work from Late Antiquity which looks to an imperial origin is the *Divisio orbis terrarum*, traditionally dated to the fifth century CE.[71] It opens with the assertion that the world is divided into three parts and, moreover, that it was Augustus 'who first displayed this fact by means of chorography'.[72] Here, once again, we have the exposition of the world linked to an age of conquest and expansion. More importantly, we are reminded of chorography's potential for pictorial representation: the contents of the world were to be looked at, as well as merely recorded. The text of the *Divisio* presents some distances, but it is far more concerned with topographic boundaries; that is, the physical characteristics which would have defined the various provinces. The descriptions are systematic: each cardinal point (starting with east and west, followed by north and south) is associated with an ocean, a sea, a river, a mountain range or a desert which acts to delimit the region in question. The possibility for displaying this gathered information in pictorial form – as well as the imperial subtext of such chorographic undertakings – is made apparent in the twelve lines of verse found at the end of the *Divisio*. The verses refer to an 'exceptional work' commissioned by Theodosius, 'in which the substance of the world is represented, in which the surface of the sea, the mountains, the rivers, harbours, straits and cities are marked'.[73] The verses also give us a small

---

[71] *Geographi Latini Minores*, Riese (ed.), 15–20. Other editions, based on different manuscripts, exist, notably that of P. Schnabel in *Philologus*, 90 (1935), 432–40. The dating comes from a reference to the fifteenth consulship of Theodosius. If we read this as Theodosius II it would place the composition after 435 CE. See Dicuil, *Liber de Mensura*, 24. The association with Theodosius II is commonly accepted but has been challenged; see N. Lozovsky, 'Maps and panegyrics: Roman geo-ethnographical rhetoric in Late Antiquity and the Middle Ages', in *Cartography in Antiquity and the Middle Ages: Fresh Perspectives New Methods*, R. J. A. Talbert and R. W. Unger (eds.) (Leiden: Brill, 2008), 173, n. 13.

[72] *Quem divus Augustus primus omnium per chorographiam ostendit. Geographi Latini Minores*, Riese (ed.), 15.

[73] *Hoc opus egregium, quo mundi summa tenetur, aequora quo, montes, fluvii, portus, freta et urbes signantur ... Geographi Latini Minores*, Riese (ed.), 19–20.

insight into the creation of the work, claiming that one person wrote while the other painted.[74]

Like Agrippa's map, the chorography of Theodosius would have been a statement of imperial knowledge, but it would have also been an advance on previous works. The verses claim that some of the measurements came from an 'old monument' – possibly Agrippa's map – but that some were improved upon for the new version. Several centuries later, the Irish geographer Dicuil, working in the court of Charlemagne, would claim to value the measurements of Theodosius over those of Pliny, suggesting that Pliny's figures have been corrupted over time by careless scribes.[75] More importantly, the chorography of Theodosius makes explicit once again the connection between the collection process and the pictorial representation. However, while the *Divisio* suggests that the collected information was later transformed into its pictorial counterpart, there is also evidence to suggest that chorographic maps could, in turn, have functioned as the source for textual expositions of the world. The *Excerpta eius sphaerae* is another fragmentary geography from the fifth century which is thought to be related to the *Cosmographia Iulii Caesaris*, and which may have been copied from a text by Julius Honorius.[76] About Honorius we know little: the student who has copied the text refers to him as a wise teacher (*magister peritus*) and he is referred to as a reliable source by Cassiodorus.[77] Whether or not the surviving text provides an accurate version of Honorius, it does inform the reader that the text should not be separated from its visual accompaniment.[78]

It is not unreasonable to suggest that the *Excerpta* could have taken its information from a cartographic source. It provides us with a list of prominent topographical landmarks – seas, islands, mountains, towns and rivers – but makes little attempt to situate them within relation to one another. Of the various features, it is only rivers which are discussed in detail: the text informs us where they begin, how far they travel (in miles) and where they end. It seems improbable that such a text could form the basis for a pictorial document, but it is easy to imagine the information

---

[74] *dum scribit pingit et alter.* An argument could be made for reading this as 'one drawing (i.e. outlines) and the other colouring (or adding details)'.

[75] Dicuil, *Liber de mensura*, 2–3.   [76] *Geographi Latini Minores*, Riese (ed.), 24–55.

[77] Ibid. 55 and Cassiodorus, *Inst.* I.xxv.1. For discussion on the transmission of Honorius, see Thomson, *Ancient Geography*, 381; Nicolet and Gautier Dalché, 'Quatre sages', 157–218; N. Lozovsky, *'The Earth is Our Book': Geographical Knowledge in the Latin West ca. 400–1000* (Ann Arbor: University of Michigan Press, 2000), 16–17; A. D. Lee, *Information and Frontiers* (Cambridge University Press, 1993), 83–4.

[78] *Hic liber exceptorum ab sphaera ne separatur. Geographi Latini Minores*, Riese (ed.), 55.

having been extracted from such a document. The form and content of such a potential late antique cartography is suggested in a panegyric by Eumenius from the third/fourth century, who desires that the children in his native Autun should have access to a portico map so that they might see not merely the lands and seas, but also the names of all places, the source and destination of the rivers, and the shape of the coastline as determined by the ocean.[79] Eumenius imagines the map as having a primarily educational function – it will allow people to learn place names – but he is also conscious of the imperial subtext: the portico map will provide a representation of the cities, peoples and nations which have been collected into the Roman world.

Between Eumenius, the *Excerpta* and the verses in the *Divisio*, we may start to arrive at some conclusions as to what the chorographic map may have contained. There would have been isolated topographic features, such as cities and towns, as well as what we might call linear features – forests, rivers, mountain ranges and coastlines – which would have delimited the boundaries of various regions. Although this information would have been situated within some tripartite organization of the *oikoumene*, it would not necessarily have been the *oikoumene* as envisaged by Eratosthenes or Ptolemy. Indeed, if the portico was the preferred location for such maps – as it was for those of Agrippa and Eumenius – the shape of *oikoumene* may well have stretched to fit the space and to allow more room for details.

Our major surviving pictorial artefact from antiquity – the Peutinger map[80] – would seem to conform quite well to the evidence for chorographic representation.[81] Its extraordinary elongation suggests a shape well-suited to a portico, and the contents presented within are familiar from the descriptions we've examined. We have a rough, very abstracted outline of the *oikoumene* created by oceans, seas and gulfs; within that space, there are mountain ranges, forests, the sources and destinations of

---

[79] *omnium cum nominibus suis locorum situs spatia intervalla descripta sunt, quidquid ubique fluminum oritur et conditur, quacumque se litorum sinus flectunt, qua vel ambitu congit orbem vel impetu inrumpit Oceanus.* Eumenius, *Pro Inst. Schol.* 20: *XII Panegyrici Latini*, ed. R. A. B. Mynors (Oxford: Clarendon Press, 1964), 242–3.

[80] For an introduction to this unique artefact, see most recently R. J. A. Talbert, *Rome's World: The Peutinger Map Reconsidered* (Cambridge University Press, 2010), which contains an extensive bibliography of earlier scholarship.

[81] The date of the Peutinger map, or rather, the archetype on which it is based, is the subject of much debate. For two very different perspectives, see P. Arnaud 'L'origine, la date de rédaction et la diffusion de l'archetype de la Table de Peutinger', *Bulletin de la Société Nationale des Antiquaires de France* (1988), 302–21; B. Salway, 'The nature and genesis of the Peutinger Map', *Imago Mundi* 57 (2005), 119–35; E. Albu, 'Imperial geography and the medieval Peutinger Map', *Imago Mundi* 57 (2005), 136–48.

rivers, and major cities.[82] Although we cannot, with any certainty, connect the Peutinger map to the imperial cartographies of Agrippa or Theodosius, the map does at least offer one possible model for late antique chorography. Ptolemy would almost certainly not have approved of the way that the map chooses to render the shape of the *oikoumene*, but as a document designed to display the contents of the entirety of the known world, the Peutinger map may well have suited the needs of its late antique viewers.

Towards a cartography of the imagination

In the first century BCE, the differences between chorography and geography may have been minimal. While both could exist as either a textual or a pictorial artefact, the tradition of geography may have concerned itself more with apprehending the true shape of the world, while chorography would have sought to catalogue its contents. Although chorography would have been associated with a high level of detail, this would not have precluded the possibility of representing a large area. Certainly the idea of world chorography does not seem to have been unfamiliar to the early imperial Romans. If chorography was a representation of known space – that is, space whose contents had been collected and recorded – then the 'world chorographies' described by Strabo and Vitruvius, and possibly as presented by Agrippa, may have become both possible and desirable during the initial period of imperial expansion. The geographer – that is, the person whose job was to create a representation of the world – would no longer have needed the mathematical basis in order to hypothesize the form and content of the world. The contents would have been experienced and recorded, and, from this empirical basis, an image of the world could take its shape.

When later writers suggest that Augustus was the first person to display the world by means of chorography, what they perhaps mean is that he was the first person to create a map large enough to hold everything which was, at that time, known about the world: all the rivers, all the mountains, all the cities. While the inclusion of so many details may have been precisely the value of such an artefact, the attempt to organize so

---

[82] The one element not mentioned in our sources, but present in the Peutinger map, is the road network. However, given the textual difficulties presented by the names along the roads, a reasonable argument could be made that the Peutinger map is, essentially, a chorographical representation of the world onto which roads and stations were later added, perhaps from a collection of itineraries. If we imagine the Peutinger map as consisting essentially of two independent layers – the chorography and the itinerary – it becomes much easier to arrive at a credible dating for the document as a whole.

much information may have compromised the formal accuracy which the Greek mathematical geographers had established. Ptolemy's criticisms of chorography may therefore have originated from a desire to reassert a mathematical basis of geography, which he saw to be on the decline. For him the absolute accuracy of the world's shape was more important than the pictorial representation of its perceived contents. However, while his assertions may eventually have come to define chorography in the early modern period – perhaps by virtue of being the only classical definition available – his assessment may not have reflected the reality of his own times.

Chorography, as it was practised in the Roman world, was not the process of translating the shape of the world onto an easily apprehended surface, but rather the collection and arrangement of significant features into a pictorial lexicon of the world's contents. Its accuracy came from the authority of experience and its accessibility from the representation of what was already understood, by consensus, to be true. So long as the individual elements are placed in some plausible relation to one another, the form of the world may reasonably take its shape around what is known or thought to be there. By extension, chorography could have provided a flexible framework for the inclusion of features whose existence could not be proven but which nonetheless remained a certainty for the people of the time. The Garden of Eden, for instance, is a common feature on medieval maps, but it is a topographical feature for which Ptolemaic coordinates would be impossible to find. Untethered from the constraints of projection, chorography's emphasis on the representation of perceived reality may have allowed space for the imagination to enter the cartographic process.

In this way, the chorographic approach to apprehending the world – that is, the approach which favours the transformation of a collection of textual sources into a subjective pictorial assessment – may have provided the perfect platform for cartographic developments in the medieval period. Certainly we may understand T–O maps and *mappaemundi*, which emerged in the centuries to come, as a continuation of a chorographic tradition which became dominant in Late Antiquity. In these later examples, an increasingly diagrammatic form allows for a more complete elaboration of content both known and believed. Although they may perhaps fall short of Ptolemy's standards, chorographic maps would not have seemed in any way inaccurate to their viewers. Indeed, by conforming to, rather than trying to dictate, what existed, chorographies may have seemed far more accurate than the abstract Ptolemaic shapes which bore little relation to that which could be perceived and understood by the observer on the ground.

Chorography, as a practice, may have outlasted the word itself. Based on the available evidence – mostly from the commentaries of Eustathius of Thessalonica – there is reason to suggest that both the text of Ptolemy and the concept of chorography were still understood in the Byzantine East throughout the medieval period. We cannot, alas, make similar claims for the Latin West, where the word may have disappeared completely sometime after the ninth century. The disappearance of the word, however, should not imply the disappearance of the practice. Without a strong Ptolemaic tradition in the Latin West there would have been no basis on which to differentiate between chorography and geography; and with no separate word to describe it, the classical tradition of chorography may thus have been subsumed into the broader geographical approach of the medieval period.

In the fifteenth century, when Ptolemy's teachings once again became widely known, so too were his particular biases about chorography rediscovered. Ptolemy's objective approach to geographical knowledge would have appealed to the rational spirit of the early modern period, while the more subjective, perceptual approach of chorography – an approach which had informed world-scale cartographic conception since imperial Roman times – would have been relegated to precisely the kind of local-area cartography to which the Ptolemaic definition had restricted it. The biases of Ptolemy seem to have ensured that chorography, as a classical tradition of geographical knowledge, has received little in the way of scholarly consideration. Furthermore, the idea that Ptolemy's texts were 'lost' and remained unknown for several centuries has fuelled the notion of a decisive break in geographical knowledge between the classical and medieval periods. In attempting to construct an alternative model of chorography based on non-Ptolemaic sources, we may better understand how this unique classical tradition may have acted as a direct precursor to the geographical mindset of the medieval world.[83]

[83] The author wishes to thank James Howard-Johnston and Veronica della Dora who read and commented on various drafts; Keith Lilley for his editorial guidance; James Ambrose and Patrick Wadden who offered insights and assistance; and finally, Andrew Merrills and Camille Serchuk for their helpful comments on the initial version of the chapter.

2 Geography and memory in Isidore's *Etymologies*

Andy Merrills

> I could not even speak of mountains or waves, rivers or stars, which are things that I have seen, or of the ocean, which I know only on the evidence of others, unless I could see them in my mind's eye, in my memory, and with the same vast spaces between them that would be there if I were looking at them in the world outside myself.[1]

It is difficult to overstate the importance of Isidore of Seville to the development of medieval geographical thought, and as a model for wider intellectual understanding of the period.[2] Ernst Curtius famously termed Isidore's *Etymologies or Origins* 'the basic book of the entire Middle Ages', and historians of geography and cartography would certainly concur with this assessment.[3] Slowly compiled during the 620s and 630s, Isidore's

---

[1] Augustine, *Confessiones*, William Watts (ed. and tr.), Loeb, 2 vols. (Cambridge, Mass.: Harvard University Press: 1989), X.8: *quod haec omnia cum dicerem, non ea videbam oculis, nec tamen dicerem nisi montes et fluctus et flumina et sidera quae vidi, et Oceanum quem credidi, intus in memoria mea viderem spatiis tam ingentibus quasi foris viderem: nec ea tamen videndo absorbui quando vidi oculis* (R. S. Pine-Coffin (tr.), *Saint Augustine: Confessions* (Harmondsworth: Penguin, 1961), 214).

[2] The research of Jacques Fontaine is fundamental on Isidore's *Etymologies*; see esp.: *Isidore de Séville et la Culture Classique dans l'Espagne Wisigothique*, 3 vols. (Paris: Études Augustiniennes, 1959); and the collected papers in *Tradition et actualité chez Isidore de Séville* (London: Ashgate, 1988). See now esp. Mark E. Amsler, *Etymology and Grammatical Discourse in Late Antiquity and the Early Middle Ages* (Amsterdam: J. Benjamins, 1989); Bernard Ribémont, *Les origines des encyclopédies médiévales. D'Isidore de Séville aux Carolingiens* (Paris: Champion, 2001); Davide Del Bello, *Forgotten Paths: Etymology and the Allegorical Mindset* (Washington, DC: Catholic University of America Press, 2007) and the provocative (and frequently amusing) study of John Henderson, *The Medieval World of Isidore of Seville: Truth from Words* (Cambridge University Press, 2007).

[3] Ernst Robert Curtius, *European Literature and the Latin Middle Ages*, W. R. Trask (tr.) (London: Routledge, 1953), 496. The standard edition of Isidore's text is W. M. Lindsay, *Isidorus Hispalensis Episcopi: Etymologiarum sive Originum libri XX* (Oxford Classical Texts, 1911); editions of individual books with facing translations (in a variety of modern languages) are emerging gradually from Budé – see specific references in the footnotes which follow. The excellent collaborative translation of Stephen A. Barney, W. J. Lewis, J. A. Beach and Oliver Berghof, *The Etymologies of Isidore of Seville* (Cambridge University Press, 2007), has made the complete text of Isidore's work available in English for the first time.

45

great text was important both as a deep reservoir of classical learning, from which later writers could draw, and as a model for the comprehension of this vast body of material. Consequently, Isidore's presence can be detected behind virtually every major Latin geographical composition of the Middle Ages, from Hrabanus Maurus and Bede in the eighth century to the great *mappaemundi* of the thirteenth and fourteenth. But what was the function of geography within Isidore's greatest work? How did descriptions of the physical world reflect his wider ambitions for the work? And how can we explain many of the peculiar features of his varied accounts of the wider world?

'Geography' was, of course, a nebulous concept in the early seventh century, and descriptions of the physical world could appear in a variety of surprising contexts.[4] It would be perfectly possible to construct a meaningful assessment of Isidore's 'geographical' understanding through the prism provided by (for example) his substantial grammatical *Sententiae* (or *Sentences*), or his narrative *History of the Goths, Vandals and Sueves*.[5] Important modern scholarship has highlighted the significance of Isidore's *De Natura Rerum* as a work of natural history, and the sober survey of physical geography in Book XIII of the *Etymologies* has also rewarded recent attention.[6] In a period in which 'geographical' subjects were most commonly viewed in their wider theological or rhetorical context, and in which theology and rhetoric could easily assume a 'spatial' dimension in turn, it is clearly impossible to provide a definitive assessment of 'Isidore's Geography'.[7]

Isidore's methodology also frustrates any attempt to focus directly upon his 'own' geographical understanding. Many of Isidore's most important passages are borrowed from elsewhere, sometimes verbatim.

---

[4] Natalia Lozovsky, *'The Earth is Our Book': Geographical Knowledge in the Latin West ca. 400–1000* (Ann Arbor: University of Michigan Press, 2000), demonstrates this point vividly.

[5] And attempts have been made. Compare Pierre Cazier, *Isidore de Séville et la Naissance de l'Espagne Catholique*, Théologie Historique, 96 (Paris: Beauchesne, 1994), on the importance of the *patria* within the *Sententiae*; on the *laus Hispaniae* which opens the longer recension of the *History of the Goths* see A. H. Merrills, *History and Geography in Late Antiquity* (Cambridge University Press, 2005), 170–228.

[6] Isidore de Séville, *Traité de la Nature*, ed. Jacques Fontaine, Bibliothèque de l'École des Hautes Étude Hispaniques, 28 (Bordeaux: Féret, 1960) and G. Gasparotto, *Isidoro di Siviglia, Etimologie: Libro XIII* (Paris: Belles Lettres, 2004), respectively provide text and translation of, and commentary on, *De Natura Rerum* and Book XIII of the *Etymologies*.

[7] Hans Philipp, *Die historisch-geographischen Quellen in den Etymologiae des Isidorus von Sevilla* (Berlin: Weidmann, 1912–13); Richard Uhden, 'Die Weltkarte des Isidorus von Sevilla', *Mnemosyne*, Third Series, vol. 3, Fasc. 1 (1935), 1–28, provide important discussions of Isidore's 'geographical' thought.

Consequently, when the scholar peers too closely at the *Etymologies*, the coherence of the text is often obscured by the plethora of allusions and citations that spring more readily into focus, whether the provincial geography of Orosius, the natural history of Pliny and Solinus, the scriptural exegesis of Augustine, Jerome and Hegesippus, or the literary expositions of Servius and the scholiasts. The implications of this cento or patchwork effect are considerable: Isidore can often seem oddly hesitant in stating his own understanding of the world, and anachronistic in his geographical perspective. If a seventh-century reader took the *Etymologies* at face value, he or she would have been confident that Pompeii was still standing, and that the Alans, Vandals and Goths were still holed up along the Danube frontier rather than dominant in the western empire (although Isidore does note that the Goths destroyed his hometown of Cartagena in 624).[8] Concerned as he was with written authority, particularly on geographical matters, Isidore is often difficult to isolate as a creative author in his own right.

The present study focuses upon four rather peculiar 'geographical' passages which appear at different points in the *Etymologies*. It will examine the way in which Isidore structured the geographical material available to him in these passages, consider the details with which he embellished them, and suggest why he adopted these approaches. From this it will argue that Isidore did articulate a view of the wider world that was very much his own, and that he regarded this material as being important enough for the dedicated student to commit it to memory. Briefly, the four important passages are: the summary of the different regions and provinces of the world that takes up most of Book XIV; the long list of *gentes* and *nationes* (and the languages that they speak) at IX.2; the adumbration of the world's cities and their founders at XV.1; and Isidore's description of the different rivers of the world at the end of Book XIII. Taken together, these passages illustrate rather well the point that 'geography' was not a single discipline within the early medieval world, and was put to a variety of different ends; having said that, it seems clear that Isidore did regard these four long digressions as a group. As I shall discuss, the ordering strategies behind these passages, and the specific nature of the descriptions included within them, can only be appreciated when the passages are considered together. I shall argue that Isidore created a consciously layered representation of the inhabited world,

---

[8] Isid. *Orig.* XV.1.51; IX.2.89, 92–101; XV.1.67. On these anachronisms cf. Lozovsky, *'The Earth is Our Book'*, 107–8.

partly for reasons of simple clarity, but also to allow his account to be more easily memorized – a point which may have some important implications for the understanding of geographical education in the early medieval world.

In some senses, of course, I have been rather selective in identifying these four passages for close reading. While I think it is likely that Isidore regarded these long geographical digressions as a group (if not to be read side by side), it is equally clear that he wished this material to be considered in other contexts. The long enumeration of peoples and languages at the start of Book IX, for example, clearly has an important geographical implication, but must also be read as a counterpoint to the detailed investigation of common family structures that concludes the same book: in this way the encyclopaedist presented human community at several different scales, like a series of nesting dolls.[9] Similarly, the enumeration of the world's cities at XV.1 needs to be considered alongside the discussion of the types of buildings and public spaces that could be found in a typical classical or late classical city which follows: here again, global and local perspectives are juxtaposed and the specific contrasted with the abstract.[10] In Book IX and in Book XV, geography was a valuable tool with which Isidore emphasized the intimate relations between things on a micro and macro level, and was able to highlight the order inherent within God's Creation. I do not wish to obscure these important themes. But I do wish to examine the order behind Isidore's geographical thinking, and to do that, the relevant passages do require special treatment.

Much of the discussion which follows is based upon two straightforward observations. The first concerns the strangely repetitive nature of Isidore's geography – over the course of his *Etymologies* he includes four distinct surveys of the known world, each of which is ecumenical in its breadth, and each of which is structured in the same way. In itself, this presentation of a fractured – yet consistent – world geography deserves some explanation. The second is the precise form which Isidore's geographical passages take, and particularly the supplementary detail with which the writer colours his account. Many of these elements are familiar features of Isidore's epistemological method, and reflect the

---

[9] Marc Reydellet, 'La signification du Livre IX des Etymologies: érudition et actualité', in *Los Visigodos, Historia y Civilizaciòn, Actas de la semana internacional de estudios Visigóticos*, Antigüdad y cristianismo, 3 (Murcia: Universidad de Murcia, 1986), 337–50; Marc Reydellet, 'Introduction', in *Isidore de Séville: Étymologies Livre IX: Les langues et les groupes sociaux* (Paris: Belles Lettres, 1989), 1–26.

[10] Jean-Yves Guillaumin and Pierre Monat (eds. and trs.), *Isidore de Séville. Étymologies. Livre 15. Les constructions et les terres* (Paris: Press universitaires de Franche-Comté, 2004).

peculiarly verbal nature of his own geographical understanding. When considered as a group, however, in the wider context of the fragmented fourfold geography, a more specific function comes to light. It will be argued here that Isidore's geographical palimpsest was intended as a structuring device – to aid his readers' navigation through the text, and to assist with the retrieval of information – and that this is most fruitfully viewed within the context of early medieval memorial practice. The fragmented geographical accounts, extraordinarily ordered descriptive method and striking illustrative passages all suggest that Isidore composed these geographical passages in the hope that they would be committed to memory.

The geographical palimpsest: or the same world, four times over

Each of Isidore's four surveys of 'world geography' – at IX.2, XIII.21, XIV and XV.1 – is dominated by a measured and systematic description of the constituent parts of the world. Crucially, each of these chapters follows the same itinerary within these descriptions: the continents of Asia, Europe and Africa are always treated in the same order, and the constituent parts of each are outlined in the same way. The list of provinces in Book XIV provides the clear paradigm for this. Isidore starts in India, and moves westwards through Asia as far as the Holy Land and Egypt. This is then followed by a second transect running from northern Asia and Scythia, through Hyrcania and Armenia to the different territories of Asia Minor. Europe forms the second part of his geography, starting in Scythia in the north-east, and moving westwards along the Danube and Rhine frontier as far as Gallia Belgica. The second section of the European geography starts in Greece and describes the provinces found there, before moving through the Balkans into Italy, thence to Gallia and finally Isidore's home territory of Hispania. The third part of the provincial geography considers Africa, and again takes the form of two broad transects. The first is westerly, starting in Cyrenaica and moving through Byzacena, Africa Proconsularis, Numidia and the Mauretanias; the second is (more ambiguously) easterly, and discusses the regions of the interior from Garama to the Aethiopias. The provincial geography ends with a discussion of the islands of the world.

The inspiration behind this ordering can only have been the provincial geography which appears in the second chapter of Orosius' *Historiarum adversus paganos libri septem* (*The Seven Books of History against the Pagans*). Orosius' *Historia* was composed *c.* 417 CE, ostensibly as a historical crib to Augustine's *De Civitate Dei*.[11] While Augustine seems to have regarded the work of his charge somewhat coolly, Orosius' *Historia* proved to be a great success in the Middle Ages, and his succinct geographical introduction – a sober summary of Roman knowledge with much implicit imperial chauvinism consciously expunged – was enormously influential. Isidore certainly knew this *Historia* well, and borrowed extensively from it. This is most immediately evident in the capsular form of Isidore's short provincial geographies, which bear a clear family resemblance to Orosius' style, but is also apparent through the preservation of certain conspicuous anachronisms. Surprisingly, Orosius' description of Hispania, for example, employs the Augustan, rather than the Diocletianic, provincial divisions, despite the fact that the writer grew up in the region.[12] Isidore does include the more recent provincial names within his account, but is at pains to preserve Orosius' division of Hispania Citerior and Hispania Superior alongside the more up-to-date political geography.[13] But there can be little doubt that the overall geographical order of Orosius' account was the single most important influence upon Isidore's work. The systematic continental survey of Book XIV was clearly taken from the fifth-century source, as many idiosyncrasies – including the unusual position of Egypt mid-way through the geography of Asia – demonstrate clearly. Isidore did slightly tidy up the provincial geography that he found in the *Seven Books of History* (he omits the long digression on the Caucasus, for example, and places his Nilotic geography elsewhere); moreover, he supplemented this material with information taken from a variety of other sources, including Solinus, Servius and Jerome. But there can be little doubt that Orosius was the principal influence behind the ordering of the geographies of the *Etymologies*.

Crucially, Isidore retained this systematic geographical order in each of the other descriptions of the world within his *Etymologies*. The discussion of the *nationes* in Book IX starts with the inhabitants of India, and moves west through Asia to the Holy Land and Egypt; reaching the south-western limit of Asia, Isidore then turns back to the north and discusses

---

[11] Merrills, *History and Geography*, 35–99, discusses Orosius and his influence. On the form of the geographical chapter see also Yves Janvier, *La géographie d'Orose* (Paris: Belles Lettres, 1982); Jocelyn Hillgarth, 'The *Historiae* of Orosius in the early Middle Ages', in *De Tertullien aux Mozarabes II: antiquité tardive et Christianisme ancien (VIe–IXe siècles): Mélanges offerts à Jacques Fontaine*, Louis Holtz and Jean-Claude Fredouille (eds.) (Paris: Institut d'études augustiniennes, 1992), 157–70. Isidore's use of the source is discussed at length in Philipp, *Die historisch-geographischen Quellen*, 31–4.

[12] Orosius, *Historia adversus paganos libri septem*, Marie-Pierre Arnaud-Lindet (ed. and tr.), *Orose: Histoire contre les païens*, Budé (Paris: Belles Lettres, 1990–1), I.2.69–74.

[13] Isid. *Orig.* XIV.4.30.

the peoples of the far reaches of the continent before turning his attention to Europe in the same order as the provincial geography. Here, he places the northerly and frontier peoples first, then the Greeks, Italians, Gauls and Spanish *gentes*. Again, Isidore closes with the discussion of African groups, broadly following an easterly itinerary along the coast, before listing the inhabitants of the interior from west to east. We see the same pattern again in the long list of rivers in XIII.21, from the Ganges onwards, and once more with the account of the world's cities in XV.1, which starts in the Indian city of Nyssa, and ends in African Cyrene. Of necessity, some minor changes to this ordering system do occasionally appear, but these were rarely conspicuous. Isidore seems to have shared the widespread assumption that there were no cities worthy of the name in northern Asia or northern Europe and so omits these regions entirely within XV.1. More confusing is his presentation of North African cities from west to east later in the same chapter.[14] This effectively reverses the order of African geographies established in the earlier chapters, but as there are only seven cities in this list the change is relatively unobtrusive. Isidore entirely omits Africa from his riparian geography, but this too is understandable, given that Isidore regarded the Nile as an Egyptian – and hence 'Asian' – river, and knew of no other major watercourses in the north of the continent. With these anomalies explained, we are presented with a striking pattern. In each of his surveys of the world, Isidore adopts the same systematic order that he had previously employed in his provincial geography. The writer made substantial changes to his hugely varied source material in order to preserve this order, and evidently regarded consistency as an important feature of his methodology.

At this point it is worth noting briefly that, in each of his surveys, Isidore adopts a separate descriptive strategy for geographical features of particular historical, scriptural or typological importance, and that these appear rather prominently at the start of each section. The human geography of Book IX, for example, is introduced with a substantial discussion of the different *nationes* and *gentes* who traced their origins from Ham, Shem and Japeth, the three sons of Noah.[15] These peoples are listed, not in geographical order, as is the case elsewhere in the chapter, but according to their genealogical relations. Similarly, the opening section of the riparian geography of XIII.21 discusses the four rivers said to flow from the Garden of Eden, and commonly identified with the Ganges, the Tigris, the Euphrates and the Nile. A brief description of Eden itself opens the provincial survey of Book XIV, and the civic

---

[14] Ibid. XV.1.74–7.   [15] Ibid. IX.2.2–37.

geography of XV.1 includes a short account of Rome, Enoch (the first city in the Bible), Babylon and Jerusalem, before turning to the familiar itinerary of the world's towns.[16] These opening passages vary greatly in length – the description of Paradise at XIV.3.2–4 runs to fifteen lines of the Oxford Classical Text, while the discussion of the descendents of Noah takes up four pages of the same edition.

What all of these introductions have in common, however, is that Isidore happily included these exceptional features of scriptural 'geography' within his ordered geographical descriptions as well as in the opening discussions. In other words, many of these specific geographical features are discussed twice: once in their own taxonomic system, and then again within the familiar geographical representation of the world. Consequently, there is some considerable repetition even within a single chapter. In IX.2, for example, the Persians, Assyrians, Chaldeans, Syrians, Armenians, Hebrews, Indians, Bactrians, Saracens, Ethiopians, Arabians, Philistines, Gauls, Scythians, Goths, Medes, Ionians, Hispani, Thracians and Sarmatians all appear twice within the same chapter – once located by their genealogy, as the descendents of Shem, Ham or Japeth, and once by their location in the inhabited world.[17] Rome and Babylon, likewise, are both mentioned twice in XV.1. This pattern of repetition is not followed absolutely faithfully: many of the biblical groups of IX.2 could not be placed within a contemporary provincial geography, and Isidore made no attempt to do so. Similarly, neither Eden nor the rivers that flowed from it appear again within the discussions of the 'inhabited' world. But the pattern still makes for striking reading; evidently, Isidore was anxious that the geographical descriptions should be as comprehensive as possible, even at the cost of repeating himself.[18]

The originality of Isidore's repetitive geography is best illustrated by comparison with other texts in circulation at that time. Many compositions, of course, simply outlined the constituent parts of the world in a single itinerary: this was the pattern adopted by Strabo, Pliny, Pomponius Mela, Ptolemy and Orosius. Other works were compiled to illuminate the more esoteric geographical allusions in the classical poets or the Bible. Vibius Sequester's *De Fluminibus*, probably written in the

---

[16] Ibid. XV.1.3–5.
[17] Respectively: Isid. *Orig.* IX.2.3, 47 [Persians]; 3, 45 [Assyrians]; 3, 48 [Chaldeans]; 4, 50 [Syrians]; 4, 61 [Armenians]; 5, 51 [Hebrews]; 5, 39 [Indians]; 5, 43 [Bactrians]; 6, 57 [Saracens]; 10, 127 [Ethiopians]; 14, 49 [Arabians]; 20, 58 [Philistines]; 26, 104 [Gauls]; 27, 62 [Scythians]; 27, 89 [Goths]; 28, 46 [Medes]; 28, 77 [Ionians]; 29, 109 [*Hispani*]; 31, 82 [Thracians]; 32, 93 [Sarmatians].
[18] This is particularly well illustrated by the repetition of material on the Saracens at IX.2.6 and IX.2.57.

fourth or fifth century, comprises a compendious list of the rivers, springs, groves, marshes, mountains and peoples which appear in Virgil and other classical poets.[19] Jerome's *De Situ et Nominibus Locorum Hebraicorum Liber* and his translation of Eusebius' *Onomasticon* provide a similar function for readers of the Old Testament.[20] Crucially, both of these texts take their structure from the texts that they comment upon, and order specific geographical entries alphabetically by name. In neither case is the reader encouraged to reflect upon the relative location of the places discussed.

The only obvious antecedent for Isidore's peculiarly fragmented geography is the *Liber Memorialis* of Lucius Ampelius, which was probably composed in the fourth century.[21] This is a short, but wide-ranging text which includes a geographical crib as well as diverse information on astrology, secular history and political institutions. Like Isidore, Ampelius lists a handful of peoples, mountains and rivers according to rough geographical location, and his rough grouping of these features into Asian, European and African examples also anticipates the seventh-century text.[22] But Ampelius' geography is haphazard; many specific features are misplaced and there is no obvious order to the short lists of features within each continent. Whereas Isidore was careful to adopt a consistent itinerary in his ordering of material, then, Ampelius simply lumped his together in ungainly lists. Ampelius may have eschewed the alphabetized abstraction of many other geographical writers, but his work lacked the consistency (and the detail) of Isidore's grand project.

This, then, is the first broad observation that I would like to make about Isidore's treatment of geographical material within the *Etymologies*: the text does not just contain one geographical survey, it contains four, and the writer was at pains to present these in as regularized a manner as possible. As I noted earlier, the fact that Isidore broke up his 'geography' may be explained in part by his desire to make other ontological connections through the structuring of his different books, and this is certainly important. But the consistency of Isidore's geographical ordering

---

[19] Vibius Sequester, *De fluminibus*, P. G. Parroni (ed.), *Vibii Sequestris: De Fluminibus, Fontibus, Lacubus etc.* Testi e documenti per lo studio dell'antichità, 11 (Milan: Istituto Editoriale Cisalpino, 1965).

[20] Jerome, *De Situ et Nominibus Locorum Hebraicorum Liber*, E. Klostermann (ed.), *Die Griechischen Christlichen Schriftsteller Gesamtbestand* (Leipzig: J. C. Hinrichs, 1904).

[21] L. Ampelius, *Liber Memorialis*, Marie-Pierre Arnaud-Lindet (ed. and tr.), *L. Ampelius: Aide-Mémoire (Liber Memorialis)*, Budé (Paris: Belles Lettres, 1993). On the dating see Félix Racine, 'Literary geography in Late Antiquity' (Unpubl. PhD thesis, Yale University, 2009), 62–3.

[22] L. Ampelius, *Lib Mem.* VI.1–15.

demonstrates that he recognized the strong association between the provinces, peoples, rivers and cities of the world, and reveals that this separation cannot have been accidental. The author was keen to ensure that each of his surveys of the world was ordered alike, and that this order broadly reflected the relative position of different regions, even when this entailed substantial manipulation of his sources and even the repetition of material which appeared in other taxonomic schemes. What results then, may either be regarded as a rather elegant geographical palimpsest, in which a coherent image of the world is built up only gradually, or as a bizarrely fragmented geography.

Scintillating details: filling in the gaps in the geographical descriptions

My second observation concerns the specific details with which Isidore fleshed out these four skeletal geographical surveys. For every *gens*, province, river or city included within his account, Isidore included a supplementary nugget of information, whether on the origin of the geographical feature, a historical event related to it or something more unusual. These asides are rarely very long, but are frequently colourful, and often allude to information found in the other descriptions. To illustrate this, we may compare the different passages relating to Persia (broadly defined) in each of the four chapters:

The Persians were named after King Perseus, who crossed into Asia from Greece and there dominated the barbarian nations with heavy and prolonged fighting. Right after his victory he gave his name to the conquered people. Before Cyrus, the Persians were an ignoble people and considered of no rank among the nations of the area.[23]

The Choaspes is a river in Persia, so named in their language because it has amazingly sweet waters, such that the Persian Kings claimed the drinking water from it for themselves for a distance that the river runs between Persian riverbanks. Some people think that the River Cydnus in Cilicia originates from this river.[24]

Persia reaches to the Indus in the east, in the west it has the Red Sea, in the north it touches Media, in the south-southwest Carmania, which is connected to Persia and in which the most renowned city of Susa is located. Persia is the

---

[23] Isid. *Orig.* IX.2.47: *Persae a Perseo rege sunt vocati, qui e Graecia Asiam transiens, ibi barbaras gentes gravi diuturnoque bello perdomuit, novissime victor nomen subiectae genti dedit. Persae autem ante Cyrum ignobiles fuerunt, et nullius inter gentes loci habebantur.* Barney et al. (tr.), *Etymologies*, 194.

[24] Isid. *Orig.* XIII.21.15: *Choaspis Persarum fluvius, vocatus eorum lingua quod miram aquae dulcedinem habeat, adeo ut Persidis fluit, sibi ex eo pocula vindicarunt. Ex hoc amne quidam Cydnum Ciliciae fluvium derivari existimant.* Barney et al. (tr.), *Etymologies*, 281.

birthplace of the magical arts. The giant Nebroth went there after the confusion of the tongues, and taught the Persians to worship fire, for in those regions everyone worships the sun, which is called *El* in their language.[25]

Perseus, son of Adea, founded the city of Persepolis, capital of the realm of Persia, very famous and stuffed with riches. Persia (*Persida*) was also named from him. The Parthians also founded Ctesiphon in Parthia in emulation of the city of Babylon. They say that Memnon's brother established the city of Susa in Persia. It was named Susa because it overlooks the River Susa. The royal palace of Cyrus is there, distinguished by its white and variegated stone, with golden columns and panelled ceilings and jewels, even containing a replica of the sky embellished with twinkling stars and other things beyond human belief.[26]

These passages illustrate Isidore's general approach well. In each, a short aside of historical, linguistic or natural historical interest is included to add substance to the geographical survey and lend some depth to what would otherwise be an undifferentiated parade of place names. As we might expect, there is some repetition between different books – the connection between Perseus and Persia is drawn twice, and the prominent city of Susa also reappears – but read together, these passages form a composite (if superficial) image of the history and geography of Persia.

Readers who are already familiar with Isidore's epistemological method will be aware of the explanation for many of these asides, of course.[27] Indeed, anyone who has ever cracked the spine of the *Etymologies* will know that Isidore's broader methodology depended upon supplementary discussions of precisely this kind. To simplify only slightly, the *Etymologies* as a whole was an exposition of the proposition that in order to understand the essence of a thing (*res*), one needs to understand the 'force' (*vis*) of its name (*nomen*) – a process best approached through etymological enquiry. As Isidore put it (in a passage cribbed more or less directly from Cicero by way of Quintilian), knowing that the word *flumen* comes from *fluendum* or (more spuriously) that *homo* comes from *humus*,

---

[25] Isid. *Orig.* XIV.3.12: *Persida tendens ab ortu usque ad Indos, ab occasu Rubrum mare habet, ab aquiline vero Mediam tangit, ab austro Carmaniam, quae Persidae adnectitur, quibus est Susa oppidum nobilissimum. In Persida primum orta est ars magica, ad quam Nebroth gigans post confusionem linguarum abiit, ibique Persas ignem colere docuit. Nam omnes in illis partibus solem colunt, qui ipsorum lingua El dicitur.* Barney et al. (tr.), *Etymologies*, 286.

[26] Isid. *Orig.* XV.1.8–10: *Persepolim urbem caput Persici regni Perseus †Adeae† filius condidit famosissimam confertissimamque opibus; a quo et Persida dicta est. Ctesiphontem quoque Parthi apud Parthiam condiderunt in aemulationem Babyloniae urbis. Dicta autem Susis quod inmineat Susae fluvio. Ibi est regia Cyri, lapide candido et vario cum columnis aureis et lacunaribus gemmisque distincta, continens etiam simulacrum caeli stellis micantibus praesignatum, et cetera humanis mentibus incredibilia.* Barney et al. (tr.), *Etymologies*, 301.

[27] On this methodology cf. esp.: Ribémont, *Les origines*, 39–80; Del Bello, *Forgotten Paths*, 104–13; Amsler, *Etymology and Grammatical Discourse*, 133–72.

helps the reader understand the nature of a 'river' or of 'mankind', respectively, and secures those images of a flowing watercourse or an earthbound mortal in his or her mind.[28] 'One's insight into anything is clearer', Isidore states, 'once the etymology [of its name] is known'.[29] In many cases, Isidore was evidently satisfied that the *vis* of these words assumed a deep allegorical significance, which could aid the understanding of God's Creation on the most profound level. Knowing that *homo* came from *humus* was, for example, a reminder of the terrestrial nature of mankind on multiple levels. Likewise, to know that Eve (*Eva*) might be cognate with 'life' (*vita*), 'calamity' (*calamitas*) or woe (*vae*) was significant both in recalling the Fall as a historic event and the (supposed) characteristics of all later women.[30] To this end Isidore privileged certain etymologies which were derived from Hebrew, Greek and Latin, and which might thus be bearers of eternal truth to the careful linguist-exegete.

Other etymologies had less arcane significance, derived as they were from human activity, rather than the eternal truth of Scripture. Isidore recognized that individuals in the past very commonly lent their names to things, either as a result of historical circumstance, or through more or less arbitrary human action. A man, Isidore reminds us, might name his slaves or other possessions according to passing fancy, and analogous whimsy had perhaps shaped the historical vocabulary of Latin as a whole.[31] In such cases, the etymologist may turn to history or to aetiology, rather than to grammar, in order to trace the origins of a particular word, and thus seek its significance in the human past. This is a practice that Isidore adopts extensively throughout the *Etymologies*. It is with good reason, then, that an alternative title for the text was the *Origines*, even among its earliest readers: Isidore's is a book about origins, as much as etymologies, about the (often spurious) histories of 'things' as much as the discussion of their names.[32]

Isidore also recognized that geography could play a role in shaping the language, and identified several different ways in which words might reflect their particular contexts:

Other words derive their names from names of places, cities [or] rivers. In addition, many take their names from the languages of various peoples, so that

---

[28] Isid. *Orig.* I.29.1; cf. Quint. *Inst.* I.6.28.
[29] Isid. *Orig.* I.29.2: *Omnis enim rei inspectio etymologia cognita planior est.* Barney et al. (tr.), *Etymologies*, 55.
[30] Isid. *Orig.* VII.6.5–6.   [31] Ibid. I.29.2–3.
[32] Cf. Braulio's introduction to the text, and the entry for aetiology (*aetiologia*) at ibid. II.21.39.

it is difficult to discern their origin. Indeed, there are many foreign words unfamiliar to Latin and Greek speakers.[33]

The categories of Isidore's account here are familiar: places, cities, rivers and languages are, of course, the subjects of the four long geographical descriptions that have already been discussed. The different passages relating to the geography of Persia illustrate several of these different forms of etymology (or aetiology) rather well. Isidore includes a euhemeristic account of King Perseus, his conquest of the East and the naming of Persia and Persepolis after himself. The city of Susa takes its name from the nearby River Susa, while the Choaspes is said to be a derivative of the local word for 'sweetness'. In this small handful of passages, then, we can identify names that are said to have been shaped by historical circumstance, by nearby cities and rivers and by the local language. While the aetiologies themselves would rarely pass muster among modern historians or historical linguists as explanations for the actual origins of the names in question, they do at least fulfil Isidore's promise that the audience will learn something of the things themselves from discussion of their names, and give some sense of the mutual reinforcement that the multiple 'world geographies' provide.

But what of the descriptive passages that do not relate directly to Isidore's aetiological method? If the author's principal concern was to explain the nature of things through the origins of their names, what was the purpose of the brief narrative of Perseus' victories in the East? Why include the superfluous detail on the royal control over the waters of the Choaspes? Or the longer discussion of Nebroth and Persian superstition? Or even the vivid description of the splendours of Persian palaces? Such incidental details certainly render Isidore's fragmented geographies more vivid – indeed, Persia stands out with particular brilliance in each of the descriptions of the world – but they provided neither a systematic portrait of the regions they purport to describe, nor a consistent explanation for the names of the geographical features themselves. We might assume that this material was included for interest alone – Isidore was nothing if not esoteric, and geography was well known in antiquity for its literary delights and curiosities – but the regularity with which these additional passages were included in the text suggests an underlying method.

---

[33] Ibid. I.29.5: *Alia quoque ex nominibus locorum, urbium [uel] fluminum traxerunt vocabula. Multa etiam ex diversarum gentium sermone vocantur. Vnde et origo eorum vix cernitur. Sunt enim pleraque Barbara nomina et incognita Latinis et Graecis.* Barney et al. (tr.), *Etymologies*, 55.

Learning the world

Having identified several idiosyncrasies of Isidore's geographical passages, it seems appropriate to pause and take brief stock of the likely function of these passages for their audience. In the classical and late antique world, geographical knowledge was frequently valued as an aid to the appreciation of the great poets of the past. It was for this reason that scholiasts like Servius included geographical addenda within their commentaries, and that writers like Vibius Sequester and Lucius Ampelius in the fourth century compiled lists of different places mentioned in the writing of Virgil, Ovid and their successors.[34] Among Christian writers, information about the physical world was still regarded as a necessary part of a rounded rhetorical education, but was also seen as an important prolegomena to the study of Scripture.[35] Augustine had advocated the study of secular natural history as a necessary foundation for biblical exegesis in a variety of his writings; Cassiodorus seconded this suggestion in his *Institutiones* – an annotated bibliography intended for the intellectual improvement for the monks in his foundation at Vivarium.[36] While Cassiodorus certainly recommended texts which ranged far beyond the territorial confines of Scripture – Latin translations of Ptolemy and Dionysius Periegetes were included on his reading list – this interest in the world beyond the Holy Land complemented the post-Apostolic vision of the entire world as the setting for the final stages of Christian history, and hence as something worthy of study in its own right.[37]

Isidore evidently valued this knowledge of the world and its parts, and it seems reasonable to assume that his decision to devote such a large proportion of his *Etymologies* to human, political, physical and civic geographies was justified by the sort of pedagogy articulated by his predecessors.[38] Moreover, as we have seen, Isidore's conviction that words might take their origins from a variety of historical, geographical and linguistic sources also meant that knowledge of the world was fundamental to his own epistemology. Motivated by the need to compile

---

[34] For more on this material, see now Racine, 'Literary geography'.

[35] For an elegant summary of this position, see Lozovsky, *'The Earth is Our Book'*.

[36] See the comments and bibliography in the recent edited translation of James W. Halporn and Mark Vessey, *Cassiodorus: Institutions of Divine and Secular Learning*, Translated Texts for Historians, 42 (Liverpool University Press, 2004).

[37] Cassiodorus, *Institutiones*, R. A. B. Mynors (ed. and tr.) (Oxford University Press, 1937), I.25. Lozovsky, *'The Earth is Our Book'*, 15–20.

[38] Lozovsky, *'The Earth is Our Book'*, 20–2; Bernard Ribémont, 'On the definition of an encyclopaedic genre in the Middle Ages', in *Pre-Modern Encyclopaedic Texts: Proceedings of the Second COMERS Congress, Groningen 1–4 July 1996*, Peter Binkley (ed.) (Leiden: Brill, 1997), 49–50.

a comprehensive body of geographical material, Isidore effectively created a substantial reference work – a text in which the provincial geography of Orosius was complemented with scattered material from a variety of other sources, including histories, literary commentaries and poetry as well as Scripture and its exegesis. This was certainly how Isidore's text was employed in the centuries that followed, and it always provided a valuable repository of information.

Within this context, the peculiar fragmentation of Isidore's geography, combined with the consistency of ordering which he follows across all four chapters, can be read on the most immediate level as a structural device – a straightforward strategy by which the reader may navigate his or her way through the geographical sections to retrieve a specific piece of information as painlessly as possible. At the risk of over-emphasizing an obvious point, such a strategy could only be effective if the geographical order employed was used consistently throughout the text: when the structuring principles of a particular geographical description are understood, navigation through a text is exceptionally easy, when they are not it is effectively impossible. In the text as it stands, a reader of the *Etymologies* who was searching for specific information about Persia (say) could rapidly find the appropriate section of each geographical book, once he had learned that Persia appeared towards the beginning of the Asian section of the description, after the geography of India, Parthia and Assyria and before the account of Mesopotamia and Chaldea. Admittedly, the reader interested in more than one feature of Persian geography would have to skip between different sections of the *Etymologies*, but the adoption of the same ordering strategy in each would minimize this inconvenience.

This provides us with a first, provisional conclusion: Isidore regarded geography as appropriate for the organization of material. In itself, this is not altogether surprising. After all, Isidore employed a staggering variety of different taxonomic systems to order the varied information within his *Etymologies*. In Books I–V, for example, Isidore organizes his discussion of disciplines by their position in the classical curriculum; in Book XII animals are arranged by their genus; in Book X adjectives are listed alphabetically (and within that schema by certain moral associations); at the end of Book V, and at the start of Book IX, chronology is used to order information; Books VII and VIII are structured in part around the assumed hierarchies of heaven (and hell), and at one stage in Book XVI gemstones are arranged according to their colour.[39] Isidore evidently felt confident that his audience would be sufficiently familiar with the

---

[39] Isid. *Orig.* XVI.7–12.

geography of the world to arrange four of his chapters in this way, but this was only one among several taxonomic systems that the writer employed.

This functions readily enough as a textual retrieval device within the *Etymologies*: once the reader has become accustomed to Isidore's structural scheme, navigation through it is exceptionally straightforward. But Isidore did not compose his *Etymologies* for the modern reader, and the peculiarly repetitive geographies of the *Etymologies* work still better as a mnemonic structure than they do as an aid to textual navigation. Like all writers (and readers) of this period, Isidore worked within a complex hermeneutic system that combined textual reference and embedded memory. As numerous didactic and mnemonic asides within the *Etymologies* make clear – and as numerous references to memory training in his other compositions confirm – his most famous work should not be regarded as a straightforward reference text, intended to be plucked off the shelf when an obscure point of information was required. Instead, it should be considered within the context of a fuller absorption of information – as the starting point or frame of reference for the memorization of information. If we view his geographical chapters from this perspective, a number of points become rather clearer.[40]

Memory and geography

The precise role of memory and memorization within early medieval education remains a complex study, despite (or perhaps because of) the ground-breaking recent work of a number of scholars.[41] Broadly, it seems likely that the 'art' of memory training, which was the subject of some discussion in the first century BCE, had become rather passé by the time that Quintilian was writing in the first century CE, and it was to remain so until the central Middle Ages at the earliest.[42] In the interim, memory and memorization remained a central feature of elementary education, and indeed of all intellectual activity, but it seems to have been viewed as a *praxis* rather than an *ars* – that is, something to be attained through constant practice in the scriptorium or study, rather

---

[40] Cf. Mary Carruthers, *The Craft of Thought: Meditation, Rhetoric and the Making of Images. 400–1200* (Cambridge University Press, 1998), 156–9; Mary Carruthers, *The Book of Memory: A Study in Medieval Memorial Culture*, 2nd edn. (Cambridge University Press, 2008), 139–40 and 218–20.

[41] Carruthers, *The Craft of Thought* and *The Book of Memory* are central here. Cf. also Kimberley Rivers, 'Memory, division, and the organization of knowledge in the Middle Ages', in *Pre-Modern Encyclopaedic Texts*, Binkley (ed.), 147–58 for a bold application of these arguments to the medieval encyclopaedic tradition more generally: an approach which has influenced the position adopted in the present chapter.

[42] For discussion see Carruthers, *The Book of Memory*, 92.

than through formal training.[43] That said, Quintilian and his successors certainly recognized methods by which speeches, literary passages or lists of specific items might be committed to memory, and Mary Carruthers has demonstrated how far such philosophies affected manuscript use within this period.

To simplify greatly, the most important themes in medieval discussions of memorization were the importance of order and familiarity within the structure of a text that was to be committed to memory (the form), and of vivid visual or verbal cues within the specific points of information to be memorized (the content). As we have seen, the characteristics of Isidore's geographical descriptions – at once ordered, repetitive and strikingly vivid – map quite closely onto these desiderata. The form in which Isidore presented his geographies would seem to suggest that he wished his audience to commit this information to memory.

The idea that information could be ordered spatially within the memory was a commonplace within classical and medieval writing on the subject.[44] In the *Rhetorica ad Herennium*, for example, the guide to the *ars memoriae* composed in the 80s BCE, the student anxious to train his memory is encouraged to imagine the different items to be recalled within a familiar environment, typically a room or villa; by 'moving through' this mental landscape and noting where each thing is 'placed', the student could expect to recall a complex list of things without fear of omission or confusion in order.[45] By the first century CE, such practices appear to have been widespread, along with other spatial metaphors for memorial practice.[46] Other examples are manifold: Quintilian implies that the memory of the trained rhetor might be imagined as a forest, with some areas more frequently accessed than others, but with everything in its proper place.[47] Adopting a different metaphor, Augustine variously referred to his own memory as 'fields or a spacious palace', and as 'vast cloisters' – a world that must still be conceptualized (and navigated) in spatial or topographical terms.[48]

In a sense, therefore, the students and scholars of the early medieval world would have had an intimate familiarity with different forms of

---

[43] Carruthers, *The Craft of Thought*, *passim*, explores this contrast.

[44] For a general introduction, see Carruthers, *The Craft of Thought*, 11–22.

[45] *Rhetorica ad Herennium*, Harry Caplan (ed. and tr.), Loeb (Cambridge, Mass.: Harvard University Press, 1954), III.29.37. And cf. Cicero, *De Oratore*, E. W. Sutton and H. Rackham (ed. and tr.), Loeb, 2 vols. (Cambridge, Mass.: Harvard University Press, 1948), II.350–60.

[46] Carruthers, *The Book of Memory*, 89–93.

[47] Quintilian, *Institutio Oratoria*, Donald A. Russell (ed. and tr.), Loeb, 5 vols. (Cambridge, Mass.: Harvard University Press, 2002), V.10.20–22.

[48] Augustine, *Conf.* X.8: *campos et lata praetoriae; aula ingenti*. Pine-Coffin (tr.), 215.

spatial cognition: ideally this would have been how *all* of their thoughts were organized. But there is clearly a considerable difference between the intimate spaces of the human memory, and the ordered geographies of the wider world, of the kind that Isidore laid out in his *Etymologies*. Isidore intended his long lists of provinces, peoples, rivers and cities to be learned and memorized: he could not have expected them to be rooted already in the minds of his audience. The abstracted spaces of memory were distinct from the concrete places of the world which the apt pupil was to learn by heart. Nevertheless, if the different parts of the world were to be committed to memory by an attentive reader, the form in which Isidore presents this material seems particularly attractive.

It is worth stressing here that the peculiar fourfold geography of the *Etymologies* may well have been intended to underscore this mnemonic. On the most immediate level, this approach has the virtue of repetition – by the fourth circuit of the world, the diligent reader or listener is already starting to appreciate the underlying spatial structure behind Isidore's organization of material. The division of geographical information into four separate chapters has the further advantage of keeping each individual entry relatively brief. The reader is provided with only one or two pieces of information on the human, physical or political geography of the world, and is not overwhelmed with a great mass of undifferentiated narrative or description. Isidore's geographies are essentially detailed lists, and lend themselves readily to the memory.

The specific descriptions of peoples, regions and places also fit closely with ancient and medieval views of what made things memorable. As the *Rhetorica ad Herennium* notes:

if we see or hear something exceptionally base, dishonourable, extraordinary, great, unbelievable or laughable that we are likely to remember for a long time ... We ought, then, to set up images of a kind that can adhere longest in the memory. And we shall do so if we establish likenesses that are as striking as possible; ... if we assign to them exceptional beauty or singular ugliness; if we dress some of them with crowns or purple cloaks, for example, so that the likeness is more distinct to us; or if we somehow disfigure them, as by introducing them stained with blood, or soiled with mud, or smeared with red paint, so that its form is more striking ... for that, too will ensure our remembering them more vividly.[49]

---

[49] Rhet ad Her. III.22.35–7: *si quid videmus aut audimus egregie turpe, inhonestum, inusitatum, magnum, incredibile, ridiculum, id diu meminisse consuevimus ... Imagines igitur nos in eo genere constituere oportebit quod genus in memoria diutissime potest haerere. Id accidet si quam maxime notatas similitudines constituemus ... si egregiam pulcritudinem aut unicam turpitudinem eis adtribuemus; si aliquas exornabimus, ut si coronis aut veste purpurea, quo nobis notatior sit similitudo; aut si qua re deformabimus, ut si cruentam aut caeno oblitam aut rubrica delibatam inducamus quo magis insignita sit forma ... nam ea res quoque facie tut facilius meminisse valeamus.* Caplan (tr.), 219–21.

Here, the anonymous author of the *Rhetorica* is discussing images to aid with the memorization of the different parts of a speech, but *mutatis mutandis* he might equally be defining Isidore's short geographical vignettes. The different sketches of Persia discussed above demonstrate how the geographies of the *Etymologies* could be striking, splendid, horrific and gaudy in precisely the manner advocated by the *Rhetorica*. In other passages physical peculiarity or vivid (and violent) historical detail helped to secure a place in the memory:

> The Gauls (*Galli*) are named for the whiteness of their bodies, for in Greek milk is called Gala. Whence the Sibyl speaks of them thus, when she says of them 'Then their milk-white necks are circled with gold.'[50]
>
> Carrhae, a city of Mesopotamia beyond Edessa, was founded by the Parthians. A Roman army was once slaughtered there, and its general Crassus was captured.[51]

Word-play and puns were also valued by theorists of memory as aids to recollection, and Isidore's etymologizing (and pseudo-etymologizing) naturally lent itself to this. The point is illustrated most vividly by one of the more improbable inclusions within Isidore's text. Midway through his enumeration of Italian cities in XV.1, Isidore includes an unlikely entry:

> Pompeii was founded in Campania by Hercules, who as a victor had led a triumphal procession (*pompa*) of oxen from Spain.[52]

What is significant here is not that Isidore included a city which had been destroyed more than 500 years earlier, (although this is surprising), but rather that the description which he offers bears no explanatory function whatsoever. Isidore follows Servius and (more directly) Solinus in his statement that Pompeii was founded by Hercules.[53] Within Isidore's sources, the association provides a colourful aside; in Isidore's geography it is best read as an *aide-mémoire*. The tradition of Hercules' triumphal procession has no direct mythological or historical relation to the foundation of Pompeii. More significantly the homonym *pompa/Pompeii*,

---

[50] Isid. *Orig.* IX.2.104: *Galli a candore corporis nuncupati sunt. Γάλα enim Graece lac dicitur. Unde et Sibylla sic eos appellat, cum ait de his: Tunc lacteal colla | auro innectuntur.* Barney et al. (tr.), 198.

[51] Isid. *Orig.* XV.1.12: *Carra civitas Mesopotamiae trans Edessam condita a Parthis, ubi quondam Romanus est caesus exercitus, et Crassus dux captus.* Barney et al. (tr.), 301. Isidore's source for this passage is not clear.

[52] Isid. *Orig.* XV.1.51: *Ab Hercule in Campania Pompeia, qui victor ex Hispania pompam boum duxerat.* Barney et al. (tr.), 304.

[53] *Servianorum in Vergilii carmina commentariorum*, Arthur Stocker and Albert Travis (eds.) (Oxford University Press, 1965), *Aen.* VII.662. For discussion, see Guillaumin and Monat, *Isidore de Séville. Étymologies. Livre 15*, 35.

which clearly motivated the addition, has no explanatory or etymological function. This triumph occurred in Spain, and Isidore claims no direct historical link to the foundation of the city in Campania. Here it is clear that Isidore's word-play has no deeper function than to provide a memorable image to secure Pompeii within the minds of his audience. *Pompa* is related to *Pompeii* as an aid to memory, not as a basis for etymologizing.

Each point of geographical reference, then, whether a people, language, river, province or city, is provided with a context of some kind: in some cases, the details adduced helped to explain the name, in others they simply secured the name in the memory. When all of these accounts are taken together, the audience to the *Etymologies* is thus presented with a fourfold mnemonic model for the world. Once these details were absorbed by the reader, Isidore's fragmented geography would be immaterial: the student could move readily enough from one register to another, by navigating through his well-ordered memory, rather than by flipping between the pages of a codex.

Isidore's *Etymologies* is a baffling text, and its geographical sections pose peculiar puzzles of their own. As a product of the early Middle Ages, Isidore seems not to have regarded 'geography' as a discrete subject, worthy of study in its own right, and yet clearly he recognized that the spatial order of the world might be a useful tool in ordering his vast encyclopaedic project, and represented an important area of knowledge that might be learned. The *Etymologies* presents relative geographical location both as something to be learned, and as an *aide-mémoire* in its own right. To this end, he adopted a systematic geographical order across four prominent books of his *Etymologies*, and included a diverse array of information and images that were intended to make his geographies memorable. There is a hint here that Isidore did not simply intend his geographical passages to be used as a crib for the reading of Scripture, or even for the greater appreciation of the pagan poets, as his predecessors had done. Instead, he would seem to be presenting the world on its own terms. Isidore's is a fragmented geography, and is frequently a confusing one, but it is memorable. And that may well have been precisely what he intended.[54]

[54] I am grateful to Keith Lilley for the invitation to talk at the 'Mapping Medieval Geographies' Conference in 2009 at UCLA, and to all of the delegates for their helpful and stimulating advice, particularly Oliver Berghof, Natalia Lozovsky, Helmut Reimitz, Camille Serchuk, Amanda Power and Jesse Simon. I would also like to thank Jen Baird for reading through versions of this chapter in draft, and for her many valuable comments.

3 The uses of classical history and geography in medieval St Gall

Natalia Lozovsky

The Benedictine abbey of St Gall, founded in the early seventh century in what is now Switzerland, functioned as a major centre of learning for more than one thousand years. The peak of the cultural and political influence of St Gall fell in the period between the ninth and eleventh centuries, when the abbey was famed for its scholarship and art and favoured by Carolingian and Ottonian rulers.[1] An excellent library was built in that time. Unlike most medieval libraries whose books and records were dispersed over centuries, the core of St Gall's medieval collection, which includes medieval catalogues and books, is still fortunately preserved in the Stiftsbibliothek St Gallen. Charters contained in the Abbey Archive document the abbey's territorial growth and its economic and social networks. Annals and chronicles composed at St Gall narrate the history of the abbey, reflecting medieval attitudes to time and space.[2] The wealth of sources surviving from St Gall provides unique

---

[1] Werner Vogler, 'Historical sketch of the Abbey of St. Gall', in *The Culture of the Abbey of St. Gall: An Overview*, ed. James C. King and Werner Vogler (Stuttgart: Belser Verlag, 1991), 9–28, esp. 12–18.

[2] For a survey of the history of the St Gall library, see Johannes Duft, *The Abbey Library of Saint Gall: History, Baroque Hall, Manuscripts*, trans. James C. King and Petrus W. Tax (St Gall: Verlag am Klosterhof, 1985). For descriptive catalogues of St Gall manuscripts, see Gustav Scherrer, *Verzeichniss der Handschriften der Stiftsbibliothek von St. Gallen* (Halle: Buchhandlung des Waisenhauses, 1875); Albert Bruckner, *Scriptoria medii aevi Helvetica: Denkmäler schweizerischer Schreibkunst des Mittelalters*, 3 (Geneva: Roto-Sadag, 1938); Beat Matthias von Scarpatetti, *Die Handschriften der Stiftsbibliothek St. Gallen, 1/4: Codices 547–669: Hagiographica, historica, geographica, 8.–18. Jahrhundert* (Wiesbaden: Harrassowitz, 2003). A significant number of St Gall manuscripts are also preserved in the Zurich Zentralbibliothek, with some manuscripts scattered in the libraries of Europe and North America. For more information and bibliography, see Anna A. Grotans, *Reading in Medieval St. Gall* (Cambridge University Press, 2006), 51, n. 10. An edition of St Gall annals can be found in *Monumenta Germaniae Historica: Scriptores* (hereafter 'MGH SS') 1. For the most recent editions of the *Casus Sancti Galli*, see Ratpert, *St. Galler Klostergeschichten (Casus sancti Galli)*, ed. and trans. Hannes Steiner, *Monumenta Germaniae Historica: Scriptores rerum Germanicarum in usum scholarum* (hereafter 'MGH SRG') 75 (Hanover: Hahnsche Buchhandlung, 2002); Ekkehard IV, *Casus Sancti Galli (St. Galler Klostergeschichten)*, ed. and trans. Hans F. Haefele (Darmstadt:

information for considering how scholars at this important centre approached historical and geographical studies. What were the ways and institutional settings in which St Gall scholars studied these subjects? What techniques did they use in order to understand classical texts written several hundred years earlier? What goals did they pursue beyond learning the language and mastering the information? To answer these questions, the present chapter will largely draw on the evidence provided by a glossed manuscript of Orosius, produced at St Gall. Several generations of St Gall monks read and studied this magisterial account, in which the fifth-century author laid out his vision of how human history developed in its spatial context.

St Gall, with its rich library and a favourable climate for the pursuit of learning, was a good place for such studies. History and geo-ethnography were very well represented among the numerous holdings in religious and secular subjects that the library accumulated between the ninth and eleventh centuries. The texts ranged from those going back to antiquity to early medieval compositions and often combined treatment of historical events with descriptions of lands and peoples. The library owned books on world history (including such late antique texts as the chronicle of Eusebius–Jerome and Orosius' *Seven Books of History against the Pagans* and the ninth-century world chronicle by Frechulph of Lisieux). Histories of different periods and regions included the Latin translation or, rather, retelling of Flavius Josephus' *Jewish Wars*, a work originally written in the first century CE. The library also had numerous lives of the saints, as well as annals and monastic chronicles produced at St Gall. Furthermore, St Gall scholars could consult books which primarily focused on geo-ethnography and contained historical information, such as the works by Roman writers Solinus and Julius Honorius. They could also turn to the fascinating and challenging text by Aethicus Ister originally composed in the seventh century and to the *Liber monstrorum*, an account of monstrous races going back to Late Antiquity. St Gall scholars also had at their disposal late antique and early medieval encyclopaedic texts, such as Martianus Capella's *On the Marriage of Philology and Mercury*, Isidore's *Etymologies*, and Bede's *On the Nature of Things*.[3] The broad range and impressive number of manuscripts, each

Wissenschaftliche Buchgesellschaft, 1980); *Continuatio 2* and Conrad of Fabaria, MGH SS 2, 148–83; Cristian der Kuchimaister, *Nüwe casus monasterii Sancti Galli*, ed. Eugen Nyffenegger (Berlin: de Gruyter, 1974). An English translation of the *Casus* by Emily Albu and Natalia Lozovsky is in progress. For charters, see *Chartularium Sangallense*, vols. 1–12, ed. Otto P. Clavadetscher (St Gall: Thorbecke, 1982–2012).

[3] For medieval catalogues from St Gall see Gustav Becker, *Catalogi bibliothecarum antiqui* (Bonn: Fr. Cohen, 1885); *Mittelalterliche Bibliothekskataloge Deutschlands und der Schweiz*,

one a product of considerable expense and effort, testify to a strong interest in these subjects. Excerpts from classical sources preserved in some codices, as well as glosses and diagrams added by St Gall scholars to classical texts, provide more specific evidence of how intensely history and geography were studied.

A text composed by a St Gall monk Notker Balbulus (the Stammerer) in the late 880s demonstrates the uses and rewards of such studies. In his *Deeds of Charlemagne (Gesta Caroli Magni)*, Notker the Stammerer collected tales and anecdotes about Charlemagne, the founder of the Frankish empire, for Charlemagne's great-grandson Charles the Fat. In this text, Notker applied his knowledge of the classical geographical tradition in a historical context and organized his account according to geographical principles.[4] He started with stories that took place in the west and proceeded eastward, gradually broadening his geographical perspective ultimately to include the faraway lands and peoples. Not only did the geographical scheme provide the Carolingian author with a convenient way to arrange historical anecdotes – its scope also underlined Notker's intention to celebrate the might of the Frankish empire. Drawing on Roman models, Carolingian imperial ideology proclaimed that the new imperial people, the Franks, not only inherited their dominion over the world from the Romans, but also expanded it. The studies of Roman history and geography by Carolingian scholars supported contemporary imperial claims. Notker's skilful employment of geography in a historical

vol. 1, ed. Paul Lehmann (Munich: C. H. Beck, 1918); Erwin Rauner, 'Notker des Stammlers "Notatio de illustribus viris", Teil 1: Kritische Edition', *Mittellateinisches Jahrbuch* 21 (1986), 34–69. On the importance of history at St Gall, see Ernst Tremp, Karl Schmuki and Rudolf Gamper, *Geschichte und Hagiographie in Sanktgaller Handschriften: Katalog durch die Ausstellung in der Stiftsbibliothek St. Gallen (2. Dezember 2002–9. November 2003)* (St Gallen, Verlag am Klosterhof, 2003); Peter Ganz, 'Geschichte bei Notker Labeo?', in *Geschichtsbewusstsein in der deutschen Literatur des Mittelalters/Tübinger Colloquium 1983*, ed. Christoph Gerhardt et al. (Tübingen: Niemeyer, 1985), 1–16; Michael I. Allen, 'Bede and Frechulf at Medieval St Gallen', *Beda Venerabilis: Historian, Monk and Northumbrian*, ed. L. A. J. R. Houwen and A. A. MacDonald (Groningen: Forsten, 1996), 61–80; J. M. Clark, *The Abbey of St. Gall as a Centre of Literature and Art* (Cambridge University Press, 1926). To my knowledge, geographical studies at St Gall have not been examined in detail; for a brief discussion, see Grotans, *Reading*, 76. The catalogue of an exhibition held at St Gallen usefully lists and illustrates the most interesting medieval maps from the St Gall library: *Karten und Atlanten: Handschriften und Drucke vom 8. bis zum 18. Jahrhundert: Katalog zur Jahresausstellung in der Stiftsbibliothek St. Gallen (3. März bis 11. November 2007)* (St Gallen: Verlag am Klosterhof St. Gallen, 2007).

[4] As the recent editor of Notker's *Gesta* Hans Haefele has demonstrated in his introduction: *Taten Kaiser Karls des Grossen [von] Notker der Stammler*, ed. Hans F. Haefele, MGH Scriptores rerum Germanicarum, nova series (hereafter 'MGH SRG n.s.'), 12 (Berlin: Weidmann, 1959), xviii–xxiii. Haefele has also agreed with earlier scholars that Notker may have used a *mappamundi*.

narrative shows his intimate familiarity with the contents and ideological uses of classical geographical learning.[5] Like other early medieval authors who used geography in a historical context to promote similar goals, Notker found inspiration and models in the classical and late antique tradition.[6]

Among authoritative works that informed Notker's approach may have been the *Seven Books of History against the Pagans* written *c*. 415–18 by Paulus Orosius, a Christian historian and theologian.[7] In his work, inspired by St Augustine, Orosius presented a Christian interpretation of the history of the world from the Creation to approximately 416 CE. Orosius viewed the rise and decline of kingdoms and empires of the world as stages in the unfolding of the divine plan in space and time, as providential movement from paganism to Christianity to the eventual salvation. The divine plan, as Orosius understood it, included not only people but also the earth they populated. Geography thus played an important role in Orosius' historical vision. He began his history with a long chapter that described the locations of the known world in an essential summary of the classical geographical learning. A spatial organization underlies his historical narrative, and descriptions of places occur throughout his book. Orosius' universal history enjoyed a wide circulation in medieval Europe, and his geographical introduction became a very popular source of geographical knowledge in its own right.[8]

---

[5] Ibid. xxi–xxii. For a reflection of Carolingian imperial ideology on geographical writings see Emily Albu, 'Imperial geography and the medieval Peutinger Map', *Imago Mundi* 57 (2005), 136–48; Natalia Lozovsky, 'Roman geography and ethnography in the Carolingian empire', *Speculum* 81 (2006), 325–64; and Patrick Gautier Dalché, 'Représentations géographiques savantes, constructions et pratiques de l'espace', in *Construction de l'espace au Moyen Age: pratiques et représentations: XXXVIIe Congrès de la Société des historiens médiévistes de l'enseignement supérieur public, Mulhouse, 12–4 juin 2006* (Paris: Publications de la Sorbonne, 2007), 13–38, esp. 15–20. On Notker's ideas of the Carolingian world empire see, most recently, Simon Maclean, *Kingship and Politics in the Late Ninth Century: Charles the Fat and the End of the Carolingian Empire* (Cambridge University Press, 2003), esp. 154.

[6] For illuminating discussions of the late antique and classical tradition see A. H. Merrills, *History and Geography in Late Antiquity* (Cambridge University Press, 2005) and Katherine Clarke, *Between Geography and History: Hellenistic Constructions of the Roman World* (Oxford: Clarendon Press, 1999). Rosamond McKitterick has demonstrated how often Carolingian history books incorporated geographical ideas, *History and Memory in the Carolingian World* (Cambridge University Press, 2004), esp. 13, 75, 226, 232.

[7] Orosius, *Historiarum adversus paganos libri VII*, ed. Karl Zangemeister, Corpus Scriptorum Ecclesiasticorum Latinorum [hereafter CSEL] 5 (Vienna: apud C. Geroldi filium, 1882); *Histoires contre les païens*, ed. and trans. Marie-Pierre Arnaud-Lindet, vols. 1–3 (Paris: Belles Lettres, 1990–1); Orosius, *The Seven Books of History against the Pagans*, trans. Roy J. Deferrari (Washington, D.C.: Catholic University of America Press, 1964).

[8] At least 249 manuscripts survive from the Middle Ages: Lars Boje Mortensen, 'The diffusion of Roman histories in the Middle Ages: a list of Orosius, Eutropius, Paulus

An Orosius codex that was produced at St Gall in the late ninth century gives us an insight into how St Gall scholars studied this important source of classical historical and geographical learning. St Gall Stifts-bibliothek MS. 621 contains the full text of Orosius' *History against the Pagans* accompanied by numerous corrections and glosses (mainly Latin and some Old High German) written in several hands, which have been dated from the late ninth to the eleventh century. While the earlier handwriting remains anonymous, the eleventh-century hand has been identified as that of Ekkehard IV (*c.* 980–*c.* 1060).[9] Ekkehard IV, a student of the famous St Gall teacher and scholar Notker the German and a formidable scholar in his own right, left his annotations in many manuscripts preserved at St Gall. He also wrote the second part of the St Gall chronicle (*Casus Sancti Galli*) and a number of poems.[10] The annotations in St Gall 621 range from simple synonyms, aimed at explaining Orosius' Latin, a foreign language to most St Gall monks, to longer comments that address broader concepts and ideas. These comments reflect their compilers' learning strategies and opinions, revealing how they perceived the uses of history and geography.

Diaconus, and Landolfus Sagax Manuscripts', *Filologia Mediolatina* 6/7 (1999–2000), 101–200. The geographical chapter was sometimes transmitted separately from the rest of Orosius' history, as witnessed by the following manuscripts: Albi, Bibliothèque municipale, MS 29, s. VIII, Spain or Septimania; and Munich, Bayerische Staatsbibliothek, MS Clm. 396, s. IX. For these and other examples, see Mortensen, 'Diffusion'.

9 This is probably the codex mentioned in the list of books that belonged to Hartmut (abbot of St Gall from 872 to 885), Ratpert, *Casus*, 228; *Mittelalterliche Bibliothekskataloge*, 1, 87. This manuscript has attracted much attention from scholars. Zangemeister used it in his edition (Zangemeister, 'Praefatio', xx–xxi). For the description of the manuscript see Scherrer, *Verzeichniss*, 202; Bruckner, *Scriptoria*, 114; Scarpatetti, *Handschriften*, 219–21; and Heidi Eisenhut, 'St. Gallen, Stiftsbibliothek, Codex 621 (Beschreibung)', in CESG – Codices Electronici Sangallenses, online at http://www.cesg.unifr.ch/virt_bib/handschriften.htm. The codex is available in digital form on the same website. J. N. C. Clark studied Ekkehard's annotations and published a selection: 'The annotations of Ekkehart IV in the Orosius MS., St Gall 621', *Bulletin Du Cange* 7 (1932), 5–35. Heidi Eisenhut has produced the full critical electronic edition of the glosses from this codex (http://www.orosius.monumenta.ch/index.php) and an extensive study of the Orosius gloss tradition, *Die Glossen Ekkeharts IV. von St. Gallen im Codex Sangallensis 621*, Monasterium Sancti Galli 4 (St Gall: Verlag am Klosterhof, 2009), excerpts available on the same website. The earlier gloss hand has been variously dated from the late ninth to the eleventh century (see above descriptions).

10 On Ekkehard IV, see E. Dümmler, 'Ekkehart IV. von St. Gallen', *Zeitschrift für deutsches Altertum* 14 (1869), 1–73; Hans F. Haefele, 'Ekkehart IV. von St. Gallen', *Die deutsche Literatur des Mittelalters: Verfasserlexikon*, ed. Wolfgang Stammler et al., vol. 2 (Berlin: de Gruyter, 1980), 455–65; and Haefele's introduction in his edition of Ekkehard's *Casus*. For poems, see *Der Liber Benedictionum Ekkeharts IV.: nebst den kleinern Dichtungen aus dem Codex Sangallensis 393*, ed. Johannes Egli (St Gall: Fehr, 1909).

A small map drawn by Ekkehard in the margin, next to the beginning of Orosius' geographical chapter, shows how the eleventh-century scholar effectively clarified Orosius' text (Fig. 3.1). Orosius began his geographical survey by mentioning two opinions about dividing the earth into continents and described the earth (*orbis totius terrae*) as *triquadrus*.[11] While Ekkehard made no comments on the differences in opinion concerning the continents, he provided an ingenious visual explanation of the word *triquadrus*. He drew a square map divided into three parts, labelled Asia, Europe and Africa, with the Ocean surrounding the landmass.[12] Ekkehard's pictorial gloss aimed to elucidate the term unfamiliar to his eleventh-century audience (it seems to have no exact parallel in the previous geographical tradition). This map was found helpful by modern editors as well.[13] At the same time the map may have served as a visual memorization tool.[14]

Orosius' picture of the world was classical in its main outlines, reflecting the world as seen by the Graeco-Roman tradition.[15] While carefully

---

[11] St Gall 621, 35. Orosius I.2.1: *Maiores nostri orbem totius terrae, oceani limbo circumsaeptum, triquadrum statuere.*

[12] On explanatory functions of maps in manuscripts see Patrick Gautier Dalché, 'De la glose à la contemplation: Place et fonction de la carte dans les manuscrits du haut moyen âge', in *Testo e immagine nell'alto medioevo: 15–21 aprile 1993*, Settimane di studio del Centro italiano di studi sull'alto Medioevo, 41 (Spoleto: Presso la sede del Centro, 1994), 693–771. On this map, see Konrad Miller, *Mappaemundi: Die Ältesten Welkarten*, vol. 6 (Stuttgart: J. Roth, 1898), 63, fig. 28; *Mappemondes A. D. 1200–1500*, ed. Marcel Destombes (Amsterdam: N. Israel, 1964), 12 and 46 (classified as type A-3, square maps showing the tripartite division of the earth); Jorg-Geerd Arentzen, *Imago mundi cartographica: Studien zur Bildlichkeit mittelalterlicher Welt- und Ökumenekarten unter besonderer Berücksichtigung des Zusammenwirkens von Text und Bild* (Munich: W. Fink, 1984), 50; and Martha Teach Gnudi, 'Might Dante have used a map of Orosius?', *Italica* 15 (1938), 112–19.

[13] Manuscript spellings of this word include *triquadrum*, *triquedrum* and *triquetrum*. Zangemeister used *triquetrum* in his 'Die Chorographie des Orosius', in *Commentationes philologae in honorem Th. Mommseni* (Berlin: Weidmann, 1877), 715–38, at 721, and *triquedrum* in his *editio maior* of 1882, at 9. However, in his preface to the *editio maior*, p. xxxviii, Zangemeister already corrected it to *triquadrum*. In his later *editio minor* (Leipzig: Teubner, 1889), vi), he explained his decision to use *triquadrum* by referring to Ekkehard, who, in his opinion, had correctly understood the meaning of this word as 'consisting of three squares' and represented those on his map: *triquadrum, i.e. tribus quadriis constans, quod Orosium scripsisse censeo* (cf. ed. mai., XXXVIII) *recte intellexit iam Ekkehartus IV. Is enim in codice Sangallensi p. 35 ad locum mappulam rudem delineauit fere talem* ... [a schematic drawing of the map follows]. The English translation circumvents the difficulty by rendering the phrase as 'a threefold division of the whole world' (Deferrari, 7). Gnudi, 'Might', 113, has suggested that *triquadrum* should stand for *triquetrum*, triangular (?). However, no trace of a triangular earth in ancient or medieval geographical tradition survives.

[14] For an excellent analysis of geography and memory in Isidore of Seville, see A. Merrills, ch. 2, this vol.

[15] Orosius shaped the tradition according to his intentions, but the focus of relevance had obviously shifted by the time his work was read by St Gall scholars. On the strategies and models that Orosius used in his geographical introduction, see Merrills, *History and Geography*, 70–97.

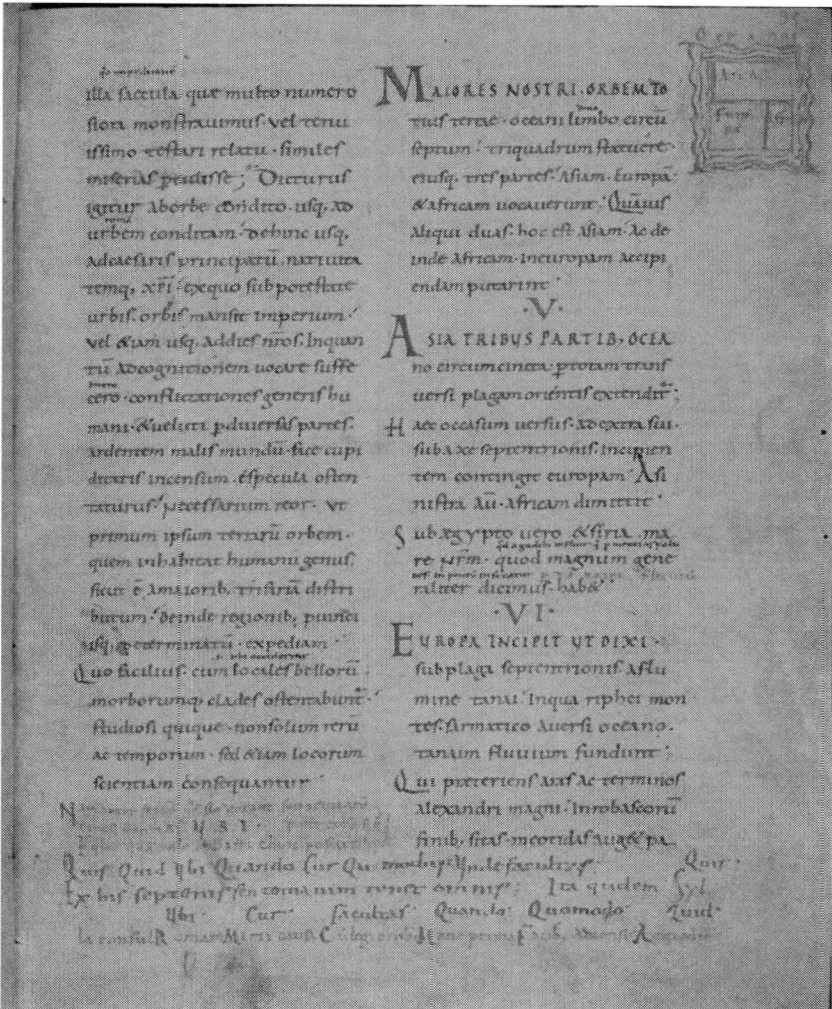

Figure 3.1 Ekkehard's 'square map' dividing the world into three parts, labelled Asia, Europe, and Africa, with the Ocean surrounding the landmass, in St Gall 621 (map is in top right-hand corner of the page). *Source:* Stiftsbibliothek St Gallen: 621, p. 35.

explaining the tradition, St Gall commentators also provided updates, focusing on places and events that lay fairly close to their home and were politically and historically significant. Thus in the geographical chapter the earlier hand supplied above the lines the current names for Moesia –

Uulgaria (Bulgaria) – and for Pannonia – Ungria (Hungary). The glosses also informed the reader that Pannonia was the region where the Ungri (Hungarians) live, and that Noricum was the place inhabited by the Baiouarii (Bavarians).[16] In a gloss that occurs later in the same codex, Ekkehard sharply criticized those in his monastery who confused the Hungarians with the Saracens: 'Hungarians now live in Pannonia, and some idiots among us are completely off the mark in calling them Hagarenes [Saracens]'.[17] He repeated the same criticism in his chronicle, the *Casus Sancti Galli*. As Meyer von Knonau has first noted, Ekkehard most likely directed his rebuke at the compilers of the St Gall Annals, who persistently referred to the Hungarians as Hagarenes and did not use the word 'Ungria' until the entry for 1030.[18] Ekkehard also noted in another manuscript, a copy of the Latin translation of Josephus, that Hungarians now lived in Pannonia.[19]

Ekkehard's efforts in getting these things right reveal his meticulous scholarly approach, as well as a special interest in the events that had directly involved St Gall. The memory of the Hungarian sack of the abbey in 926 must have been a painful one for St Gall inhabitants, and it still ran fresh in the eleventh century when Ekkehard wrote his glosses and composed his chronicle. In his *Casus*, Ekkehard told the dramatic story of how the abbey fell to the Hungarians. Whereas the glosses on the Hungarians in St Gall 621 provided a learned context for conceptualizing these contemporary barbarians, linking them to those familiar from the classical tradition, the vivid details of the enemies' customs and behaviour that Ekkehard recorded in his chronicle differ from standard classical accounts and most likely go back to the abbey's oral tradition.[20]

---

[16] St Gall, Stiftsbibbliothek 621, 41A (Orosius I.2.55) MOESIA quae nunc uulgaria; PANNONIAM quae nunc ungria; 41 B (Orosius I.2.60): *PANNONIA in qua ungri NORICUS in qua baioarii RHETIA in qua Alemanni et Rhetii curiales.*

[17] Ibid., 315 A (Orosius 7.32.14): *PER PANNONIAS in quibus nunc ungri . Quos longe a uero lapsi . Idiotae nostri quidam . nunc agarenos uocant* ... Also at 267 A (Orosius 6.21.14): *NORICI nunc baioarii . ILLIRICI . PANNONII huni . nunc ungri.* Ekkehard, *Casus* 82, 170: *Qui autem Ungros Agarenos putant, longa via errant.* I thank Emily Albu for supplying a felicitous turn of phrase for the translation of this passage.

[18] Ekkehard, *Casus Sancti Galli*, ed. Meyer von Knonau, Mitteilungen zur vaterländischen Geschichte, N.F. 5–6 (St Gall: Fehr, 1877), 298, n. 988.

[19] St Gall 626, 130 (*Hegesippi qui dicitur historiae libri v*, ed. Vincenzo Ussani, CSEL 66, II.9.41, 151): *PANNONIORUM ibi nunc ungri sunt* ... *PANNONIUS i. ungar.* This codex is a ninth-century copy of Hegesippus' Latin version of Flavius Josephus' *Jewish Wars*. The text is accompanied by annotations written by four contemporary or slightly later hands, including the hand of Ekkehard IV. For the description of the manuscript see Scherrer, *Verzeichniss*, 204; Scarpatetti, *Handschriften*, pp. 230–32; and Eisenhut, 'St. Gallen'.

[20] For the Hungarian conquests in the late ninth–early tenth century, see Charles R. Bowlus, *Franks, Moravians, and Magyars: The Struggle for the Middle Danube, 788–907* (Philadelphia: University of Pennsylvania Press, 1995), 236–67. For the sack of St Gall

The updates made by St Gall commentators formed part of their endeavour to understand the order of events in space and time, beginning with ancient history and often leading up to the present.[21] Thus when Orosius summarized the number of years that passed between the time of Ninus and Abraham and the time of Emperor Augustus, Ekkehard reminded his readers that Ninus and Abraham lived at the same time.[22] Concerning more recent history, the annotators showed a particular interest in establishing how the order of power changed in certain regions and how one ethnicity replaced another. Following up on the matter of the Hungarians, the earlier commentator picked up on Orosius' mention of the Pannonii and reported that 'now they do not exist, and the Hungarians possess their land'.[23] When Orosius in his geographical chapter mentioned the Suevi, the people who lived in the area roughly corresponding to modern southern Germany and Switzerland where St Gall was located, the earlier annotator introduced a brief history of changes in power and population: 'They [the Suevi] are called so because of [the name of] the Mountain Sueus. But those who came from these parts are now called Alemanni, from the name of the lake of Lemannus [Lake Geneva], the area around which they have subjugated by arms'.[24] When later Orosius explained that Pannonia, Noricum and Rhaetia were encircled by the Danube in the north and that the Danube also divided Gaul and Germany, the same earlier annotator added that these parts 'are populated by the Alamanni, the Alsatii, and the part of the Franks who are called the Teutoni and that the Gauls are now called the Franks'.[25] The earlier

---

by the Hungarians, see Ekkehard, *Casus* 51–5 with notes by Von Knonau and Haefele in their respective editions and *Die Ungarn und die Abtei Sankt Gallen. Magyarok és a Szent Galleni Apátság: Akten des wissenschaftlichen Kolloquiums an der Universität Eötvös Loránd Budapest vom 21. März 1998 anlässlich der Ausstellung 'Die Kultur der Abtei Sankt Gallen' im Ungarischen Nationalmuseum (21. 3.–30. 4. 1998)*, ed. György J. Csihák and Werner Vogler (Zurich: Ungar-Historischer Verein, 1999). For a discussion of St Gall glosses on barbarians, see Lozovsky, 'Roman geography', 343–4.

[21] Lozovsky, 'Roman geography'.

[22] St Gall 621, 33 B (Orosius I.1.6): *A NINO AUTEM UEL ABRAHAM contemporaneis.*

[23] St Gall 621, 268 A (Orosius VI.21.23): *PANNONIOS Hunos* [written by Ekkehard] *qui nunc nulli sunt . quorum ungri nunc terram tenent* [written by the earlier hand].

[24] Ibid., 41 A (Orosius I.2.53): *SUEUI a sueuo monte circa quem habitant sic nominati. Sed et ab ipsis egressi . alemanni nunc uocantur . a Lemanno uidelicet laco . cuius propinqua armis subegerant.* Cf. Isidore of Seville, *Etymologiarum sive Originum libri XX*, ed. Wallace M. Lindsay, 2 vols. (Oxford: Clarendon, 1911), IX.2.94: *Lanus fluvius fertur ultra Danubium, a quo Alani dicti sunt, sicut et populi inhabitantes iuxta Lemannum fluvium Alemanni vocantur;* and IX.2.98: *Dicti autem Suevi putantur a monte Suevo....*

[25] St Gall 621 42 A (Orosius, *Hist.* I.2.60): *DANUBII FONTEM ET LIMITEM QUI GERMANIAM A GALLIA INTER DANUBIUM GALLIAMQUE SECERNIT Ipsum istum limitem . alemanni Alsatii . et franchorum pars teuto*[earlier hand]*na Inc*[Ekkehard]*olunt . Nam et galli nunc franci solent uocari*[earlier hand].

annotator also pointed out that the Langobards settled down in northern Italy, and Ekkehard repeated this information later in the codex.[26] The annotators of St Gall 621 also made a special effort to reinforce the Christian contents of Orosius' history and geography. The space that Orosius described in his geographical introduction was Roman in its main outlines, with no references to Christian landmarks. For instance, Orosius never mentioned Paradise or the Holy Land, and Ekkehard supplemented his text. In a gloss above the line which mentions the island of Taprobane (Ceylon), he added: 'They say that beyond it [Taprobane] and the infinite sea lies Paradise'.[27] Ekkehard also drew a map, which appears to be the earliest extant medieval example of a regional map that focuses on the Holy Land and the surrounding areas (Fig. 3.2).[28]

The map shows Palestine, indicated as 'The Promised Land', enclosed within the space formed by the river Jordan, which originates from the confluence of two streams, the Ior and the Dan. Both the Jordan and its sources are indicated by a double line, unlike the river Nile, which is represented by a single line. A broken line running across a body of water (which is meant to represent the Red Sea but is not so named), indicates the route taken by the Israelites in their flight from Egypt. A strip of land shows the parting of the Red Sea. The only two cities named within the Holy Land are Jerusalem and Jericho. Both are marked by pictures of buildings, with Jerusalem represented by a larger and more elaborate building with a cross on top. Some features of the map correspond to the information that Orosius gives in his geographical section, while others are drawn from other sources.[29] For instance, the map shows Arabia

---

[26] The earlier annotator, St Gall 621, 156 B (Orosius IV.13.15): *MEDIOLANUM ciuitas cisalpinae galliae . quae uariis nominibus distinguebatur . nam galli bello in Italia sicut et post langobardi consederant.* Ekkehard, ibid. 166 B (Orosius IV.20.4): *INSUBRES . BOII . ATQUE CENOMANNI galli de planis italiae . ubi post langobardi.*

[27] St Gall 621, 37 A (Orosius I.2.16): *INSULA TAPROBANE ... Ultra quam . et mare infinitum . aiunt esse Paradysum.*

[28] On medieval maps of the Holy Land, see P. D. A. Harvey, 'The Holy Land on medieval world maps', in *The Hereford World Map: Medieval World Maps and their Context*, ed. P. D. A. Harvey (London: British Library, 2006), 243–51; Catherine Delano-Smith, 'Geography or Christianity? Maps of the Holy Land before A.D. 1000', *Journal of Theological Studies* 42 (1991), 142–52; and Kenneth Nebenzahl, *Maps of the Holy Lands: Images of Terra Sancta through Two Millennia* (New York: Abbeville Press, 1986).

[29] Thus I cannot entirely agree with Arentzen, who argues that the map is not derived from Orosius' text because it contains Christian elements, missing in Orosius' description, and omits most of the names important for the orientation system of the text (Arentzen, *Imago*, 51). The map is described and reproduced in Miller, *Mappaemundi*, 63 and fig. 29, who gives an erroneous date for the map, repeated by later scholars, such as Delano-Smith, 'Geography or Christianity?', 150; and James M. Scott, *Geography in Early Judaism and Christianity: The Book of Jubilees* (New York: Society for New Testament Studies Monograph Series, 2002), 243, n. 47.

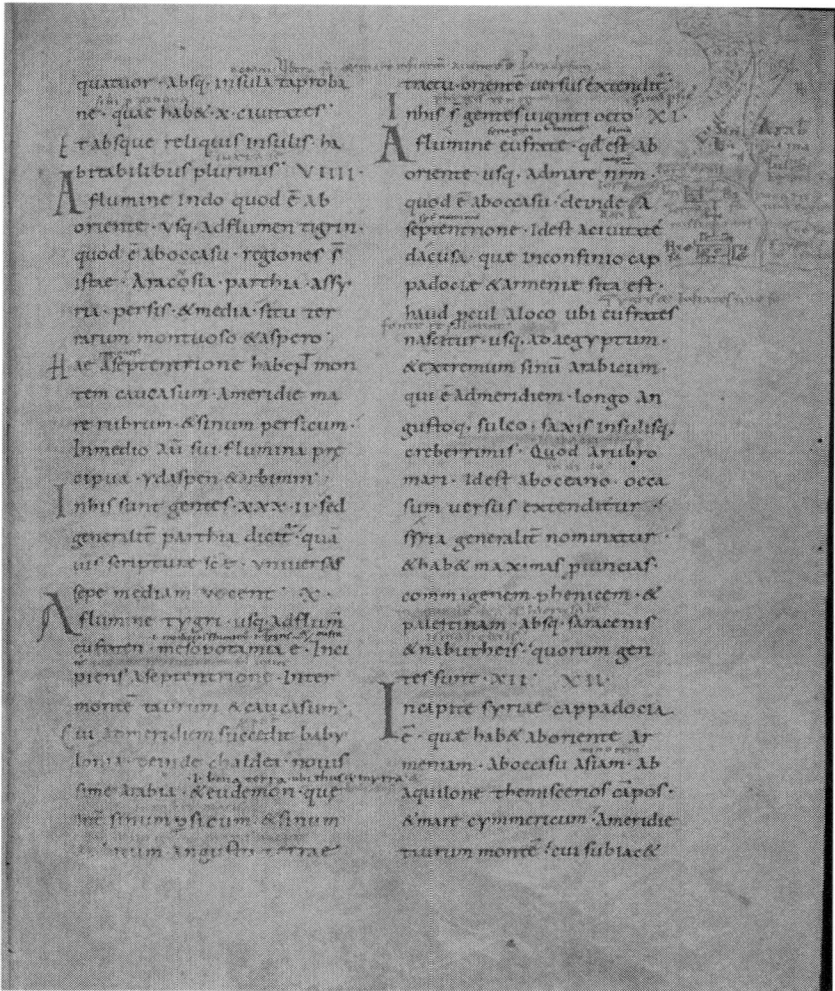

Figure 3.2 Ekkehard's map of Palestine in St Gall 621, which appears to be the earliest extant medieval example of a regional map that focuses on the Holy Land and the surrounding areas (map is in top right-hand corner of the page).
Source: Stiftsbibliothek St Gallen: 621, p. 37.

Eudemon in the same location and shape as those described by Orosius, who says that Arabia Eudemon 'extends toward the east in a narrow strip between the Persian Gulf and the Arabic Gulf'.[30] The map also follows Orosius, who reports that Superior Egypt extends in length towards the east and has the Arabic Gulf to the north and Lower Egypt to the West.[31] True to Orosius' description, the Nile runs through the length of Egypt, with Meroe shown as a large island. However, certain features do not correspond to Orosius' account. For instance, Ekkehard's map supplied the information about the sources of the Jordan, the Ior and the Dan, missing in Orosius but available from other texts, ranging from various accounts about the Holy Land to the *Etymologiae* by Isidore of Seville.[32] Showing the same persistency as in the case of the Hungarians, Ekkehard added the names of the sources of the Jordan in another manuscript.[33]

Ekkehard's map bears similarities to the way the Holy Land is depicted on some *mappaemundi*, sharing their orientation, spatial layout, and such specific features as the sources of the Jordan and the indication of the Red Sea crossing. Thus a world map, which he could consult in the St Gall library, was likely to serve as a model for Ekkehard.[34] In his *Casus*

---

[30] Orosius I.2.21: *Arabia Eudaemon, quae inter sinum Persicum et Arabicum angusto terrae tractu orientem uersus extenditur.*

[31] Orosius I.2.34: *Aegyptus superior in orientem per longum extenditur. cui est a septentrione sinus Arabicus, a meridie oceanus, nam ab occasu ex inferiore Aegypto incipit, ad orientem Rubro mari terminatur.*

[32] Accounts of the Holy Land: *Ps.-Antonini Placentini Itinerarium* 7; Theodosius, *De situ terrae sanctae* 2; Adomnan, *De locis sanctis* II.19, all edited in Corpus Christianorum Series Latina 175. Isidore, *Etym.* XIII.21.18: *Iordanis Iudaeae fluvius, a duobus fontibus nominatus, quorum alter vocatur Ior, alter Dan. His igitur procul a se distantibus in unum alveum foederatis, Iordanis deinceps appellatur. Nascitur autem sub Libano monte, et dividit Iudaeam et Arabiam; qui per multos circuitus iuxta Iericho in mare Mortuum influit.* St Gall scholars could consult some of the pilgrims' accounts and Isidore's encyclopaedia in their own library. St Gall 133, an eighth-century manuscript that may have been produced at St Gall, includes the *Antonini Placentini Itinerarium*. St Gall 231 and 232, produced at St Gall in the late ninth century, transmit the text of the *Etymologiae*. St Gall 237, an early ninth-century copy of the *Etymologiae* not necessarily written at St Gall, may have already been present at the library in the ninth century.

[33] St Gall 626, 171 (Hegesippus III.6.5, 198): *INCIPIT ENIM A FONTIBUS IORDANIS i. ior et dan.*

[34] Ekkehard's map shares similarities with such *mappaemundi* as the Albi Map (Albi, Bibliothèque Municipale MS 29, f. 57v, *c.* 730, Spain or south-western France), the Vatican Map (Vatican, Bibliotheca Apostolica Vaticana MS Vat. Lat. 6018, ff. 63v–64r, 762–77, Italy), and the map in Munich, Bayerische Staatsbibliothek MS Clm 6362, f. 74r, s. XI, Freising. Good reproductions of these maps can be found in Leonid S. Chekin, *Northern Eurasia in Medieval Cartography: Inventory, Texts, Translation, and Commentary* (Turnhout: Brepols, 2006). For maps of Palestine copied from *mappaemundi*, see Harvey, 'Holy Land', 243. Ekkehard's map also shares similarities with the so-called Jerome map of Palestine found in a manuscript of Jerome's works which was produced *c.* 1200 in Tournai (London, British Library, Additional MS 10049, f. 64v). The map of Asia drawn on the recto of the same folio (f. 64r) presents

Sancti Galli, Ratpert mentions a *mappamundi* of exquisite workmanship made by the order of abbot Hartmut, but it is not found among the extant St Gall manuscripts. None of the maps present in other St Gall codices, which Ekkehard had an opportunity to consult, is detailed enough to have served as his model.[35]

Ekkehard's map of the Holy Land provided an effective visual tool for introducing and illustrating a Christian interpretation of geographical space. Further throughout the codex, both the earlier annotator and Ekkehard also pointed out and reinforced other Christian lessons that could be drawn from Orosius' history and geography. Their reactions to Orosius' texts also give us an insight into how they perceived the place and meaning of history and geography in the world and in the system of knowledge. Following his vision of world history as a movement from the original sin to salvation that embraced both the human population and their physical environment, Orosius often emphasized connections and correlations between the two, demonstrating how the physical earth shared in the fate of humanity. Thus in interpreting the biblical story of the destruction of Sodom and Gomorra, Orosius pointed out that God punished its very soil for the sins of its people. The earlier St Gall annotator emphasized the importance of Orosius' interpretation by drawing special attention to this passage and urging the reader to mark Orosius' words: 'Note that the earth with its fruits always suffers punishment for the sins of the people'.[36]

Orosius' ideas about intimate connections between human history and the earthly space lent a deeper spiritual dimension to his appreciation of history and geography both as objects and as subjects of study. While drawing on the achievements of his predecessors who used geography in their historical works, Orosius appears to be the first writer in the Latin

---

an even more intriguing similarity with glosses from St Gall 621, updating the information on Moesia in the same way: *mesia hec est vulgaria*. Ekkehard's map and the two 'Jerome' maps may ultimately go back to the same model, a connection which merits further investigation but has not been pursued, although the update on the Jerome map of Asia has been noted (for this and the recent bibliography, see Chekin, *Northern Eurasia*, 134–5).

[35] Ratpert, *Casus*, 228: *Inter hos etiam unam mappam mundi subtili opere patravit, quam inter hos quoque libros connumeravit.* Also in *Mittelalterliche Bibliothekskataloge*, vol. 1, 87, lines 23–4. Maps surviving in St Gall 236, 89 and St Gall 237, 1 and 219, are tripartite schematic representations of the T-O and T-Y type. The scope of this chapter does not allow for a discussion of these maps, some of which have drawn considerable scholarly attention, esp. the map in St Gall 237, 1: see Miller, *Mappaemundi*, vol. 6, 57, fig. 27; Gautier Dalché, 'De la glose', 727–8; Anna-Dorothee von den Brincken, *Fines Terrae: Die Enden der Erde und der vierte Kontinent auf mittelalterlichen Weltkarten* (Hanover: Hahnsche Buchhandlung, 1992), 52.

[36] St Gall 621, 51 A (Orosius I.5.11): *TERRA QUOQUE IPSA QUAE HAS HABUERAT CIUITATES nota quod semper terra cum fructibus suis peccata hominum luit.*

tradition to have placed a long geographical description at the beginning of his world history. He provided no elaborate justifications for his decision, simply stating his wish to tell the reader about both times and places in which events happened.[37] While St Gall commentators did not elaborate on this question either, their glosses allow us a glimpse into how they envisioned the place of geographical learning in the system of education and knowledge. Ekkehard's gloss, which follows a passage in Orosius mentioning time and place, suggests that the eleventh-century scholar instinctively identified rhetoric as a possible educational context of the text discussing time and place. Picking up on the notions of 'place' and 'time' in the text of Orosius, Ekkehard explained that the question 'where?' occupies an important place among the seven *circumstantiae*, used for analysing statements. Reminding the readers about the rules of classical rhetoric used throughout the Middle Ages in order to analyse and understand texts, he recited the seven questions (*who? what? why? how? where? when? by what means?*) and demonstrated how to use them in analysing a sample sentence.[38] This impression is reinforced by glosses in which Ekkehard and the earlier commentator pointed out rhetorical figures throughout the manuscript.[39]

Could this manuscript have been used in education? Although it somewhat differs in appearance from what has been termed a 'school book', this would not preclude its use for education, probably at an advanced stage.[40] While neither history nor geography had a firmly established place in the medieval curriculum, these subjects fit into

[37] Specifying relations and connections between history and geography and understanding the role that geographical descriptions played in historiographical works seems to be a largely modern preoccupation. For a theoretical discussion, see Clarke, *Between Geography and History*, and Merrills, *History and Geography*, the latter also specifically discussing Orosius. On the place and function of Orosius' geographical introduction see also Eugenio Corsini, *Introduzione alle 'Storie' di Orosio* (Turin: G. Giappichelli, 1968), 73–83; Fabrizio Fabbrini, *Paolo Orosio: uno storico* (Rome: Edizioni di storia e letteratura, 1979), 322–7; Yves Janvier, *La géographie d'Orose* (Paris: Belles Lettres, 1982), esp. 9–10. Orosius I.1.17: *quo facilius, cum locales bellorum morborumque clades ostentabuntur, studiosi quique non solum rerum ac temporum sed etiam locorum scientiam consequantur.*

[38] St Gall 621, 35 A: *Nam inter septem quas sic uocant sententiarum circumstantias U B I . postremum non est . E quibus quidam tale Distichon posuit . Quis Quid Ubi Quando Cur Qui modus . Unde facultas . Ex his septenis sententia uim tenet omnis; Ita quidem . (Quis) Sylla consul (Ubi) Romam (Cur) Marii causa (Facultas) Cum legionibus (Quando) Mane prima (Quomodo) Facibus accensis (Quid) Aggreditur.* For discussion, see Grotans, *Reading*, 225.

[39] For instance, St Gall 621, 91 B (Orosius II.7.1-2); glosses in the earlier hand: *DELIBERATIO FUIT uerbum Rhetoricum i. consultatio . . . CIUITATEM STERNENDAM suasio de utili rethorico . . . SE PROMISSUROS suasio de honesto rhetorico.*

[40] Scholars have not yet reached a consensus about the function of glossed manuscripts in medieval education and learning. For the most recent contribution to the discussion, see Mariken Teeuwen, 'Glossing in close co-operation: examples from ninth-century Martianus Capella manuscripts', in *Practice in Learning: The Transfer of Encyclopaedic*

various niches within the liberal arts, encyclopaedias and treatises 'On the Nature of Things'.[41] Like other secular subjects, the studies of history and geography ultimately were meant to fulfil the highest goal of medieval education, which was to approach the understanding of God through learning about the created world.[42] The annotators of St Gall 621 seemed to keep multiple educational goals in mind. In their effort to make ancient history and geography comprehensible, the annotators explained a broad range of subjects, from vocabulary and grammar to theological ideas. They also reinforced the Christian lessons that studies of history and geography could provide. All these features would make the manuscript very useful in the context of education and learning.[43] This supplements our evidence about a possible place of geography among the liberal arts in St Gall's educational curriculum. According to a letter preserved in a late ninth-century collection, students would learn about 'locations of regions' along with other subjects pertaining to natural sciences while studying the subjects of the quadrivium.[44] Glosses on Orosius point at the verbal arts of the trivium (grammar, rhetoric and dialectic) as a context that could accommodate historical and geographical studies.

Knowledge in the Early Middle Ages, eds. Rolf H. Bremmer Jr. and Kees Dekker (Leuven: Peeters, 2011), 85–99. In the course of this debate, Gernot Wieland proposed a set of criteria for identifying a glossed manuscript as a classbook: 'The glossed manuscript: classbook or library book?', *Anglo-Saxon England* 14 (1985), 153-73, esp. 170. The Orosius manuscript actually fits most of these criteria. Lars Boje Mortensen, on the contrary, did not think that the extant Orosius manuscripts were used at school: 'The diffusion of Roman histories in the Middle Ages: a list of Orosius, Eutropius, Paulus Diaconus, and Landolfus Sagax Manuscripts', *Filologia Mediolatina* 6/7 (1999–2000), 101–200, at 106.

[41] For a recent overview and bibliography, see John J. Contreni, 'The Carolingian Renaissance: education and literary culture', in *The New Cambridge Medieval History*, vol. 2: *c.700–c.900*, Rosamond McKitterick (ed.) (Cambridge University Press, 1995), 709–57. On the role of geography in Christian education in the early Middle Ages see Lozovsky, *'Earth is our Book'*, ch. 4. For a brief mention of geographical information within the quadrivium as practised at St Gall, see Grotans, *Reading*, 76.

[42] For history, see Allen, 'Bede and Freculf', 68; K. F. Werner, 'Gott, Herrscher und Historiograph: Der Geschichtsschreiber als Interpret des Wirkens Gottes in der Welt und Ratgeber der Könige (4. bis 12. Jahrhundert)', in *Deus qui mutat tempora: Menschen und Institutionen im Wandel des Mittelalters: Festschrift für Alfons Becker zu seinem fünfundsechzigsten Geburtstag*, ed. E.-D. Hehl et al. (Sigmaringen: J. Thorbecke, 1987), 1–31. For geography, see Lozovsky, *'Earth is our Book'*.

[43] Scholars in later centuries recognized this potential and copied the text with accompanying glosses. For analysis of the three twelfth-century manuscripts directly dependent on St Gall 621, see Eisenhut, *Glossen Ekkeharts*.

[44] *Das Formelbuch des Bischofs Salomo III von Konstanz aus dem neunten Jahrhundert*, ed. Ernst Dümmler (Osnabrück: O. Zeller, 1964), 51: *de regionum situ quaerere, de cursu planetarum uario scitari, de stellarum effectibus admirari*. For discussion and more evidence, see Clark, *Abbey*, 119–21, and Grotans, *Reading*, 76-7.

While meticulously following the established traditions of study, St Gall annotators also made what appear to be their own unique contributions. Both Ekkehard and his earlier colleague wished to make Orosius' history and geography not only understandable but also relevant for their audience and provided updates of the classical tradition. Although we cannot be certain that all glosses preserved in this codex were composed at St Gall, this appears to be the case for the updates that focus on places and peoples significant for the monastery.[45] Glosses in St Gall 621 in general reveal a high level of personal engagement with the material. In their immediacy and distinctive contemporary concerns, the St Gall annotations seem different from sets of glosses on other texts, which were produced at different places, such as those on Martianus Capella or Virgil. The ninth-century glosses on Martianus Capella's geographical account, for example, seem largely frozen in time. In an effort at what might be seen as historical preservation, the annotators carefully explain the classical picture of the world, which must have seemed antiquated even in the fifth century when Martianus wrote his encyclopaedia. The modifications of the classical picture of the world that they introduce are rare and subtle.[46] Although the St Gall annotators also often engage in 'historical preservation', they practise a different approach to the text when it mentions places close to home. When this happens, we can witness snippets of an ongoing conversation, a lively dialogue with the past.[47]

---

[45] Neither the place of composition nor the source of the St Gall glosses is entirely clear. At the end of Orosius' text, Ekkehard mentioned that he had access to two other copies of Orosius' history, which he used to correct the present manuscript. See St Gall, 621, 351: *Plura in hoc libro fatuitate cuiusdam ut sibi uidebatur male sane asscripta. Dominus Notker abradi et utiliora iussit in locis asscribi; Assumptis ergo duobus exemplaribus quae deo dante ualuimus . tanti uiri iudicio fecimus.* Some scholars have suggested that these manuscripts may have contained glosses; see, for instance, Otto Skutsch (ed.), *The Annals of Q. Ennius* (Oxford: Clarendon Press, 1985), 25–6.

[46] The tradition of Orosius glosses preserved in St Gall 621 and codices copied from it also differs from gloss traditions found in other Orosius manuscripts. Glosses from one such tradition as presented in Vatican City, Biblioteca Apostolica Vaticana, MS Vat. lat. 1650 have been edited by Olivier Szerwiniack, 'Un commentaire hiberno-latin des deux premiers livres d'Orose, *Histoires contre les paiens*', *Bulletin Du Cange* 51 (1992–93), 5–137 and 65 (2007), 165–207. For a detailed discussion of various traditions, see Eisenhut, *Glossen Ekkeharts*. For geography and history in Martianus Capella commentaries, see Lozovsky, '*Earth is Our Book*', 113–38 and Natalia Lozovsky, 'Perceptions of the past in ninth-century commentaries on Martianus Capella', in *Carolingian Scholarship and Martianus Capella*, ed. Mariken Teeuwen and Sinead O'Sullivan, Cultural Encounters in Late Antiquity and the Middle Ages (CELAMA), 12 (Turnhout: Brepols, 2011), 123–45.

[47] David Ganz has discussed the dialogue with the past in which an earlier St Gall scholar, Notker Balbulus, was engaged; see 'Humour as history in Notker's *Gesta Karoli Magni*', in *Monks, Nuns, and Friars in Mediaeval Society* (Sewanee: University of the South Press, 1989), 171–83, esp. 180. The glossators' immediate reactions to Orosius' text are not

This deep engagement in conversation with the classical tradition provided St Gall scholars with a framework for interpreting and context-ualizing their own times. When reading Orosius, Ekkehard often noted the role of fortune and misfortune in the rise and fall of kingdoms and rulers, and his contemplation of this topic informed Ekkehard's own historical work.[48] One of the main themes of his *Casus*, declared in the introduction and followed throughout the text, is the interplay of fortune and misfortune in the history of his monastery.[49] Hans Haefele, the most recent editor of Ekkehard's chronicle, has argued that it was the reading of Boethius that had inspired Ekkehard's fascination with the vicissitudes of *fortuna*.[50] Ekkehard's notes in the Orosius manuscript demonstrate that it was also his serious study of Orosius' history that provided the St Gall scholar with inspiration and concrete examples of how *fortuna* affected peoples and kingdoms.

★★★

In the period between the ninth and the eleventh centuries, St Gall scholars actively pursued the studies of history and geography. They copied and excerpted authoritative texts, produced glosses and diagrams, and used their learning in writing their own works. Annotations in St Gall 621 show how the glossators explained Orosius by employing traditional strategies of textual scholarship and addressing contemporary concerns. These glosses, possibly used in education, complement our

---

limited to their updates of geo-ethnographical information and deserve further study. For more examples, see Eisenhut's edition of St Gall glosses.

[48] Ekkehard shared this interest with the earlier annotator: St. Gall 621, p. 72 B (Orosius II.4.2): *PARIQUE SUCCESSU CRUDELITATIS fortunio* [written by an earlier hand] *uel magis infortunio* [written by Ekkehard]; p. 92 A (Orosius II.17.6): *OCCISO ALCIBIADE AUSPICANTUR hoc infortunio portendunt* [earlier hand]; p. 142 A (Orosius IV.6.9): *CYRI REGIS PERSARUM GESTA SUNT apud cartaginem . cuius infortunia praelibamus* [Ekkehard]; p. 172 B (Orosius IV.22.4): *AURI ARGENTIQUE METALLIS SUPPLEUERUNT uide urbem ditissimam . tam infortunatam . Ut numquam in diuitiis spem ponas* [Ekkehard]; p. 236 B (Orosius VI.7.7): *ARREPTA NAUICULA ultro per fortunam caesaris* [Ekkehard]; p. 255 B (Orosius VI.16.7): *CUM SUBITO UERSUS IN FUGAM sine ui . fortuna caesaris* [earlier hand].

[49] In the beginning of his *Casus*, Ekkehard writes: 'we have undertaken the difficult task of relating some of the fortunate and unfortunate events [*cum infortuniis tradere fortunia*] at the monastery of Saints Gall and Otmar' (*Casus, Preloquium*, p. 16). On *fortunia et infortunia* as the leading theme of Ekkehard's *Casus*, see Hans F. Haefele, 'Untersuchungen zu Ekkehards IV. Casus Sancti Galli. 2. Teil', *Deutsches Archiv* 18 (1962), 120–70.

[50] Hans F. Haefele, 'Zum Aufbau der Casus Sancti Galli Ekkehards IV', in *Typologia litterarum: Festschrift für Max Wehrli*, ed. Stefan Sonderegger et al. (Zurich: Atlantis, 1969), 155–66, esp. 160. See also the critique in Ernst Hellgardt, 'Die Casus Sancti Galli Ekkeharts IV. und die Benediktsregel', in *Literarische Kommunikation und soziale Interaktion: Studien zur Institutionalität mittelalterlicher Literatur*, ed. Beate Kellner et al. (Frankfurt: Lang, 2001), 27–50, at 34.

evidence about the place of geography in St Gall's curriculum. Studies of history and geography provided St Gall scholars with information and inspiration. The classical tradition inspired Notker Balbulus, who used a geographical framework to organize his account of the deeds of Charlemagne, thus embedding recent history in a frame of classical scholarly and ideological references. Ekkehard's in-depth study of Orosius' text informed his approach to writing the history of his monastery.

Even the small portion of St Gall materials discussed above shows their exceptional promise. St Gall evidence gives us a rare opportunity to observe how historical and geographical learning developed and changed at one centre throughout the Middle Ages and beyond. While new discoveries and approaches have radically changed the old negative picture of medieval geographical knowledge, an alternative story of how it was learned, taught and used still remains to be told. St Gall materials provide abundant evidence for writing such a new story by allowing us to observe traditional strategies and immediate reactions of St Gall scholars, to witness the joy and pride they took in learning, and to reconstruct how they used ancient history and geography to understand their place within the physical world and human history.[51]

---

[51] I thank Emily Albu, who read this chapter in several drafts, and Kathy Lavezzo for their comments. All errors and infelicities that remain are my own.

4   The cosmographical imagination of Roger Bacon

Amanda Power

In the summer of 1264, according to Roger Bacon, the planet Mars held sway over human affairs. Its fiery influence irradiated the heavens and the earth below, arousing people to fury and discord. It generated a terrifying comet that sped through the skies 'like iron rushing toward a magnet', igniting wars and conflict across the lands that lay beneath its path. Fighting was still going on in England, Spain, Italy and other countries some two or three years later when Bacon was writing a series of treatises on the reform of learning at the request of Pope Clement IV.[1] Caught up in the effects of this cosmological disturbance, Bacon's own family suffered financial ruin during the English wars when they supported Henry III against his barons. In the chaos, he had lost touch with them.[2] Yet, in Bacon's view, people with the right knowledge of both geography and astrology could have anticipated the effects of the celestial spheres on the sublunary world and anyone could exercise their free will to resist them. The comet of 1264 brought such devastation because the population of the Latin West, and particularly its leaders, were ignorant of these sciences, unprepared and therefore susceptible to the torrents of disharmony pouring through the atmosphere. 'Oh, how great an advantage might have been obtained for the Church of God', Bacon wrote,

if the qualities of the heavens in those times had been understood beforehand by scholars, prelates and princes, and the knowledge had been used to engender a fixed resolve for peace. For then such a great slaughter of Christians would have been avoided, and so many souls would not have been sent to hell.[3]

---

[1] *Sicut ferrum currit ad magnetem*, Roger Bacon, *Opus maius*, J. H. Bridges (ed.), 3 vols. (Oxford: Clarendon Press, 1897–1900), IV.iv.16, 1:385.

[2] Roger Bacon, *Opus tertium*, in *Fr. Rogeri Bacon Opera Quaedam Hactenus Inedita*, J. S. Brewer (ed.), Rerum Britannicarum Medii Aevi Scriptores [Rolls Series], 15 (London, 1859), 3–310, ch. 3, 16.

[3] *O quanta utilitas ecclesiae Dei potuisset procurari, si coeli qualitas istorum temporum fuisset praevisa a sapientibus, et praelatis et principibus cognita, et pacis studio mancipata! Nam non fuisset tanta Christianorum strages nec tot animae positae in infernum*, Bacon, *Opus maius*, IV. iv.16, 1:386.

In his account of the causes of the wars in the mid-1260s and his confident belief that greater understanding would enable humans to exercise some control over their environment and circumstances, Bacon trod an uneasy path through a moral – and, as will be seen, cosmographical – quagmire. Two points should be emphasized. First, it is obvious that knowledge of geography was crucial, for the influence of the comet was felt in certain parts of Europe, and it was there that wars broke out. In order to determine the effects of the heavens on specific places and peoples, the astronomer needed to be able to pinpoint them with mathematical precision. It was imperative that the Latins acquire the necessary information and develop the appropriate expertise to produce accurate maps of the earth. These could then be used in conjunction with astronomical observations, and the meaning given to the movements of the heavens by the discipline of astrology, so that the whole functioning of the cosmos could be seen and understood.

Secondly, there was the issue of action and moral responsibility. There was no doubt in Bacon's mind that Mars was the cause of the discord, operating on an unwary population that was swayed, unknowingly, to unusually aggressive behaviour by external forces. Yet however antithetical to contemporary theological thought this might have sounded, it had important implications within a Christian world-view.[4] The emotions unleashed by the influence of Mars led people to behave with such violence that, when they were killed, they were damned for eternity. It was apparent that the effect of planetary influences on humans could decide the fate of souls if it was not combated. Bacon remained within the bounds of orthodoxy in his insistence that free will could have and should have been exercised. He simply maintained that humans left in ignorance of cosmic forces, and acting according to the promptings of fallen human nature, were at a great disadvantage. They were in effect being deprived of the opportunity to exert their free will. The salvation of souls was the business of the Church; consequently, the Church had a responsibility to accumulate knowledge of the cosmos and to act on it.

---

[4] The relationship between astronomy/astrology and orthodoxy was often far more ambiguous than the stark public denunciations of the former by medieval intellectuals would suggest. On this point, see V. Flint, 'World history in the early twelfth century: the "Imago Mundi" of Honorius Augustodunensis', in *The Writing of History in the Middle Ages: Essays Presented to Richard William Southern*, R. H. C. Davis and J. M. Wallace-Hadrill (eds.) (Oxford: Clarendon Press, 1981), 211–38; and more generally, V. I. Flint, *The Rise of Magic in Early Medieval Europe* (Oxford: Clarendon Press, 1991); J.-P. Boudet, *Entre science et nigromance: Astrologie, divination et magie dans l'Occident médiéval (XIIe–XVe siècle)* (Paris: Publications de la Sorbonne, 2006).

Bacon's thought can seem idiosyncratic at first sight, and this example is no exception. Yet the passage occurs in a work that was written at the request of the Pope by a Franciscan friar and respected intellectual who had undertaken years of research, thought deeply and was deadly serious.[5] The old assumption that astrology was invariably regarded with suspicion has, in any case, been challenged in recent decades by research demonstrating that skilled practitioners of astrology and other occult arts were valued at both royal and papal courts.[6] The *Opus maius* and its two accompanying treatises contained complex and radical ideas about ways in which Christians might, through the embrace of new kinds of learning, be better able to carry out God's commands. Bacon was at the forefront of contemporary efforts to absorb and react to the intellectual stimulus provided by growing engagement with the philosophical heritage of Greeks, Muslims and Jews. His treatises for the Pope were designed to provide powerful arguments in favour of doing so. He was certainly aware that aspects of what he was suggesting might be resisted in some quarters and he framed his discussion carefully so that his readers' natural apprehensions might be allayed. Despite his palpable caution, there is no indication that he feared papal censure for his proposals; rather, he expected that once Clement had understood them, he would endorse them and ask for further information.[7]

These points need to be emphasized in order to counterbalance the dominant assumption of much Bacon historiography that he was a marginal, suspect and defensive figure.[8] The evidence for his putative 1277 condemnation is uncertain, but we know without doubt that his works were received and read at the papal court and that they were

---

[5] For a more detailed defence of this view of Bacon, see A. Power, 'The remedies for great danger: contemporary appraisals of Roger Bacon's expertise', in J. Canning, M. Staub and E. King (eds.), *Knowledge, Discipline and Power: Essays in Honour of David Luscombe* (Leiden: Brill, 2011), 63–78.

[6] See: A. Paravicini-Bagliani, *Medicina e scienze della natura alla corte dei papi nel Duecento* (Spoleto, 1991); S. J. Williams, 'The early circulation of the Pseudo-Aristotlian *Secret of Secrets* in the West: the papal and imperial courts', *Micrologus* 2 (1994), 127–44.

[7] See Bacon, *Opus maius*, IV.iv.16, 1:403.

[8] The *Chronica XXIV Generalium Ordinis Minorum* (compiled 1369, partly from earlier fragments) recorded that Bacon was condemned by his order in 1277 for certain 'suspected novelties'. Even if this somewhat lurid chronicle is an accurate account of events that took place some years before it was written, it has been allowed to colour assessments of Bacon's standing in the late 1260s to a problematic extent. A man suspect in 1277 may also have been suspect in 1267 – but if so, why was the Pope asking for his opinion? The traditional acceptance of the *Chronica* account was seriously questioned by Lynn Thorndike, 'Roger Bacon and experimental method in the Middle Ages', *Philosophical Review* 23 (1914), 271–92. It has been revived by discussions, such as P. L. Sidelko, 'The Condemnation of Roger Bacon', *Journal of Medieval History* 22 (1996), 69–81.

disseminated widely over the following decades. In addition to his writing on mathematics and optics, his discussions of occult arts and magic gained him a reputation for great learning and mastery over the natural world.[9] In view of this evidence that contemporaries and subsequent generations found Bacon's writings – as he intended – both stimulating and sufficiently orthodox, perceptions of Bacon's writings and their reception may need to be modified. Furthermore, it provides yet another example to support the view that intellectuals of the past did take magic and occult arts very seriously and did not necessarily distinguish them from disciplines we would now call 'scientific'.[10] It has been argued that past cultures can be examined productively where they seem most strange to us.[11] With this in mind, it is worth considering whether Bacon's views on the way that the Church should use astrological learning to defend against the malign influences of comets can provide a way into a world whose differences from our own can be deeper, more various and more deceptive than they often seem within the familiar narratives of the past.

In this chapter, I want to consider the ways in which approaches to geography and mapmaking espoused by an individual might fruitfully be

[9] See D. C. Lindberg (ed.), *Roger Bacon's Philosophy of Nature: A Critical Edition, with English Translation, Introduction, and Notes, of* De multiplicatione specierum *and* De speculis comburentibus (Oxford University Press, 1983), xxv; D. C. Lindberg, 'Lines of influence in thirteenth-century optics: Bacon, Witelo and Pecham', *Speculum* 46 (1971), 66–83, at 73–5; A. P. Bagliani, 'Storia della scienza e storia della mentalità: Ruggero Bacone, Bonifacio VIII e la teoria della "prolongation vitae"', in *Aspetti della Letteratura Latina nel secolo XIII: Atti del primo Convegno internazionale di studi dell' Associazione per il Medioevo e l'Umanesimo latini (AMUL) Perugia 3–5 ottobre 1983*, C. Leonardi and G. Orlandi (eds.) (Florence: Società Internazionale per lo Studio del Medioevo Latino, 1985), 243–80; L. DeVun, *Prophecy, Alchemy and the End of Time: John of Rupecissa in the Late Middle Ages* (New York: Columbia University Press, 2009), esp. 80–9, 134–6; A. Power, 'A mirror for every age: the reputation of Roger Bacon', *English Historical Review* 121 (2006), 657–92.

[10] Arguments about the importance of situating intellectuals in their cultural context have been made repeatedly – notably in the case of Issac Newton – but have still not fully penetrated discussions of medieval 'science' and scholarship. B. J. T. Dobbs, *The Foundations of Newton's Alchemy: Or, 'The hunting of the greene lyon'* (Cambridge University Press, 1975); D. C. Lindberg, 'Medieval science and its religious context', *Osiris*, 2nd ser. 10 (1995), 60–79; Alister Chapman, John Coffey and Brad S. Gregory (eds.), *Seeing Things Their Way: Intellectual History and the Return of Religion* (University of Notre Dame Press, 2009).

[11] For example R. Darnton, *The Great Cat Massacre and Other Episodes in French Cultural History* (London: Penguin Books, 2001); P. E. Dutton, *Charlemagne's Mustache and Other Cultural Clusters of a Dark Age* (New York: Palgrave Macmillan, 2004). There is debate over whether it is helpful to focus on a presumed inherent 'alterity' of the medieval period, see: P. Freedman and G. M. Spiegel, 'Medievalisms old and new: The rediscovery of alterity in North American medieval sudies', *American Historical Review* 103 (1998), 677–704, and discussions arising from it. See also J. J. Cohen (ed.), *The Postcolonial Middle Ages* (London: Palgrave, 2000).

looked at as the product of the active imagination of that individual.[12] Historians of later medieval cartography and geography are well aware of the sources of information available to medieval authors and thus the basic architecture of their collective imagination. Medieval cosmographers are, in this mode of analysis, essentially passive recipients of second-hand information, distinguishable from one another chiefly through their varying critical powers and access to material. Less investigated has been the agency demonstrated by the same authors: how and – more importantly – *why* they used their sources as they did in the context of their own lives. How did those sources shape and interact with the experience, educational background, personal or institutional objectives, immediate preoccupations, curiosity and desires of people existing within a particular place and time?[13] The nature of medieval texts on geography and cosmography makes such an investigation a difficult undertaking, yet the specificity of each text must be borne in mind as much as possible. Long-standing approaches to the history of ideas, perhaps especially of geography, have often had a flattening effect, prioritizing continuities and developments in thought over immediacy and individual context.[14] This has tended to obliterate a sense of the authors as individuals who were using conventional material deliberately for ends very particular to the needs of the moment as they perceived them.

Roger Bacon is an unusually rich subject for the exploration of these issues for several reasons. He is a well-known figure who wrote a great deal on a wide range of subjects. We possess enough autobiographical

---

[12] 'Imagination' is a category of thought frequently invoked by historians, but less often defined by them. See J. Le Goff, *The Medieval Imagination*, A. Goldhammer (tr.) (Chicago University Press, 1988), 1–17. Imagination and *mentalité* are closely related, although Le Goff made a distinction between them (e.g. p. 3).

[13] This has, perhaps, been investigated most effectively in the case of Columbus. See V. I. J. Flint, *The Imaginative Landscape of Christopher Columbus* (Princeton University Press, 1992). More recently, see A. J. Hingst, *The Written World: Past and Place in the Work of Orderic Vitalis* (Notre Dame: University of Notre Dame Press, 2009) and D. Birkholz, ch. 10, this vol.

[14] Although the last two decades have seen important reconceptualizations in the field, the base narrative remains fundamental. It provides a teleological context for the thought of individual authors, often at the expense of their wider contemporary contexts. It traces the evolution of European geographical knowledge from an early medieval stagnation dependent on classical texts through an accelerating expansion of horizons towards the 'age of discovery'. See the classic works: G. H. T. Kimble, *Geography in the Middle Ages* (London: Methuen, 1938); J. K. Wright, *The Geographical Lore of the Time of the Crusades: A Study in the History of Medieval Science and Tradition in Western Europe*, 2nd edn. (New York: Dover, 1965); and more recently, J. R. S. Phillips, *The Medieval Expansion of Europe*, 2nd edn. (Oxford University Press, 1998). For a thorough reappraisal of the nature and content of early medieval geography see N. Lozovsky, *'The Earth is Our Book': Geographical Knowledge in the Latin West ca. 400–1000* (Ann Arbor: University of Michigan Press, 2000).

fragments to discern some of the personal history behind his public writings.[15] More generally, he lived at a particularly vital moment for geographical and cosmographical reflection. He was of the first generation in the Latin West that had available to them the bulk of the newly translated Greek and Arabic texts touching on the subject. At the same time, Latin Christians were participating in the intricacies of the Byzantine and Muslim worlds of the Near East and learning rapidly from their early contacts with the vast Mongol empire.[16] In writing about these more distant regions, Bacon had the benefit of the reports of John of Plano Carpini and William of Rubruck, among others.[17] He was conscious of the fact that geographical knowledge was in dramatic flux from both high theoretical and practical perspectives, as, in his view, it had not been for centuries. Finally, he was a member of the highly influential Franciscan order and thus greatly interested in the conversion of non-Christians. One of the most crucial factors in Latin reactions to external stimuli was the distinctive nature of contemporary spirituality. The thirteenth century was a high point of papal power and ambition. It was also the time when followers of men like Francis of Assisi sought to infuse the Church and the laity with a profoundly apostolic sense of Christianity.

[15] Although it should also be noted that virtually all the directly autobiographical statements occur in the problematic context of self-justification or complaints to Clement IV.

[16] For Bacon's views on Muslims, see J. D. North, 'Roger Bacon and the Saracens', in *Filosofia e scienza classica, arabo-latina medievale e l'età moderna: Ciclo di seminari internazionali (26–27 gennaio 1996)*, G. F. Vescovini (ed.) (Fédération Internationale des Instituts d'Études Médiévales. Textes et études du Moyen Age, 11. Louvain-la-Neuve, 1999), 129–60. For the impact of Latin involvement in the Byzantine world on Bacon, see A. Power, 'The importance of Greeks in Latin thought: the evidence of Roger Bacon', in *Shipping, Trade and Crusade in the Medieval Mediterranean: Studies in Honour of John Pryor*, R. Gertwagen and E. Jeffries (eds.) (London: Ashgate, 2012), 351–78. On Muslim–Christian interactions, see R. L. Euben, *Journeys to the Other Shore: Muslim and Western Travellers in Search of Knowledge* (Princeton University Press, 2006). On the intellectual and cultural impact of the Mongols, see F. Schmieder, *Europa und die Fremden: Die Mongolen im Urteil des Abendlandes vom 13. bis in das 15. Jahrhundert*, Beiträge zur Geschichte und Quellenkunde des Mittelalters, 16 (Sigmaringen: Thorbecke, 1994); P. Jackson, *The Mongols and the West, 1221–1410* (Harlow: Pearson Education, 2005); A. Ruotsala, *Europeans and Mongols in the Middle of the Thirteenth Century: Encountering the Other*, Humaniora Series, 314 (Helsinki: The Finnish Academy of Sciences, 2002). More generally, see J. H. Bentley, *Old World Encounters: Cross Cultural Contacts and Exchanges in Pre-modern Times* (Oxford University Press, 1993).

[17] For the impact of William's account on Bacon's text see: J. Charpentier, 'William of Rubruck and Roger Bacon', *Geografiska Annaler* 17 (1935), supplement: *Hyllningsskrift tillägnad Sven Hedin*, 255–67; M. Gueret-Laferté, 'Le voyageur et le géographe: L'insertion de la relation de voyage de Guillaume de Rubrouk dans l'*Opus Majus* de Roger Bacon', *Perspectives médiévales* (1998), supplément 24, *La Géographie au Moyen Âge: Espaces pensés, espaces vécus, espaces rêvés*, 81–96.

Many people, Bacon among them, felt that the world was oppressively full of unbelievers, and Christians were few.[18] They remembered the evangelism of the early Church – how the apostles had taken the word of God to every corner of the world – and they felt a great responsibility to do the same again. To such people, knowledge of geography was not merely a matter of academic or even spiritual interest – it was an urgent moral necessity. If they were ignorant or indifferent about the rest of the world, they might be called to account for it by God.

The elements of imagination

Thinking more closely about the constitutive elements of the way in which Bacon imagined the cosmos, the world and the Church's role within it, what factors determined the individual shape, limitations and character of his geographical thought? In the spaces of his writing, the contours of his existence and consequently his imagination can be glimpsed. Despite his travel, Bacon seems to have lived a comparatively restricted and possibly sedentary life, although he was never specific about the limits of his travels. He crossed the English Channel several times and knew Oxford and Paris well. He had gone far and wide, he felt, in his efforts to consult other scholars and to search for manuscripts.[19] Yet the places that he had actually seen with his own eyes were only a fraction of the whole broad world that he could conjure up in his mind's eye and write about with authority and a measure of conviction. He inhabited a physical landscape that expanded over the years as he travelled about, but that always had a frontier where familiarity and experience blended into what was essentially an imagined world based on different kinds of hearsay. His memories of places he had seen and known gave way on every side to a wealth of words, to descriptions mundane or fantastic, to careful calculations and projections, to outlines of continents drawn on parchment. His direct experience of people and places was his own and unique to him, but his knowledge of the vast regions of earth where he had never been was dependent on the results of other kinds of inquiry. As he put it: 'everyone can describe the details of the place where they were born, but needs to be taught about foreign parts by others'.[20]

---

[18] P. Biller, *The Measure of the Multitude: Population in Medieval Thought*, new edn. (Oxford University Press, 2003), esp. 217–49. On the evangelical activities of the Church, see: J. Richard, *La Papauté et les missions d'Orient au Moyen Age (XIIIe–XVe siècles)*, 2nd edn., Collection de l'Ecole française de Rome, 33 (Rome: Ecole française, 1998).

[19] Bacon, *Opus tertium*, chapter 17, 59.

[20] *Quilibet potest loca natalis soli describere, et per alios de locis extraneis edoceri.* Bacon, *Opus maius*, IV.iv.16, 1:304.

What he could know intimately, however, was the world around and, importantly, above him. Much emphasis is laid on travel and textual sources in the study of medieval knowledge of the world. It is too easy to forget that in those days the nights were dark and the air clear. Bacon was a man who observed closely and reflected extremely carefully and persistently on what he saw. Scattered through his writings are references to the details of his existence: the play of light in morning dewdrops or the plumage of birds; sunlight entering rooms through cracks; the colours of a garden full of plants on a bright day; the size of the autumn moon swimming into view on the horizon; the effects of humidity and perspective on the night skies; the movements of the planets; the puzzling scintillation of the stars.[21] Reading his writings is to discover, casually among the more formal elements, a relationship with nature and with the cosmos that was profoundly visceral. Wherever he stood, he was at the still, small, heavy centre of the universe, encircled by the lower elements and, above them, the revolving layers of the heavens. Sun, moon, planets and stars were in motion around him; he saw and knew them. Standing beneath them, he felt, as he put it, that: 'the vastness of [celestial] things rouses us to reverence for the Creator'. As Scipio looked down, Bacon looked up and saw an immensity that showed him God. The celestial spheres inspired him with poignant yearning. As he wrote to the Pope in justification of astronomy: 'we believe that we will live corporeally and eternally in the heavens – so nothing should be known by us more than the heavens; nothing should be so desired'.[22]

This sort of evocation of Bacon's daily experience could be extended, but the point is that it was a great deal easier for him to study the heavens than the places of the earth. He valued certainty and precision; mathematics was his passion. He distrusted descriptive narrative sources, although their impact on his imagination was considerable – especially, it must be stressed, that of scriptural and patristic material. His mind wandered ardently through the terrain of the Holy Land,

---

[21] Bacon, *De multiplicatione specierum*, in *Roger Bacon's Philosophy of Nature: A Critical Edition, with English Translation, Introduction, and Notes, of 'De multiplicatione specierum' and 'De speculis comburentibus'*, D. C. Lindberg (ed.) (Oxford University Press, 1983), 156, 57–60, 61; D. C. Lindberg (ed.), *Roger Bacon and the Origins of Perspectiva in the Middle Ages* (Oxford: Clarendon Press, 1996), I.v.1, 60, 61; III.ii.4, 312, 13–16, 17; II.iii.7, 232, 33.

[22] *Rerum magnitudo excitet nos ad reverentiam creatoris; credimus nos fore mansuros corporaliter in coelo et perpetue. Quapropter nihil deberet tantum sciri a nobis sicut coelum, nec aliquid in humanis tantum desiderari.* Bacon, *Opus maius*, IV.iv.16, 1:180–1. For intellectual background to this sensibility, see: P. Ellard, *The Sacred Cosmos: Theological, Philosophical, and Scientific Conversations in the Twelfth-Century School of Chartres* (Chicago: University of Scranton Press, 2007).

a place that he never visited, but knew and loved through the words of others. Nothing, he felt, could be more important to the Christian than the most intimate understanding of all the meanings embedded in descriptions of that environment steeped in the sacred past.[23] Yet he felt that only astronomy could give sure results and the reliable information so essential to the future of Christendom. It becomes, then, less surprising that he would look upwards for his knowledge of the world in which he lived.

There were other reasons for this preference, which lie in his education. Numerous accounts of his intellectual formation exist, so a brief sketch will suffice here.[24] He was trained in the new university in Oxford and then, in the 1240s, was among the first to lecture in Paris on the natural philosophy of Aristotle, which had been banned there previously. Towards the end of the decade, he left the university world and embarked on twenty years of private study. He learned Greek, some Hebrew and a little Arabic; he read extensively in Greek and Muslim science and philosophy; he pursued arcane material, especially the Pseudo-Aristotelian *Secretum Secretorum*, and the study of astrology, alchemy and some forms of magic. As has already been suggested, these approaches to natural philosophy were at once somewhat unconventional and more widespread than theologians cared to admit. It is vital, however, to recognize that Bacon believed such studies to be more, rather than less, advanced and capable of providing rational and reliable information about the fabric of the universe. His desire was to draw closer to the ultimate wisdom of God and he thought this combination of disciplines would take him there. It was probably no coincidence that it was also during these years, possibly in the late 1250s, that Bacon entered the Franciscan order. He gave no direct indication of his reasons for doing so, but his internalization of many aspects of the Franciscan mission seems evident from the purpose and content of his writings. He also made it clear that he thought the moral perfection of the religious life was the *sine qua non* of the pursuit of wisdom.[25] He had been a Franciscan for a decade when, between *c.* 1266 and *c.* 1268, he produced his *Opus maius*

---

[23] Bacon, *Opus maius*, IV.iv.16, 1:183–7.
[24] J. Hackett, 'Roger Bacon: his life, career and works', in *Roger Bacon and the Sciences: Commemorative Essays*, J. Hackett (ed.), Studien und Texte zur Geistesgeschichte des Mittelalters, 57 (Leiden, 1997), 9–23; Lindberg, 'Introduction', in *Roger Bacon's Philosophy of Nature*, xv–xxvi; G. Molland, 'Roger Bacon (*c.*1214–1292?)', in *Oxford Dictionary of National Biography*, H. C. G. Matthew and Brian Harrison (eds.) (Oxford University Press, 2004), http://www.oxforddnb.com/view/article/1008 (accessed 30 November 2009).
[25] This thesis is defended in A. Power, *Roger Bacon and the Defence of Christendom* (Cambridge University Press, 2012).

and related works at the urgent request of the Pope.[26] In them, he addressed the attainment of wisdom and its application to all human concerns. These he grouped into four categories: the running of the Church, the ordering of the affairs of the faithful, the conversion of unbelievers, and the defence of Christendom against those who could not be converted. It must be stressed that everything he wrote about geography and cartography was framed by these objectives. His was a text which subordinated abstract information to a clear set of purposes.

It should be evident from this account of his intellectual formation that when Bacon thought about the world beyond Christendom, he had a great range of material at his disposal and that it was not easily brought together into a coherent picture. In the first place, he had the conventional scriptural, patristic and classical material fundamental to Latin geographical thought. These texts had been used for centuries to supply routine descriptive and anecdotal material about distant places and people to a wide variety of ends. However, they did not really offer Bacon any kind of methodological or critical framework beyond the mingled notions of revelation, authority and the 'reliable witness' around which they were respectively structured. The information that they supplied was also, as Bacon sometimes recognized, dated. It was difficult to integrate it with the far more dynamic theoretical and mathematically based treatises that had become available through new translations in the course of recent decades.

Aristotle's natural philosophy was at the core of this second corpus, but his work had been much expanded and refined in various ways by generations of Muslim intellectuals.[27] The principle on which Aristotle's system rested was that everything was caused, or moved, by something else. The whole cosmos was in perpetual motion, perpetual generation and decay. The lower motions, those on and of the earth, were determined by the motions of the higher, superlunary spheres. The chain ended with the 'unmoved mover' – for monotheists: God. Bacon, as many Christian philosophers before him, borrowed this model to supplement the vaguer cosmography of the Scriptures. He began by outlining a theoretical conception of the nature and functioning of the universe in which every element of the cosmos existed within a hierarchy of

---

[26] *Et per tuas nobis declares litteras quae tibi videntur adhibenda remedia circa illa, quae nuper occasione tanti discriminis intimasti*, 'Epistola Clementis Papae IV ad Rogerium Baconem', in *Fr. Rogeri Bacon Opera Quaedam Hactenus Inedita*, J. S. Brewer (ed.), Rerum Britannicarum Medii Aevi Scriptores [Rolls Series], 15 (London, 1859), 1.

[27] On this process, see: D. Gutas, *Greek Thought, Arabic Culture: The Graeco-Arabic Translation Movement in Baghdad and Early 'Abbāsid Society (2nd–4th/8th–10th Centuries)* (London: Routledge, 1998).

interlocking causal relationships. He called this the 'multiplication of *species*', by which he meant the process through which the force or influence (*species*) of one thing operated on another.[28] The *species* moved in straight lines between objects, either being absorbed on reception or reflected or refracted depending on the nature of the object. This enabled the relations between things to be expressed in a series of mathematical principles which would then allow further relations to be projected and the natures of things to be understood, at least in general terms. He hoped that a proper comprehension of this process would enable future scholars to map out the likely effects of the heavens on the earth, providing at least a notional picture of the probable nature of places and peoples.

The most contemporary elements in his thought, however, were the stories told by men returning from Louis IX's disastrous crusade in Egypt and, above all, the two reports written by Franciscans who had travelled among the Mongols: the papal envoy John of Plano Carpini and the missionary William of Rubruck.[29] Both reports showed conventional geography in flux. As regions were always known in descriptive geographies and travel narratives by their peoples, cities, religions and products, the Mongol invasions had changed the face of the known world, devastating cities, destroying and replacing populations, and destabilizing the long hegemonies of particular faiths. The old picture of the world was unsettled by a new ambiguity and potential for rapid alteration. Bacon wondered whether the imminence of the apocalypse could be detected in all this turbulence. Above all, he sensed opportunities.

His efforts to decide what to do with this distinctly incompatible combination of sources and ideas informed and enlivened his approach to geography. He told the Pope that two possibilities were open to him. He could use a theoretical approach to hypothesize the physical and human geography of the world on the basis, chiefly, of astronomy infused with elements of what we would today call astrology. Alternatively, he could collate classical, scriptural and modern descriptions of peoples and places to provide as exhaustive and up-to-date picture of the world as possible.[30] Both proved, in practice, to present their own difficulties.

---

[28] His writings on *species* drew heavily on Arabic material, particularly the writings of Al-Kindi. See Lindberg, *Philosophy of Nature*, xxxv–lxxi.

[29] Bacon claimed to have a range of informants, but there is no direct evidence for his use of reports by other envoys to the Mongols, although they were accessible to others. See M. Guéret-Laferté, *Sur les routes de l'Empire mongol: Ordre et rhétorique des relations de voyage au XIIIᵉ siècles*, Nouvelle Bibliothèque du Moyen Âge, 28 (Paris: Honoré Champion Éditeur, 1994), esp. 285–8.

[30] Bacon, *Opus maius*, IV.iv.16, 1:304–6.

Recognizing this, he offered each in succession and highlighted their inadequacies to Clement as a demonstration of the urgent need to improve Latin knowledge of the wider world.

An apostolic cosmography

The urgency was the product of a range of considerations, but in particular the evangelical imperative embedded in the wider mission of the Church and the specific mission of the friars. The connection between geography and evangelism was elaborate but critical. Bacon wrote: 'place is the beginning of our being, just like a father ... And we see that everything varies in accordance with the different places of the world, not only in nature, but in the customs of men'.[31] He believed this to be true in the case of whole peoples and of every individual. The heavens laid their specific and indelible impression on a child at the moment of birth, giving it particular inclinations in morals, learning, languages, occupation and much else. Yet, he emphasized, 'he will be able to change himself through free will, the grace of God, the temptation of the Devil, and good or bad advice, especially when he is young'.[32] Following sources such as Ptolemy's *Almagest* and Abu Ma'shar, he believed that there were six main religions in the world, and could only be six, just as there were six planets. The planets, in particular conjunctions, had and imparted the qualities of the religions. Venus in conjunction with Jupiter, for example, signified the lasciviousness of Islam, and Saturn, with its slow movements, the antiquity of the Jews.[33] Bacon, like his authorities, was nevertheless insistent that the planets did not compel the will, which remained free, but they did affect the body, so that: 'the soul united to the body is excited strongly and influenced effectively'.[34] It was this and little more, he insinuated, that caused people to adhere to the religions that held sway in the regions of their birth.

---

[31] *Sed locus est principium generationis, quemadmodum et pater, ut dicit Porphyrius. Et nos videmus, quod omnia variantur secundum loca mundi diversa non solum in naturalibus, sed homines in moribus.* Bacon, *Opus maius*, IV.iv.5, 1:138. Bacon had probably borrowed the quotation from Grosseteste, but it came originally from Porphyry.

[32] *Licet poterit se mutare per libertatem arbitrii, et per gratiam Dei, et per tentationem diaboli et per bonum aut malum consilium, maxime a juventute.* Bacon, *Opus maius*, IV.iv.5, 1:139.

[33] On these stereotypes see: J. J. Cohen, 'On Saracen enjoyment: some fantasies of race in late medieval France and England', *Journal of Medieval and Early Modern Studies* 31 (2001), 113–46; E. Zafran, 'Saturn and the Jews', *Journal of the Warburg and Courtauld Institutes* 42 (1979), 16–27.

[34] *Anima corpori unita excitatur fortiter et inducitur efficaciter.* Bacon, *Opus maius*, IV.iv.16, 1:267.

This was not a token deference to the orthodox position on determinism versus free will. It was, rather, the whole point. For Bacon, the idea that morals, in particular, were the result of impartial external conditioning was the cause for immense optimism. He wrote that human diversity was not therefore due to 'difference of the rational soul'.[35] If moral and religious dispositions were not inherent, then individuals and whole nations of *infideles* could be made conscious of the fact and drawn towards the Christian truth. Rational argument, based in a philosophical demonstration of the truth of Christianity, was the most obvious approach, and the one that Bacon most clearly elaborated. His suggestions to this end have received a fair amount of attention, and are not hard to follow.[36] The same is not true of his other, more occult approach, which has, perhaps unsurprisingly, not been investigated in the context of Franciscan mission.[37]

In several places, Bacon gave cautious indications that he thought occult arts might profitably be employed to 'improve the morals' of people as well as to defend Christendom from its enemies. This was dangerous territory, so he took refuge behind stories from the Scriptures and other authorities, particularly Aristotle. In the *Secretum Secretorum*, which he persisted in attributing to Aristotle, it was related that when Alexander the Great encountered peoples with evil morals, he wrote to Aristotle to ask what he should do with them. Aristotle replied: 'If you can alter their atmosphere [*aer*], allow them to live; if not, kill them all'. Bacon, evidently missing the force of the remark, commented: 'Oh, how enigmatic is the response, yet how full of the power of wisdom! For

---

[35] *Diversitatis animae rationalis*. Bacon, *Opus maius*, IV.iv.16, 1:250.

[36] Influential discussions are: E. R. Daniel, *The Franciscan Concept of Mission in the High Middle Ages* (Lexington: University of Kentucky Press, 1975), esp. 55–66; R. W. Southern, *Western Views of Islam in the Middle Ages* (Cambridge, Mass.: Harvard University Press, 1962), 52–61; B. Z. Kedar, *Crusade and Mission: European Approaches towards the Muslims* (Princeton University Press, 1984), 177–80; J. Tolan, *Saracens: Islam in the Medieval European Imagination* (New York: Columbia University Press, 2002), esp. 225–9.

[37] For existing work on this aspect of Bacon's thought, see: D. Bigalli, *I Tartari e l'Apocalisse: Ricerche sull' escatologia in Adamo Marsh e Ruggero Bacone* (Florence: La Nuova Italia, 1971), esp. 168–90; J. Hackett, 'Astrology and the search for an art and science of nature in the thirteenth century', in *Ratio et Superstitio: Essays in Honor of Graziella Federici Vescovini*, G. Marchetti, V. Sorge and O. Rignani (eds.), Textes et études du Moyen Âges, 24 (Louvain-la-Neuve: Brepols, 2003), 117–36; J. Hackett, 'Aristotle, *Astrologia*, and controversy at the University of Paris (1266–1274)', in *Learning Institutionalized*, Van Engen (ed.), 69–111; J. Hackett, 'Roger Bacon on Astronomy-Astrology: the sources of the *Scientia Experimentalis*', in *Roger Bacon*, Hackett (ed.), 175–98; H. M. Carey, 'Astrology and Antichrist in the later Middle Ages', in *Time and Eternity: The Medieval Discourse*, G. Jaritz and G. Moreno-Riaño (eds.), International Medieval Research, 9 (Turnhout, 2003), 515–35.

Aristotle understood that, following a change of the atmosphere, which contains celestial forces, the morals of men are changed'. He went on:

Aristotle intended that Alexander should change the quality of the atmosphere of those peoples for the better, so that ... they could be inspired towards virtuous morals without going against their free will, since every nation is influenced towards its morals by its own atmosphere, which contains the forces of the stars that are above the heads of the inhabitants, and in accordance with the signs or planets dominant over the particular regions.[38]

In effect Bacon was arguing that, since everyone was subject to arbitrary celestial influences, changing the nature of the influences for the better did not affect the freedom of the inhabitants of such places. It did not prevent them from exercising free will to resist the improved influences, but it meant that those who remained passive – as the population had done previously – would be far more likely to find salvation.

How, then, did he propose to go about altering atmospheres? This brings us to an important element in the medieval imagination, or cosmological sensibility: the porous quality of the terrestrial sphere. The influence of the celestial bodies, through the multiplication of *species*, could have major effects on what was below. These were not the only influences at work, for the power of the Christian supernatural was also palpable in daily life. Agents of God, demons, unearthly spirits and saints had the power to influence nature. Bacon accepted their roles in a matter-of-fact fashion, enthusiastically recounting commonplace variants on the tales in which demons were defeated by the power of the Eucharist.[39] It was therefore natural enough that he should look to saints and exceptionally virtuous people to exercise the sort of influence over nature that was implied in 'Aristotle's' advice. He disapproved of spells and incantations when used by magicians and old women, but the rational soul of a wise and holy man was the most powerful thing after God and the angels and, as he wrote, 'can have a great potency for altering the things of this world'.[40] Avicenna, he said, taught that: 'the soul that is

---

[38] *Si potes alterare aerem ipsorum, permitte eos vivere; si non, interfice omnes. O quam occultissima responsio est, sed plena sapientiae potestate! Nam intellexit quod secundum mutationem aeris, qui continet coelestes virtutes, mutantur mores hominum ... Voluit ergo quod Alexander in bonum mutaret qualitatem aeris illarum gentium, ut ... excitarentur ad honestatem morum sine contradictione liberi arbitrii; sicut quaelibet natio excitatur ad proprios mores per aerem proprium habentem virtutes stellarum quae sunt super capita hominum, et secundum quod signa vel planetae dominantur singulis regionibus.* Bacon, *Opus maius*, IV.iv.16, 1:393.

[39] Rogeri Baconis, *Moralis Philosophia*, E. Massa (ed.) (Turin: Thesaurus Mundi, 1953), IV.iii.1, 225–6.

[40] *Possunt habere magnam virtutem alterandi res mundi hujus.* Bacon, *Opus maius*, IV.iv.16, 1:397.

pure and clean from sins is able to change the universe and the elements'.[41] The implications of this are fascinating, and can be explored in various ways. Here, the point is Bacon's provocative analysis of the fabric of the universe: its power to condition the environment, lives and belief systems of humans, and its capacity to be manipulated by the right kind of people. It seems highly likely that the learned and saintly men whom Bacon had in mind were Franciscans and Dominicans who had been properly trained in the occult arts and were working with the grace of God.

It must have been obvious for some time that, for any of this to work in practice, the most precise geographical knowledge was required. Although it ought to have been possible to describe the nature of the world by pure extrapolation from a study of the heavens, this could only be useful to the interests of Christendom if all this information could be mapped on to the known cities and regions of the world. As Bacon put it after his discussion of astronomy: 'it is crucial that we consider the diversity of the regions of the earth; how each region changes over time; and how different things in the same region receive different influences at the same time. But we cannot know all this unless we can clarify the size and shape of the habitable earth and its zones [climata]'. He went on: 'Since these zones and the famous cities in them cannot be perfectly envisaged through verbal description alone, a map must be used to make them clear to our senses'.[42]

Bacon's map has intrigued historians of cartography for its promise of innovation. It has not survived, nor are there any reliable visual indications in the manuscript tradition to show how it might have looked.[43] His approach to mapping undoubtedly owed much to Ptolemy of Alexandria's *Almagest*, which had been translated into Latin in the twelfth century. The *Almagest* used a coordinate system to describe the locations of superlunary bodies in the celestial spheres. Bacon applied the same method to the terrestrial realm using the available tables of latitudes,

---

[41] *Anima sancta et munda a peccatis potest universale et elementa alterare.* Bacon, *Opus maius*, IV.iv.16, 1:403.

[42] *Necesse est considerare quae sit diversitas regionum mundi, et quomodo eadem regio in diversis temporibus variatur, et quomodo res diversae ejusdem regionis diversas recipiunt passiones in eodem tempore. Sed haec sciri non possunt, nisi quantitatem et figuram habitabilis terrae et climata ejus distinguamus; quoniam haec climata et civitates famosae in eis non possunt evidenter percipi sermone, oportet quod figura sensui ministretur.* Bacon, *Opus maius*, IV.iv.16, 1:288, 295–6.

[43] He described it at *Opus maius*, IV.iv.16, 1:294–300. The question has been examined carefully. See D. Woodward, 'Roger Bacon's terrestrial coordinate system', *Annals of the Association of American Geographers* 80 (1990), 109–22; D. Woodward and H. M. Howe, 'Roger Bacon on Geography and Cartography', in *Roger Bacon*, Hackett (ed.), 199–222.

probably those of al-Khwārizmī, which had been translated into Latin by Adelard of Bath.[44] The visual relationship between his map and contemporary maps is unclear. It may have been a somewhat hybrid affair. He described it as containing written descriptions of places on the same parchment as the map, a standard approach to cartographical depictions. Much of the material that he wished to include was generic and designed to provoke spiritual reflection as much as to supply fresh material about distant lands. This, too, was in keeping with the usual purposes of the extant *mappaemundi*.[45] At the same time, his method required a level of spatial accuracy more characteristic of portolan charts than *mappaemundi* (a word that he did not use to describe his map). The important point, as far as his innovation is concerned, is that he understood the need to use precise coordinates of longitude and latitude to indicate the positions of places. He informed Clement that in making his map he had been obliged to depend on inaccurate tables because the longitude and latitudes of many cities and regions had not yet been established among the Latins, and patronage was imperative if they were ever to be so.[46] He wrote optimistically:

Once we have acquired accurate information, we should be able to understand under which stars each place is found, how far it is from the path of the sun and the planets, and by which planets and [astrological] signs each is dominated. All these elements engender the different qualities of places. If they could be grasped, people would be able to unravel the nature of their influences on everything on earth, since everything on earth derives its nature and properties from the effects of its location.[47]

[44] Ptolemy, *Almagest*, G. J. Toomer (tr.) (London, 1984), VII.4, 339–40. This is an outline, and the following books of the *Almagest* explained and developed the concepts. There is debate over which tables of latitude Bacon used. On Adelard's, see: C. Burnett, 'Adelard of Bath and the Arabs', in *Recontres de Cultures dans la philosophie médiévale: traductions et traducteurs de l'antiquité tardive au XIVe siécle*, J. Hamesse and M. Fatton (eds.) (Université catholique de Louvain, 1990), 89–107, esp. 95–107. On the tables more generally, see: G. J. Toomer, 'A Survey of the Toledan Tables', *Osiris* 15 (1968), 5–174; G. Saliba, *A History of Arabic Astronomy: Planetary Theories during the Golden Ages of Islam* (New York University Press, 1994), 53.

[45] See D. Woodward, 'Medieval mappaemundi', in *The History of Cartography Volume 1: Cartography in Prehistoric, Ancient, and Medieval Europe and the Mediterranean*, J. B. Harley and D. Woodward (eds.) (University of Chicago Press, 1987), 286–370.

[46] Bacon's greatest objection to the Toledan tables was that they had defined 'east' and 'west' too arbitrarily (*Opus maius*, IV.iv.16, 1:298–300). They would also have allowed him to include no more than a mere sixty cities on his map. See Toomer's Table of Geographical Co-ordinates in Toomer, 'A survey of the Toledan Tables', 134–5. Contemporary reports took a long time to be incorporated into maps. I. Baumgärtner, 'Weltbild und empire: die Erweiterung des kartographischen Weltbilds durch die Asienreisen des späten Mittelalters', *Journal of Medieval History* 23 (1997), 227–53.

[47] *Sed haec certificatio . . . tunc enim sciremus sub quibus stellis est quilibet locus, et quantum a via solis et planetarum, et quorum planetarum et signorum loca recipiant dominium, quae omnia*

As he had not yet acquired the information he needed, he was forced to return to traditional methods of describing the world. These could not be jettisoned because it was too important to have such information as there was. As I have indicated, his sources were conventional, but updated in the areas explored by crusaders and by travellers in Mongol lands. The result was a meandering and eclectic itinerary of the habitable world.

From the perspective of cosmography or geography this was not a satisfactory text. It was, importantly, probably not meant to be. It existed for another purpose: namely, to indicate directions for future research. For us the cosmographical imagination that underlies the way he wrote about geography, cartography and astronomy may be of greater interest than the information that he supplied. In the first place, his ardent sense that the whole world might be and must be encompassed, discovered, changed for the better was both remarkable and yet quite characteristic of the ambitions of his order and of the thirteenth-century Church. What we see in Bacon's writings is the impact that this apostolic hunger had on a responsive, individual imagination. Bacon's sense of the cosmos was not fragmentary; not the patchy, clichéd picture which faded out into monstrous races around the edges – quite the reverse. Wrapped in the rotating celestial spheres – which he loved to contemplate because he loved God – the earth was given unity, life and character, for good or ill, by the imprint of the heavens. Yet amid these influences, humans remained free, and because they were free, they could be brought to salvation, and live for eternity beyond the stars. The task of the Church – one that it had neglected since the days of the first apostles – was to bring this salvation, in any way permissible, to the people who lived in moral darkness beyond the borders of Christendom. Bacon's cosmos was a place that demanded dynamic action. His imagination is a witness to that of his times, in its careful methods of evaluation and its bold assumptions; its awkwardness between intellectual traditions; its diverse use of geographical ideas and, perhaps above all, its quickening energy.

---

faciunt diversas complexiones locorum: quae si scirentur, possit homo scire complexiones omnium rerum mundi et naturas et proprietates quae a virtute loci contrahunt. Bacon, *Opus maius,* IV.iv.16, 1:30.

5   Reflections in the Ebstorf Map

Cartography, theology and *dilectio speculationis*

Marcia Kupfer

Measuring just over three and a half metres square, the celebrated Ebstorf Map is one of the most imposing works of medieval cartography on record. The full-size, hand-coloured reconstruction on parchment, produced ten years after its destruction, drives home its suprahuman scale (Fig. 5.1).[1] Our knowledge of the giant *mappa mundi* depends on witnesses to the original artefact between its discovery, c. 1830 at the women's convent on the Lüneburg heath in Lower Saxony, and its loss in Allied bombing at Hanover on 8–9 October 1943. Hartmut Kugler's 2007 monumental critical edition and atlas launches a new phase of research with nothing less than a veritable archaeology of the map's transmission over the course of its modern lifespan.[2] In navigating an archive of transcriptions and reproductions, Kugler not only mines it for data pertinent to the map's physical status, cartographic content and artistic character. Equally importantly, he also elucidates the extent to which the pre-war documentation is intrinsically compromised, a screen between the original and us.

To come to grips with the Ebstorf Map is to reckon as much with the philological objectives and intellectual priorities of nineteenth-century scholarship as with the technical limitations of contemporary photo-mechanical processes. The visual record is especially fraught. Left with facsimiles derived from successive manipulations of hand-doctored collotypes, we see the Ebstorf Map through a glass very darkly indeed. Why not, then, turn this predicament into an opportunity? To be sure, our necessary reliance on copies of flawed copies to gain some insight into an artwork that we can no longer know directly is vexing. Yet it puts us in a position to appreciate analogous issues of loss and mediation that the

---

[1] The dimensions of the Ebstorf Map are known to have been superseded by two other medieval *mappaemundi*, both, like the Ebstorf, no longer extant: Marcia Kupfer, 'The lost mappamundi at Chalivoy-Milon', *Speculum* 66 (1991), 540–71; Marcia Kupfer, 'The lost Wheel Map of Ambrogio Lorenzetti', *Art Bulletin* 78 (1996), 286–310.

[2] Hartmut Kugler, *Die Ebstorfer Weltkarte*, 2 vols. (Berlin: Akademie Verlag, 2007); hereafter cited as Kugler followed by volume and page number.

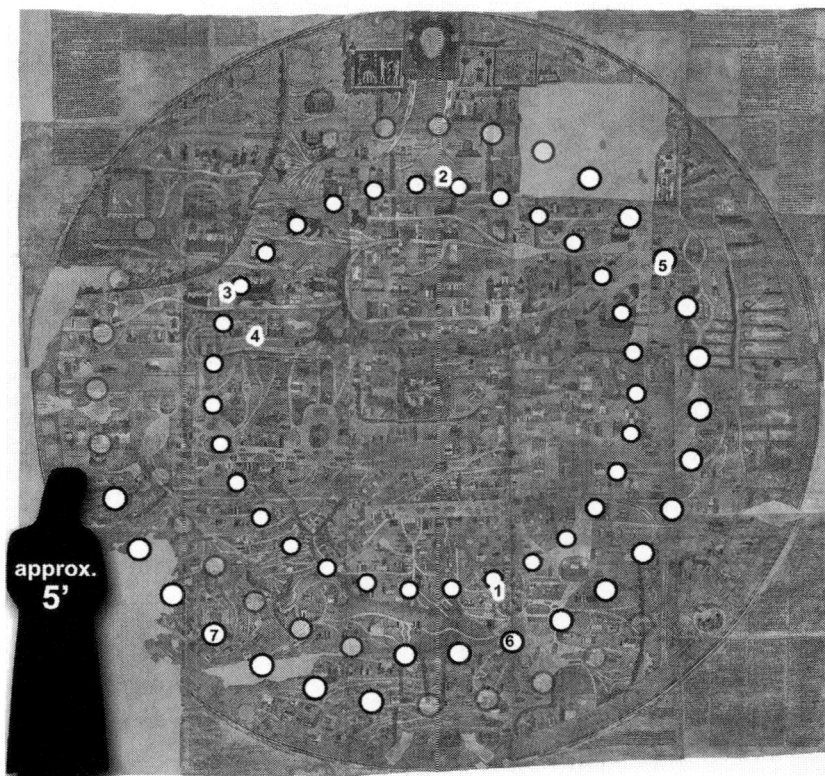

Figure 5.1 Author's schematic of the Ebstorf Map created in
Photoshop by Asa S. Mittman. The base layer is a scan of the map's
digital reconstruction in Hartmut Kugler, ed., *Die Ebstorfer Weltkarte*,
2 vols. (Berlin, 2007). The overlays show the size of the Map relative to
the height of viewers (fourteenth-century nuns) and suggest the map's
radial construction. The rings represent the progressive evangelization
of the world in three stages: apostles; the patron saint of Ebstorf cloister;
Ebstorf. The numbers indicate sites discussed in the text: 1. Rome,
2. Tomb of St Thomas, 3. Tomb of St Bartholomew, 4. Tomb of
St Philip, 5. Thebes, 6. Abbey of St Maurice at Agaune, 7. Ebstorf
cloister.

map's cloistered audience also confronted – on a spiritual plane. How
to recover the divine image at the root of human identity (Gen. 1:26–28)
but obscured by the Fall? How to approach God, his countenance
withdrawn from our gaze, if not through contemplation of his
vestigia in the sensory world? The map, produced *c.* 1300 for the

Benedictine nunnery of Ebstorf, had a special role to play in setting its readers on the right path.[3]

For a start, the map's unique pictorial structure reflects theological and ecclesiological commitments. It insists on God's immanence and omnipresence, while marshalling at the same time a discourse of place to advance a vision of universal participation in the Church as Christ's body. Moreover, the map's cartographic agenda relates to its communal function, as revealed by its self-professed claims. Along the top of the map, a line of paragraphs covers such matters as the celestial sphere, paradise, the earth and its divisions, and the days of Creation. Following this cosmological 'prologue', which concerns things in themselves (*res*), comes a passage in the far right corner on the representation that is the map. It proceeds from a definition of terms and a historical sketch to a statement of purpose:

Mappa dicitur forma. Inde mappa mundi id est forma mundi. Quam Julius Cesar missis legatis per totius orbis amplitudinem primus instituit; regiones, provincias, insulas, civitates, syrtes, paludes, equora, montes, flumina quasi sub unius pagine visione coadunavit; que scilicet non parvam prestat legentibus utilitatem, viantibus directionem rerumque viarum gratissime speculationis dilectionem.

[*Mappa* means image. Hence a *mappa mundi* is an image of the world. Which Julius Caesar, having sent legates throughout the breadth of the whole world, first procured; regions, provinces, islands, cities, sandy coasts, marshes, flat expanses [of seas or plains], mountains, and rivers he brought together, as it were, for viewing on a single page. It offers to readers no small utility, to wayfarers direction and the pleasure of the most pleasing sight (*gratissime speculationis dilectionem*) of things along the way.][4]

Love of speculation is what the map is for.

More than any other example of the cartographic genre to which it belongs, the Ebstorf Map brings into focus the motivating force of two

---

[3] Jürgen Wilke, *Die Ebstorfer Weltkarte*, 2 vols. (Bielefeld: Verlag für Regionalgeschichte, 2001) (hereafter cited as Wilke), builds a compelling, multi-pronged case for the Map's origin *c*. 1300 in the environs of Lüneburg–Verden, conclusions endorsed and further reinforced by Kugler 2:32–7, 61–9 (cf. esp. 62–3 and 69) and Hartmut Kugler, 'Die Ebstorfer Weltkarte ohne Gervasius von Tilbury', in *Kloster und Bildung im Mittelalter*, Nathalie Kruppa and Jürgen Wilke (eds.) (Göttingen: Vandenhoeck & Ruprecht, 2006), 497–512. For another point of view, see most recently Armin Wolf, 'The Ebstorf *Mappamundi* and Gervase of Tilbury: the controversy revisited', *Imago Mundi* 64 (2012), 1–27.
[4] Kugler, 1:42, no. 7/1, 2:86, no. 7/1. The translation is mine, with thanks to Jeffrey Hamburger for the felicitous turn of phrase 'pleasure of the most pleasing sight'. My thinking about this passage is indebted to Cornelia Herberichs, '... *quasi sub unius pagine visione coadunavit*: Zur Lesbarkeit der Ebstorfer Weltkarte', in *Text – Bild – Karte: Kartographien der Vormoderne*, Jürg Glausner and Christian Kiening (eds.) (Freiburg im Breisgau: Rombach Verlag, 2007), 201–17, esp. 212–14.

complementary Pauline verses and holds them in tension. To the challenge of *videmus nunc per speculum in enigmate* and the promise of *tunc autem facie ad faciem* (1 Cor. 13:12), it responds with Romans 1:20: 'For the invisible things of him from the creation of the world are clearly seen, being understood by the things that are made (*invisibilia enim ipsius a creatura mundi per ea quae facta sunt intellecta conspiciuntur*)'. Jeffrey Hamburger has surveyed the exegetical career of Romans 1:20, explaining its significance to medieval practices of speculation that embraced 'the mirror as a metaphor, not for the transience of experience and the limits of human knowledge, but on the contrary, for the beauty of nature and the possibilities of sensory perception'.[5] This perspective complements Patrick Gautier Dalché's discussion of how the *mappa mundi* tradition instantiated visual tropes at play in monastic contemplation.[6] The reading of the Ebstorf Map that follows draws upon these critical insights. By the later Middle Ages, pictorial images, things of human artifice, might serve as vehicles by which the embodied soul ascended from and through divine reflections in the Creation to attain a foretaste of the *visio dei*. But only a map can in the process also answer two questions: where is God, where is Ebstorf?

W/here

The Ebstorf Map conjoins a teeming representation of the inhabited world, its three parts standing for the whole *orbis terrarum*, with a cruciform arrangement of Christ's body parts. Head, hands and feet emerge at the cardinal points as if piercing the landmass, yet remain confined to the earthly sphere. Discrete, severed members present themselves, yet at the same time belong centripetally to a single body, reconstituted at the site of Christ's resurrection in Jerusalem, *umbilicus mundi*. There, the figure of the triumphant Christ is rotated ninety degrees off axis to

---

[5] Jeffrey Hamburger, 'Speculations on speculation: vision in the theory and practice of mystical devotion', in *Deutsche Mystik im abendländischen Zusammenhang: neu erschlossene Texte, neue methodische Ansätze, neue theoretische Konzepte. Kolloquium Kloster Fischingen 1998*, Walter Haug and Wolfram Schneider-Lastin (eds.) (Tübingen: Max Niemeyer Verlag, 2000), 353–408, passage quoted at 355; Jeffrey Hamburger, 'Idol curiosity', in *Curiositas: Welterfahrung und ästhetische Neugierde in Mittelalter und früher Neuzeit*, Klaus Krüger (ed.) (Göttingen: Wallstein Verlag, 2002), 21–58.
[6] See Patrick Gautier Dalché, 'De la glose à la contemplation: place et fonction de la carte dans les manuscrits du haut Moyen Âge', in Patrick Gautier Dalché, *Géographie et Culture: La représentation de l'espace du VIe au XII siècle* (Aldershot: Ashgate, 2006), ch. 8, esp. 749–64; Patrick Gautier Dalché, 'Pour une histoire des rapports entre contemplation et cartographie au moyen âge', in *Les Méditations Cosmographiques à la Renaissance*, Frank Lestringant (ed.), Cahiers V. L. Saulnier 26 (Paris: Presses de l'Université Paris-Sorbonne, 2009), 19–40.

face north.[7] The world is coextensive with Christ's body, of which the head, significantly, takes the form of the *vera icon*.[8]

As Roman relic, the Veronica preserves the earthly trace of the ascended Lord even as miraculous image, it captures the effigy that mortals may discern only *per speculum in enigmate*.[9] Specific reference to this Pauline verse had attached to its veneration since 1216.[10] The circle-in-square of the Veronica crowns the enormous circle-in-square of the framed map. Thus the world mirrors the divine image through which it was generated and the art object, the relic.[11] *Sudarium* and *mappa* are cognates, too, in that the denotative sense 'handkerchief' or 'napkin' endures even as the connoted graphic representation takes precedence and functions independently: a defining image, replicated in other media, subsumes the originary cloth support.[12] The one displays the impression of Christ's Face, the other the Creator's handiwork, his

[7]   The Jerusalem vignette includes Mount Zion perched at the south-east corner of the city walls (Kugler 1:92, no. 30/7 and 2:172, no. 30/7). This placement is roughly consistent with that found in other maps, e.g. a mid-twelfth-century map of Palestine (London, British Library, MS Additional 10049, f. 64v) and the Psalter Map. The topology indicates that the city itself is oriented on axis; it is only the *figure* of the risen Christ that faces north. On the iconographic background, function and rotation of the image of the resurrected Christ in the Ebstorf Map, see Marcia Kupfer, 'The Jerusalem effect: rethinking the centre in medieval world maps', in *Visual Constructs of Jerusalem*, Bianca Kühnel, Galit Noga-Banai and Hanna Vorholt (eds.), forthcoming.

[8]   The Holy Face in the Ebstorf Map conforms to the Byzantine schema of the Mandylion as exemplified in the Novgorod and Laon icons, the latter identified with the Latin Veronica by 1249; this Veronica type appears *c.* 1280 in the Hours of Yolande of Soissons (New York, Pierpont Morgan Library, M 729, f. 15r). The flanking Alpha and Omega, however, recalls Matthew Paris's rendition of the Veronica in part one of his *Chronica Majora* (Cambridge, Corpus Christi College, MS 16, f. 49v); see *Il volto di Cristo*, exh. cat., Giovanni Morello and Gerhard Wolf (eds.) (Milan: Electa, 2000), cat nos. III.13, 97–99, IV.2, 169–71 and IV.4, 172–3.

[9]   The Veronica ('the true image'), an image of Christ, was a cloth relic kept in the church of St Peter's, Rome, referred to in eleventh- and twelfth-century documents.

[10]   See Jeffrey Hamburger, *The Visual and the Visionary: Art and Female Spirituality in Late Medieval Germany* (New York: Zone Books, 1998), 317–82; *The Holy Face and the Paradox of Representation: Papers from a Colloquium held at the Bibliotheca Hertziana, Rome and the Villa Spelman, Florence, 1996*, Herbert L. Kessler and Gerhard Wolf (eds.) (Bologna: Nuova Alfa Editoriale, 1998); *Il volto di Cristo*, 103–210; Christiane Kruse, *Wozu Menschen malen: Historische Begründungen eines Bildmediums* (Munich: W. Fink, 2003), 269–306; *L'immagine di Cristo dall'acheropita alla mano d'artista: dal tardo medioevo all'età barocca*, Christoph L. Frommel and Gerhard Wolf (eds.) (Vatican City: Biblioteca Apostolica Vaticana, 2006), esp. 91–165.

[11]   On the operation of formal similitude in structuring how medieval maps are read, see Ingrid Baumgärtner, 'Erzählungen kartieren: Jerusalem in mittelalterlichen Kartenräumen', in *Projektion-Reflexion-Ferne: Räumliche Vorstellungen und Denkfiguren im Mittelalter*, Sonja Glauch, S. Köbele and U. Störmer-Caysa (eds.) (Berlin: De Gruyter, 2011), 193–223.

[12]   Patrick Gautier Dalché, 'Les sens de mappa (mundi): IVe–XIVe siècle', *Archivum Latinitatis Medii Ævi* 62 (2004), 187–202.

imprint in nature. The physical world is but a *mappa* on which God drew the *forma* and painted all the *res*.[13]

Christine Ungruh has proposed an ingenious typology of the Christ figure as a conflation of the Veronica with the *Volto Santo*, the Lucca crucifix attributed to Nicodemus.[14] In focusing on iconographic relationships, however, she sidelines the most important aspect of the linkage between both images and the Map. Especially relevant to the incorporation of the Lucca figure in the *mappa mundi* is the legend, reported by Gervasius of Tilbury, that Christ's body left an imprint on a cloth spread to cover his nudity while he hung on the cross; the cruciform imprint then served as the divine model for Nicodemus's carving. Medieval viewers like Gervasius insisted on the equivalence of the *Volto Santo* and the Veronica, among other miraculous images of Christ (including the cloth of Edessa).[15] The map can refer its cruciform Christ figure to the full-body cloth imprint that lies behind the *Volto Santo*, even as it directly 'quotes' the Roman Veronica. The double genealogy reflexively glosses the cartographic enterprise described in the framing text-block. No less than pictorial 'copies' that stand in for image-traces of Christ impressed on linen, the map is a surrogate for the world 'brought together for viewing on a single page'. *Mappa mundi* is a simulacrum of the fabric of Creation in which the *vestigia dei* (lit. God's footprints) are immanent.

Ebstorf is the only medieval map to depict the figure of Christ entirely inscribed *within* the bounds of the *oikumene*. Scholars agree on the visual facts of the singular representation, but differ on how to interpret them. 'The world *is* Christ's body or at least symbolizes it', said Armin Wolf first in 1957, and most scholars have since reprised this perception even as they have qualified aspects of his analysis or rejected his conclusions.[16] However, Barbara Bronder, writing in 1972, and Kugler more recently dispute any suggestion that the map pictorially identifies Christ with the world.[17] They appeal to unspecified

---

[13] On this conceit, see now Kruse, *Wozu Menschen malen*, 137–73.

[14] Christine Ungruh, 'Paradies und *vera icon*: Kriterien für die Bildkomposition der Ebstorfer Weltkarte', in *Kloster und Bildung im Mittelalter*, 301–29; on the geometricity of the Ebstorf Face, see her astute observation, 308 n. 24.

[15] Gervase of Tilbury, *Otia Imperialia: Recreation for an Emperor*, S. E. Banks and J. W. Binns (eds.) (Oxford: Clarendon Press, 2002), 598–607.

[16] Armin Wolf, 'Die Ebstorfer Weltkarte als Denkmal eines mittelalterlichen Welt- und Geschichtsbildes', *Geschichte in Wissenschaft und Unterricht* 8 (1957), 204–15; Jörg-Geerd Arentzen, *Imago Mundi Cartographica: Studien zur Bildlichkeit mittelalterlicher Welt- und Ökumenekarten unter besonderer Berücksichtigung des Zusammenwirkens von Text und Bild* (Munich: W. Fink, 1984), 267–74; Wilke, 1:72–3.

[17] Barbara Bronder, 'Das Bild der Schöpfung und Neuschöpfung der Welt als orbis quadratus', *Frühmittelalterlichen Studien* 6 (1972), 188–210, esp. 209, n. 88; Kugler,

'theological grounds' as justification for bringing the image into line with normative iconographic templates that instead extol God's majesty. Their alternative solutions stem from a common philological method that denies the validity of the idiosyncratic trait and corrects it against a canonical reading. Clarifying what distinguishes the Map's overarching pictorial conception, or better, what it simultaneously invokes and renounces, is therefore in order.

The Ebstorf Map deploys the widely disseminated schema of the *syndesmos* figure, to use Anna Esmeijer's descriptive term for the cruciform pose assumed by Christ's members.[18] By virtue of its all-encompassing breadth, length, height and depth (Eph. 3:18), the erect stature and outstretched arms take on cosmogonic significance. Christ *syndesmos* proliferated in images that diagram the nested dynamics of the *machina universitatis* enveloped within and bound together by a towering God. The Ebstorf Map, however, eschews the depiction of a transcendent Christ-Logos who *surpasses* the world. He neither embraces the circumference of the *orbis terrarum* as in the second of two miniature maps in a mid-thirteenth-century English Psalter, nor stretches beyond it, as in a map of *c.* 1300 accompanying a section of the *Historia Britonum* (Fig. 5.2).[19] For Bronder, the weight of convention, Scripture and exegesis vitiates Ebstorf's exception to the rule; she argues that the relative size, or extent, of the Christ figure has no bearing on meaning. On her reading, the Ebstorf Map reduces to an illustration of the phrase inscribed at Christ's left hand, *terram palmo concludit*, 'He holds the earth

2: 19–21, esp. 20; Baumgärtner, 'Erzählungen kartieren', 193–4, notes the difference of opinion and leaves the question open.

[18] Anna C. Esmeijer, 'La macchina dell' universo', in *Album discipulorum: aangeboden aan J.G. van Delder ter gelegenheid van zijn zestigste verjaardag 27 Februari 1963*, Josua Bruyn (ed.) (Utrecht: Haentjens, Dekker & Gumbert, 1963), 5–15; Anna C. Esmeijer, *Divina quaternitas* (Amsterdam: Van Gorcum Assen, 1978), esp. 97–108.

[19] On the two Psalter Maps of *c.* 1265 (London, BL, Add MS. 28681, ff. 9r and 9v), see most recently Peter Barber, 'Medieval maps of the world', in *The Hereford World Map: Medieval World Maps and their Context*, Paul D. A. Harvey (ed.) (London: British Library, 2006), 15–19; on the map accompanying chapter 17 of the 'Gildas' version of Nennius's *Historia Britonum* (London, Lambeth Palace Library MS. 371, f. 9v), see Scott D. Westrem, 'Geography and travel', in *A Companion to Chaucer*, Peter Brown (ed.) (Oxford: Blackwell, 2000), 195–271, esp. 206–9; for a transcription and colour reproduction, see Leonid S. Chekin, *Northern Eurasia in Medieval Cartography: Inventory, Text, Translation, and Commentary* (Turnhout: Brepols, 2006), 72–3 and 371. Ungruh, 'Paradies und *vera icon*', 314–15, also cites the Lambeth Palace map to offset the Ebstorf composition. The distinctive terrestrial containment of the Ebstorf Christ, however, undercuts her comparison to the *Volto Santo* in so far as the sculpture was framed by the addition of the cosmic circle. The framing circle is conjoined with the Lucca crucifix in such a way that Christ's hands and feet (though not the head) extend beyond it, a feature reiterated in fourteenth-century images of the *Volto Santo* (cf. her figures 11, 12); the addition of the cosmic reference incorporates the *Volto Santo* into the normative tradition that Ebstorf rejects.

Figure 5.2 *Mappa mundi*
Source: *Historia Britonum*, London, Lambeth Palace Library, MS. 371,
f. 9v, *c.* 1300. Reproduced by kind permission of The Trustees of
Lambeth Palace Library.

in his palm'.[20] Yet notwithstanding the textual exaltation of God's
immensity and overwhelming might, the pictorial configuration emphasizes divine immersion in the earthly realm.

---

[20] On the inscription, not mentioned by Bronder, see Kugler, 1:98–9, and 2:186 no. 35/4.

The transposition of the *syndesmos* figure from extrinsic, containing force to intrinsic, embedded subject recalls the iconography of man as microcosm.[21] But the Ebstorf schema stands convention on its head: God takes the place of man, and the earthly sphere overtakes the cosmos.[22] Visual reference to the tradition through inversion creates new meaning. From the twelfth century, the representation of micro- and macrocosm fed into twin intellectual preoccupations with the humanity of God and the divinity of man. The theme thus engaged the abbess and physician Hildegard of Bingen, for example, informing her spiritual epiphanies in which the architecture of the physical universe coalesced with the history of salvation.[23] Artists adapted cosmological formulae to the Trinitarian imagery of her visions: miniatures in the deluxe Lucca manuscript of her *Liber divinorum operum* (c. 1230) feature a youthful Adam, reminiscent of the beardless Son in the Rupertsberg copy of her *Scivias* (variously dated c. 1165–before 1195).[24] The *syndesmos* posture of the fully embodied microcosm in the Lucca miniature implies the generative principle of creation and the regenerative power of the cross through which the New Adam restored the Old *ad imaginem dei*.[25] Conversely, in a late twelfth- or early thirteenth-century copy of Hrabanus Maurus's *Liber sanctae crucis*, the image of a gigantic *Christus triumphans* is embedded in gloss explicating the cosmological significance of his identity as the New Adam.[26] The Ebstorf Map pushes a Christological typology of the microcosm

---

[21] Wolf, 'Die Ebstorfer Weltkarte', first suggested the relevance of the micro- and macrocosm theme to a discussion of the Map. For an updated analysis and corrections, see Wilke, 1:73–4, 134–6, 139.

[22] As Patrick Gautier Dalché has noted, the Ebstorf schema concerns neither *minor mundus* (man) nor *maior mundus* (the cosmos), but rather God and the terrestrial world. See his review of *Kloster und Bildung im Mittelalter*, in *Francia-recensio*, 2010 no. 1, Mittelalter, at www.perspectivia.net/content/publikationen/francia/francia-recensio/2010-1/MA (accessed 24 July 2013).

[23] Barbara Newman, *Sister of Wisdom: St. Hildegard's Theology of the Feminine* (Berkeley and Los Angeles: University of California Press, 1987), 44–6, 97, 251; Barbara Newman, *God and the Goddesses: Vision Poetry, and Belief in the Middle Ages* (Philadelphia: University of Pennsylvania Press, 2003), 51–89, 138–9.

[24] Barbara Newman, 'Die visionären Texte und visuellen Welten religiöser Frauen', in *Krone und Schleier: Kunst aus mittelalterlichen Frauenklöstern*, exh. cat. (Munich: Hirmer Verlag; Bonn: Kunst- und Ausstellungshalle der Bundesrepublik Deutschland; Essen: Ruhrlandmuseum, 2005), 105–17, esp. 108–9; Robert Suckale, 'Hildegard von Bingen, *Liber divinorum operum*', in *Krone und Schleier*, 310–11, no. 198; Lieselotte E. Saurma-Jeltsch, *Die Miniaturen im 'Liber Scivias' der Hildegard von Bingen: Die Wucht der Vision und die Ordnung der Bilder* (Wiesbaden: Reichert, 1998), 93–5.

[25] For other examples of Adam *syndesmos*, see Esmeijer, *Divina quaternitas*, 100–4.

[26] Lucca, Biblioteca Statale, MS. 370, f. 121v, the figure of Christ crucified is considered a pictorial reference to the Lucca *Volto Santo*: see Ungruh, 'Paradies und *vera icon*', 323–4 (with colour reproduction); and *Il volto di Cristo*, exh. cat., Morello and Wolf (eds.), 271, cat. no. VI.2.

much further than the Hildegard miniature and pursues the cosmological implications of the crucified in a different direction than the Hrabanus Maurus miniature. In Ebstorf, the suffering, wounded body of Christ *syndesmos* is broken down into Eucharistic parts, spatialized at the scale of the *oikumene*. The members tend toward a higher unity, but one that, contrary to the aggregative macrocosm, is installed at the map's centre, in the glorious body triumphantly emerging from the Holy Sepulchre.

Like Bronder, Kugler, too, dislocates the Map's representation of Christ from the surrounding world, albeit by other means. Noting the visual pre-eminence of the Holy Face at the zenith of the central vertical axis, he aligns the map with schemes that elevate God above the cosmos. In particular, he refers the work to diagrams that chart the soul's climb through the celestial spheres and up the 'ladder of being' to arrive at the enthroned Christ outside and beyond the system.[27] He wants correspondingly to see the Holy Face, flanked by the apocalyptic letters Alpha and Omega and words *primus et novissimus*, as raised above the terrestrial world.[28] Now, medieval artists *did* develop visual strategies for representing Christ's dual nature through his bodily division between celestial and terrestrial realms, head in heaven, feet on earth, and for identifying the Holy Face with the firmament.[29] Against this background, the earthbound Veronica of the Ebstorf Map stands out saliently. The metaphor of the ladder and, concomitantly, the idea of spiritual ascent are certainly important to the theory of speculation that I will later bring to bear on the map. But so too is the work's refusal to follow through on the depiction of God's transcendence and heavenly sovereignty. The Hereford Map of *c.* 1300 and the first of the Psalter maps *do* portray a Majesty who occupies a separate, superior realm from which he presides over the world; they are consistent in this respect with the aforementioned diagrams of the soul's ascent (among other cosmological images).[30] Not Ebstorf.

---

[27] Kugler 2:20 figure 18 from Paris, BnF, MS. lat 3236 A, f. 90r. On the diagram, see Christian Heck, *L'échelle céleste dans l'art du moyen âge: une image de la quête du ciel* (Paris: Flammarion, 1997), 100–2.

[28] Miller recorded the three-word inscription in the lower corners of the field (not visible in Sommerbrodt's plates, nor mentioned by him); Kugler, 2:79–80, nos. 4/2 and 4/7.

[29] See Herbert L. Kessler, *Seeing Medieval Art* (Peterborough, Ontario: Broadview Press, 2004), 75–6 and *passim*; Herbert L. Kessler, *Spiritual Seeing: Picturing God's Invisibility in Medieval Art* (Philadelphia: University of Pennsylvania Press, 2000), 53–63; Herbert L. Kessler, 'Face and firmament: Dürer's *An Angel with the Sudarium* and the limit of vision', in *L'immagine di Cristo*, Frommel and Wolf (eds.), 143–65, esp. 152–3.

[30] For the Hereford Map, see *Hereford World Map*, Harvey (ed.). Examples of similar conceptions of God's exalted place above/outside the cosmos: a miniature in Gossuin de Metz, *Image du monde*, bound in a miscellany of *c.* 1276–77 (Paris, Bibliothèque Ste-Geneviève, MS. 2200 f. 115v), repr. *Liber Floridus* website http://liberfloridus.cines.fr/

The place of the Veronica and its eschatological framing sustain the map's interest in the modalities of God's presence in the Creation, and how to perceive it. The adjacent image of Paradise and the Fall recounts the original sin even as the Veronica's apocalyptic notation alludes to the Last Judgment. The split perspective points up, Janus-like, the qualitative difference between the vision of God possible *nunc* in the time-bound, corporeal world – i.e. mediate his traces – and the unencumbered gaze upon his Face that the blessed will enjoy *tunc* at the consummation of the *saeculum*.[31] How better to define simultaneously the current plight and ultimate goal of the map's readers, to situate them, as it were, in terms of their fallen condition yet propel them forward on their spiritual journey? I will return presently to the significance of the Veronica as a geographic marker. Suffice it to say, for the moment, that the dual temporal orientation of the Holy Face at the eastern summit of the world circle institutes the breach between the beginning and the end through which all times and places come to pass.

Even as the Ebstorf Map disperses Christ's members, assimilating them into the profusion of disparate cartographic elements, the circumference of the world circle visually binds them together. Contrary to cosmological images in which the supramundane figure of Christ-Logos effects cohesion, in the Ebstorf Map it is the inhabited world through which the enfleshed, incarnate Word coheres. The composition recasts the topos of micro- and macrocosm in terms of Christ's twice-figured appearance, dismembered at the periphery of the *oikumene* and intact at the centre. In sum, the map emphatically identifies God's body and world. It remains to investigate what is thereby accomplished and on what theological grounds.

Henri de Lubac has traced how, beginning in the mid-twelfth century and with increasing momentum over the course of the thirteenth, theologians thrashed out a sharpened, dialectical understanding of the distinct sacramental and ecclesial realities to which the Eucharist pertains.[32] According to the doctrine of real presence, the consecrated host is the *corpus verum*, the historical body born of the Virgin and expired on the

---

textes/biblio_fr.html, nos. 197, 198; catalogue entry at http://www.calames.abes.fr/pub/ms/BSGB12236, both accessed 24 July 2013; a miniature depicting the creation of light and the spheres in a mid-fourteenth-century English Book of Hours (London, BL, Egerton MS. 2781, f. 1v), Michael Camille, *Gothic Art: Glorious Visions* (New York: Harry N. Abrams, 1996), 43, figure 28.

[31] Ungruh, 'Paradies und *vera icon*', 312–14, 327.

[32] Henri de Lubac, *Corpus Mysticum: The Eucharist and the Church in the Middle Ages*, Gemma Simmonds (tr.), Laurence Hemming and Susan Frank Parsons (eds.) (University of Notre Dame Press, 2006).

cross; with this body, Christ redeemed the sins of a fallen humanity. At the same time, the sacrificial victim taken in communion spiritually binds together the faithful into the *corpus mysticum*. The application of this term to mean the body of the Church first appears in the opening paragraph of the bull, *Unam sanctam*, promulgated by Boniface VIII on 18 November 1302: 'one holy catholic Church ... which represents one sole mystical body whose head is Christ and the head of Christ is God'.[33] The Ebstorf Map must be set in the context of the movement culminating in the corporatist redefinition of the Church as a universal Christendom under papal rule: as the the bull concludes, 'it is absolutely necessary for salvation that every human creature be subject to the Roman Pontiff'.[34]

Whereas theologians and jurists aspired to analytic clarity, however, the art work achieves pictorial synthesis, reconciling the sacramental and ecclesial dimensions of Christ's body. On the one hand, the broken, fragmented figure represents the crucified body consumed in the host. The composite Christological and cartographic image projects to the ends of the earth the relationship of part equivalent to whole that operates in the sacrifice of the Mass, there not according to the theoretical principle of reflection governing the micro- and macrocosm but ontologically: in each particle of every wafer consecrated at any altar anywhere Christ's true body is fully present and whole. Because the Eucharist is a transformative power operative in individual communicants at all churches everywhere throughout Christendom, the Christ figure on the other hand also represents the integrity of the *corpus mysticum*. The substitution of the *vera icon* for the head signifies the universality of the Church under the aegis of Rome, for, as Gerhard Wolf has observed, the relic/image 'bears an ecclesiological reference within itself' and 'became a "universal symbol" of the Roman, i.e. of the universal church'.[35] The Veronica doubles as a marker for Rome literally in the role of *caput mundi*. Thus, conversely, the composite cartographic and Christological image projects the submission

[33] De Lubac, *Corpus Mysticum*, 3. *Unam sanctam ecclesiam catholicam ... Quae unum corpus mysticum repraesentat cujus caput Christus, Christi vero Deus*. I quote the English translation available from the 'Internet Medieval Sourcebook' at http://www.fordham.edu/halsall/source/b8-unam.html, accessed 24 July 2013; for the Latin text see http://www.ccel.org/s/schaff/history/6_ch01.htm#_ednref30 (at section no. 4), accessed 24 July 2013.

[34] Along these lines, see further Bildhauer, 'Blood, Jews and monsters', in *The Monstrous Middle Ages*, Bettina Bildhauer and Robert Mills (eds.) (University of Toronto Press, 2003), 75–96.

[35] Gerhard Wolf, 'From Mandylion to Veronica: picturing the "disembodied" face and disseminating the True Image of Christ in the Latin West', in *The Holy Face and the Paradox of Representation*, 153–79, esp. 177; Wolf, '"Or fu sì fatta la sembianza vostra?"', in *Il volto di Cristo*, 104, 108–10.

of the *oikumene* to the Church, not merely by rhetorical analogy, as in the topos of members subordinated to the head, but actually through the medium of the Saviour's flesh.

Apart from scattered remarks that have offhandedly, and somewhat crudely, characterized the map as a 'colossal wafer', a serious Eucharistic reading of the image has not hitherto been pursued.[36] Yet Eucharistic overtones permeate the work, most obviously crystallizing around individual pictorial elements that exerted a special devotional appeal for religious women. The Holy Face and the Holy Sepulchre from which Christ rises triumphantly are the two most salient images in the map. Given the map's enormous size, they offered miniature *Andachtsbilder* in their own right. Jeffrey Hamburger has pointed out that 'disembodied representations of the Holy Face came to stand by synecdoche for the whole of Christ's body, especially ... as present in the Eucharist'. Moreover, 'the Veronica, in keeping with its connotations of real presence, also stood for the body of the resurrected Christ'. In discussing the frequent juxtaposition of the two images, he further observes that the Veronica 'supplies historical witness to the continuing presence of Christ following the Resurrection'.[37] The axial alignment of these foci in the map delineates a comparable theme. At Ebstorf, the cartographic ground of the pictorial conjunction imbues the message of Christ's *enduring* presence, a temporal modality, with spatial extension. God's ubiquity is also at stake.

I would go further to argue that even the map's status as cartographic representation participates in its Eucharistic resonance. As Gerhard Wolf has explained, 'the *ostensio* of the Veronica is a pendant to the elevation of the host'; the visual *manducatio* of the Eucharist, the burgeoning cult of the Veronica and the institution of the feast of Corpus Christi are interrelated phenomena.[38] It is within this devotional-ritual constellation that the geographic formation of the *orbis terrarum* might also be analogized to the consecrated host. In contemporary scenes of the celebration of Mass, the elevated wafer can sometimes show a tripartite division anticipating its actual tripartite fraction; the internal marking echoes the conventional T–O scheme for the tripartite division of the *oikumene*

---

[36] Armin Wolf, 'Ebstorfer Weltkarte', in *Lexikon des Mittelalters* (Munich: Artemis Verlag, 1986), vol. 3, 1534–35: 'Die Erde erscheint auf diese Weise – wie eine riesige Oblate von doppelter Menschengrösse – als Christi Leib' (1535).

[37] J. Hamburger, 'Vision and the veronica', in J. Hamburger, *The Visual and the Visionary. Art and Female Spirituality in Late-Medieval Germany* (New York: Zone Books, 1998), 317–82, quotes at 333, 340; Kugler, 2:21–2.

[38] Wolf, 'From Mandylion to Veronica', 168 and Wolf, '"Or fu sì fatta la sembianza vostra?"', 108; on how the Volto Santo and Nicodemus legend relates to the elevation of the host, see Ungruh, 'Paradies und *vera icon*', 320.

(Fig. 5.3).[39] Although the cartographic field of the Ebstorf Map by no means reduces to a T–O scheme, what matters here is the underlying geographic doctrine of *triple partition*, acknowledged in the framing text-block and reinforced by way of a T-diagram (Fig. 5.4).[40] The iconography of the tripartite host yields insight into the map's bid to sacramentalize the Creation. The *forma* that God impressed on the fabric of the physical world at creation is that of the Eucharist to be broken and shared in communion. *Mappa mundi*, the real one in nature, bears the inherent trace of God's consecrated body just as the actual relic of the *sudarium* bears the trace of his visage.

It so happens, too, that the map programmatically recapitulates a standard liturgical allegory for the wafer's threefold fraction. The triple fraction symbolized Christ's threefold body ever since Amalarius of Metz first put forth the idea in the ninth century.[41] According to the tradition as formulated by Peter Lombard among others: 'The part that is offered and put in the chalice represents the body of Christ which has already risen' (corresponds to the vignette at map centre); 'the part that is eaten represents him *still walking on earth*' (corresponds to the head, hands and feet *contained within* the terraqueous disk; this part stands for the living); 'the part that remains on the altar until the end of mass signifies the body lying in the tomb, because until the end of the age the bodies of the saints

---

[39] Additional examples include: Psalter-Hours from Trier, *c.* 1250–74 (New York, Pierpont Morgan Library, M 94, f. 90r); Liège Psalter, *c.* 1280–90 (New York, Pierpont Morgan Library, M 183, f. 123r, Mass of St Giles illustrating Ps 97 and f. 252v illustrating the Funeral Mass for the Office of the Dead); Liège Psalter-Hours, *c.* 1285–90 (Liège, Bibliothèque de l'Université, MS 431, f. 198v, Funeral Mass for the Office of the Dead). On the thirteenth-century examples in the Trier and Liège MSS, see Judith Oliver, *Gothic Manuscript Illumination in the Diocese of Liège (c. 1250 – c. 1330)*, 2 vols. (Leuven: Peeters, 1988), vol. 1, 72–4, 96–8 and vol. 2, 259–62, 278–80, 397, pl. 88, 445 pl. 143. The images in the Morgan manuscripts can easily be accessed through CORSAIR at http://www.utu.morganlibrary.org/medren/ListOfMssWithImages.cfm. On the leaf from a fourteenth-century Hamburg missal, see *Goldgrund und Himmelslicht: Die Kunst des Mittelalters in Hamburg*, exh. cat. (Hamburg: Hamburg Kunsthalle, Dölling und Galitz Verlag, 1999), cat. no. 6, 133–5. Jeffrey Hamburger has called my attention to another instance of the tripartite host in a miniature for the Corpus Christi feast in a gradual, dated *c.* 1380, from the convent Paradies bei Soest (Düsseldorf, Universitäts- und Landesbibliothek, Gradual D 11, p. 319). The scene is that of the Last Supper. The image centres on the consecrated wafer displayed on gilded ground, surrounded by the words *hoc est corpus meum*, and juxtaposed with a chalice held up by Christ; see Jeffrey Hamburger, Susan Marti and Drew Massey, 'Medieval hypertext: the illuminated manuscript in the age of virtual reproduction', in *Text und Bild im Mittelalter*, Barbara Schellewald and Karin Krause (eds.), Sensus: Studien zur mittelalterlichen Kunst 2 (Cologne: Böhlau, 2011), 365–408.

[40] Kugler, 1:40–41, 2:84. On the semiotic function of the gap between the complexity of the cartographic representation and the T-diagram as symbol, see Herberichs, '. . . *quasi sub unius pagine visione coadunavit*', 209 and esp. 215 n. 56.

[41] De Lubac, *Corpus Mysticum*, 265–301.

Figure 5.3 Elevation of the Host, Funeral Mass for the Dead, Psalter, Liège, *c.* 1280–90
Source: The Pierpont Morgan Library, New York. MS M.183, f. 252v. Photography by Graham S. Haber, 2012.

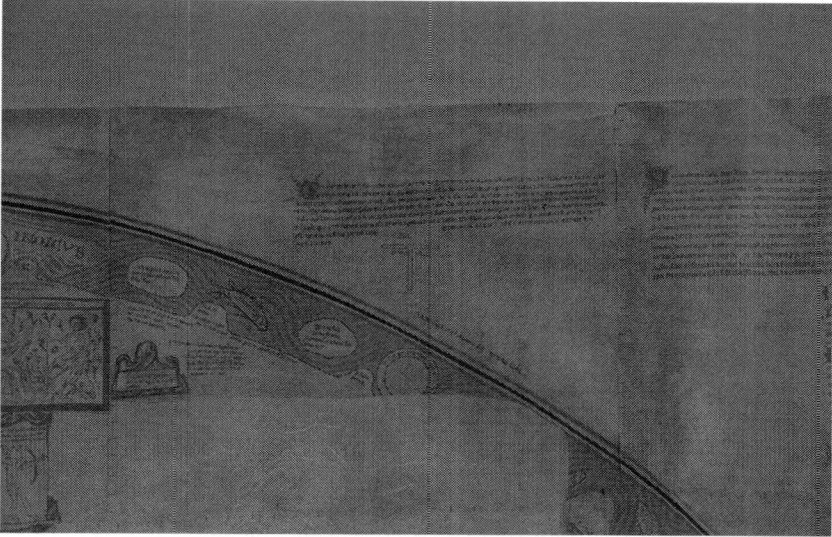

Figure 5.4 Diagram of the tripartite oikumene in the text-block framing
the Ebstorf Map, from one of the four facsimiles of the Map at original
size (3.56 x 3.58 m.) as reconstructed by Rudolf Wieneke in 1953.
Source: The Kloster Ebstorf. Photo reproduced by permission of The
Kloster Ebstorf.

will be in tombs'.[42] Between the intact and dismembered bodies of
Christ, the map proper plots saints' burials in a manner that demonstrates the place of Ebstorf Cloister within the share of the blessed
awaiting the resurrection.

A deliberate radial pattern structures the map's geographic content.[43]
Mid-way between the world's perimeter and nucleus is a latent,
second-order circle of sites locating the bodies of apostles and therefore

[42] Ibid. 283. Thirteenth-century commentaries on the triple fraction (e.g. by Bonaventure
and Thomas Aquinas) reflect the doctrine of purgatory; thus the reserved part represents
the faithful now buried, of whom some wait in purgatory for the future resurrection and
others, the saints, wait already in heaven (see 290–1).

[43] Radial organization in *mappaemundi* centred on Jerusalem is discussed by Asa Simon
Mittman, *Maps and Monsters in Medieval England* (New York and London: Routledge,
2006), 39–42. My reading of the Ebstorf Map complements and fleshes out Kugler's
observation regarding the importance of graves as an organizing principle: see Hartmut
Kugler, 'Die Gräber der Ebstorfer Weltkarte', in *In Treue und Hingabe: 800 Jahre Kloster
Ebstorf*, Marianne Elster and Horst Hoffmann (eds.) (Ebstorf: Kloster, 1997), 53–65. The
diagonal that, as he has noticed, aligns Ebstorf, Jerusalem, and the pious, Jerusalem-going
Nubians at the south-eastern edge of the world is part of the radial system I lay out. This
geometric structure amounts to what Herberichs, '... *quasi sub unius pagine*

delimiting the extent of their mission to spread the Word. The circle intersects Rome (south-west at 'five o'clock') with its cluster of famous shrines, notably St Peter's and St Paul's.[44] It passes through Thomas's burial place due east of Jerusalem (not quite 'twelve o'clock' with respect to Christ's face); a turreted structure encloses his tomb and a burning lamp hangs above. Also on the circle lies Bartholomew's tomb (north-east at 'ten o'clock'), which receives an even more monumental edifice, crowned by a bell-tower; it is coupled with Philip's smaller tomb diagonally below. Jürgen Wilke has correlated the differential importance accorded the tombs of Thomas, Bartholomew and Philip with church and altar dedications in the diocese of Verden to which Ebstorf Cloister belonged, and specifically with the relic collection of the principal Lüneburg monastery, St Michaelis.[45] Saints' bodies, then, are equally present at their faraway tombs and in their locally dispersed relics. In both the multiplicity of their fragments and the higher-order integrity of being in which all the remains everywhere participate, saints' bodies replicate the Eucharistic model of Christ's own.[46]

Beyond the apostolic ring, south-east of Jerusalem (at 'two o'clock') lies the Egyptian city of Thebes. There the only rubricated inscription in the map indicates the origins of St Maurice, patron saint of Ebstorf Cloister.[47] Maurice was martyred with his Roman legion at Agaune, where the abbey of St Maurice is located directly below Rome on the same circle as Thebes. (The label *Agaunensium regio*, however, appears further north atop a mountain range.)[48] The convent of Ebstorf itself appears as a tiny speck

---

visione coadunavit', 208, refers to as the work's 'immanente Bildlogik'. Thomas O'Loughlin, 'The view from Iona: Adomnán's mental maps', *Peritia* 10 (1996), 98–122, esp. 114–16, argues persuasively that the textual organization of Adomnán's *De locis sanctis* reflects a mental map of concentric circles according to which Jesus' message spread from Jerusalem, to the Holy Land, and westward to Rome and ultimately Iona.

[44] For inscriptions and commentary, see Kugler, 1:122–3 and 2:265–6.
[45] Wilke, 1:185–91.
[46] Peter Dinzelbacher, 'Die "Realpräsenz" der Heiligen in ihren Reliquien und Gräbern nach mittelalterlichen Quellen', in *Heiligenverehrung in Geschichte und Gegenwart*, Peter Dinzelbacher and Dieter R. Bauer (eds.) (Ostfildern: Schwabenverlag, 1990), 115–74; Caroline Walker Bynum, 'Material continuity, personal survival and the resurrection of the body: a scholastic discussion in its medieval and modern contexts', in Caroline Walker Bynum, *Fragmentation and Redemption: Essays on Gender and the Human Body in Medieval Religion* (New York: Zone Books, 1991), 239–98, esp. 280, 285; Bruno Reudenbach, 'Visualizing holy bodies: observations on body-part reliquaries', in *Romanesque Art and Thought in the Twelfth Century: Essays in Honor of Walter Cahn*, Colum Hourihane (ed.) (University Park, PA: Pennsylvania State University Press, 2008), 95–106, esp. 103–5; Derek Krueger, 'The religion of relics in Late Antiquity and Byzantium', in *Treasures of Heaven: Saints, Relics, and Devotion in Medieval Europe*, exh. cat., Martina Bagnoli, Holger A. Klein, C. Griffith Mann, and James Robinson (eds.) (Baltimore: Walters Art Museum, 2010), 5–18, esp. 8; Arnold Angenendt, 'Relics and their Veneration', ibid. 19–28, esp. 23.
[47] Wilke, 1:186; Kugler 1:82–3 and 2:125–6.
[48] Kugler, 1:132–3 no. 52/12 and 134–5 no. 53/7; 2:290, 94.

on the very north-western rim of the world sandwiched between Verden and the prominently featured Welf capital of Lüneburg.[49]

At the Ebstorf convent, three squares and an inscription indicate the graves of martyrs who, according to legend, had long ago fallen to pagans. Their cult had begun shortly before the period of the map's production to attract local pilgrimage on account of sudden miraculous flows of healing oil.[50] The confrontation between Christianity and the heathen realm in local history reiterates that which had occurred during apostolic times in the interior regions of the world. Between the two eras, St Maurice engaged in the same struggle when he travelled from Thebes, already beyond the radius attained by the apostles themselves, to the far north-west where he died protecting fellow Christians. Clearly, the Ebstorf–Lüneburg sector is a *minor mundus* that conforms to the overarching sacred order of times and places.[51] The paradigm of micro- and macrocosm, transfigured in Christological terms, governs the spiritual concordances according to which salvation history unfolds across space.[52] It is by identifying with an ever more remote centre that the convent and neighbouring religious houses in the diocese attach themselves to, indeed lodge themselves within, the macrocosm of the universal church. The incorporation of the fringe member into a unified corpus follows the centripetal model of Christ's Eucharistic and resurrected bodies.

Echoing the Lord's call to Adam, 'ubi es?', a fallen, exiled humanity cries out, 'Where is God?': 'Everywhere', answers the Ebstorf Map. But God's ubiquity is not unqualified; the work's incarnational perspective and institutional agenda together foreclose any possibility of pantheistic error.[53] As we have seen, the Word spreads forth from the centre through the evangelizing activity of saints so that, at the limits, Christ's body in the Eucharist comes to all; in the process, its superimposed 'clone', the body of the Church, attains ecumenical realization.

[49] Kugler, 1:128–9 nos. 50/7–17; 2:279–81 nos. 50/7–17.

[50] Wilke, 1:224–6; Kugler, 2:62, 64, 280 nos. 50/12 and 14.

[51] For a similar argument about the regional church as microcosm of the universal church and its history, see Jennifer O'Reilly, 'Islands and Idols at the Ends of the Earth: Exegesis and Conversion in Bede's *Historia ecclesiastica*', in *Bède le vénérable: Entre tradition et posterité*, Stéphane Lebecq, Michel Perrin and Olivier Szerwiniack (eds.) (Lille: Université Charles de Gaulle, 2005), 119–45, esp. 129.

[52] For other aspects of macro-micro spatiality, see Keith D. Lilley, *City and Cosmos: The Medieval World in Urban Form* (London: Reaktion Books, 2009).

[53] For a discussion of the tension between God's ubiquity and pantheism, see Amos Funkenstein, *Theology and the Scientific Imagination from the Middle Ages to the Seventeenth Century* (Princeton University Press, 1986), 42–57.

In tandem with this internal compositional dynamic, the map articulates a threefold hierarchy that parallels the textbook discussion of how 'God can be said to be in things', as laid out by Peter Lombard.[54] First, 'God ... is by means of His presence, power, and essence in every nature or essence without His own limitation, and in every place without circumscription, in every time without mutability'. To this category correspond the myriad creatures, places and times that fill the world circle. Second, God 'is in holy spirits and souls in a more excellent manner, namely through indwelling grace'. To this category belong the saints, 'in whom He is through grace', indicated on the Map by their graves. Describing the saints as dwelling-places of divine grace, the Master of the Sentences makes the point that 'not wheresoever He is does he dwell there, but where He does dwell' – n.b. in Ebstorf's martyrs – 'there He is'. Finally, asserts Peter, God is 'in the Man Christ in a most excellent manner, in whom the fullness of the Divinity indwells corporally'. The Map figures the pictorial correlate in terms of sacramental and ecclesial mediation: the Eucharistic body, to which the faithful gain access via the priesthood, actualizes the *corpus mysticum* of which the navel is Jerusalem and the head, Rome.[55]

From specula *to* speculum

Nothing in the antiquarian record concerning the Ebstorf Map, or in the archaeological record concerning the convent, allows us to surmise the context of the work's display. A description published in 1834 relates that the map was found in a storage room along with '*vasa sacra*, staffs perhaps used in processions, *Muttermarienbilder*, and altar coverings', and, more intriguingly, that poles were attached to it so that it could be rolled and unrolled.[56] The map's preservation amidst a collection

---

[54] Peter Lombard, *Sententiae in IV Libris Distinctae*, 3rd edn., vol. 1, pt. 2 (Grottaferrata: Editiones Collegii S. Bonaventurae ad Claras Aquas, 1971), liber 1, distinctio 37, 263–75; an English translation was available along with the Latin from the 1882 Quaracchi edition at http://www.franciscan-archive.org/lombardus/opera/ls1-37.html, accessed 7 July 2009; I quoted passages in chapters 1 and 2.

[55] For a general history of these anthropomorphic tropes, see Beat Wolf, *Jerusalem und Rom: Mitte, Nabel – Zentrum, Haupt. Die Metaphern 'Umbilicus mundi' und 'Caput mundi' in den Weltbildern der Antike und des Abendlands bis in die Zeit der Ebstorfer Weltkarte* (Bern: Peter Lang, 2010).

[56] Georg Heinrich Wilhelm Blumenbach, 'Beschreibung der ältesten bisher bekannten Landkarte aus dem Mittelalter, im Besitze des Klosters Ebstorf', in *Vaterländisches Archiv für hannoveranisch-braunschweigische Geschichte* (1834), 1–21, esp. 2. Eckhard Michael, 'Das wiederentdeckte Monument: Erforschung der Ebstorfer Weltkarte, Entstehungsgeschichte und Gestalt ihrer Nachbildungen', in *Ein Weltbild vor Columbus: Die Ebstorfer Weltkarte. Interdisziplinäres Colloquium 1988*, Hartmut Kugler (ed.) (Weinheim: VCH, 1991), 9–22, esp. 11.

of liturgical and para-liturgical objects might conceivably support its conjectural display in the church building itself where it could be encountered by pilgrims visiting the martyrs' graves.[57] Kugler explores this hypothesis, citing the example of the Hereford Map.[58] Alternatively, the Ebstorf Map may well have hung in the convent school, the existence of which is documented in 1307, and where, following Jürgen Wilke, it may have also served a pedagogical function.[59] In the end, however, Kugler sensibly concludes that the map's attachment to rollers militates against a single, permanent installation. His survey of the work's multiple functions therefore accommodates public exhibition and internal, conventual use. For the cloistered whose vows prevented actual travel to holy sites, the Map could have provided a stimulus to spiritual journeys and visitations (*peregrinatio in stabilitate*).[60]

Without limiting the audience for the map, I want nevertheless to extend appreciation of its communal function beyond the domains of schoolroom instruction on the one hand and compensatory 'imagined pilgrimage' on the other.[61] To be sure, the Ebstorf map encompassed encyclopaedia, bestiary and universal chronicle.[62] Yet over and above the various aspects through which the work collectively elaborated a *summa* of knowledge about the world, it is the expansive spread of the *mappa* – to which, by definition, all *picturae et scripturae* were tributary[63] – that put the *forma mundi* panoramically before one's eyes. In this respect, the Map's 'no small utility' exceeded imagined itineraries to specific destinations. It lay in providing physical and intellectual support for the monastic practice of speculation along much broader lines.

---

[57] See Klaus Jaitner, 'Kloster Ebstorf', *Germania Benedictina 11: Norddeutschland* (1984), 165–92, esp. 186; Hamburger, *The Visual and the Visionary*, 35–109, esp. 54–7.

[58] Kugler, 2:65; for the installation of the Hereford Map, see now Thomas de Wesselow, 'Locating the Hereford Mappamundi', *Imago Mundi* (forthcoming).

[59] Wilke, 1:257–71, esp. 270–1.   [60] Kugler, 2:14, 65–7.

[61] Hamburger, 'Vision and the Veronica', 322–3, for example, discusses how replicas of the Holy Face 'permitted an interior, proxy pilgrimage to Rome' for cloistered women. For an in-depth discussion of imagined pilgrimage as a monastic practice, see Daniel K. Connolly, *The Maps of Matthew Paris: Medieval Journeys through Space, Time and Liturgy* (Woodbridge: Boydell Press, 2009), esp. 5–89, and most recently, Kathryn M. Rudy, *Virtual Pilgrimages in the Convent: Imagining Jerusalem in the Late Middle Ages* (Turnhout: Brepols, 2011).

[62] Anna-Dorothee von den Brincken, 'Mappa mundi und chronographie: Studien zur "imago mundi" des abendländischen Mittelalters', *Deutsches Archiv für Erforschung des Mittelalters* 24 (1968), 118–86. Uwe Ruberg, 'Die Tierwelt auf der Ebstorfer Weltkarte im Kontext mittelalterlicher Enzyklopädik', in *Ein Weltbild vor Columbus*, 319–46; Margriet Hoogvliet, *Pictura et scriptura: textes, images et herméneutique des mappae mundi (XIIIe–XVIe siècles)* (Turnhout: Brepols, 2007), 163–8.

[63] Hoogvliet, *Pictura et scriptura*, 23–9.

Gautier Dalché has explained that the relationship between the *mappa mundi* genre and the Benedictine tradition of speculation developed out of their mutual engagement with the topos of the cosmic vision, adapted from Platonic and Stoic philosophies to Christian ends.[64] The world map became the graphic homologue of the celestial prospect available to God in which privileged contemplatives might momentarily share. Already Orosius had deployed the conceit of aerial observation to unfold the geographical tableau, a verbal *mappa mundi*, which opens his great world *History*. The imaginary view from a watchtower (*specula*) set up a moralizing vantage point from which to survey, and concomitantly reflect upon, the sorry course of human affairs, prelude to Christ's coming. Gregory the Great spiritualized the character of the *speculatio* afforded to the lofty gaze. He recounted how St Benedict, absorbed in prayer, once beheld from a tower's highest window the vastness of the entire world encompassed within a single ray of light; divine illumination had so dilated his soul that, carried beyond the world, he saw its true size and (in)significance.

Beginning in the early Middle Ages, detailed world maps variously marshalled the format and architecture of the codex to simulate for their monastic readers the *extra mundum* perspective that the rapt visionary casts over a minuscule world. One of the earliest preserved examples, dating from the second half of the eighth century (Albi, Bibliothèque municipale MS. 29, f. 57v), introduces the geographical description excerpted from Orosius' history. The map's distance from the reader combined with its full-page size recreates the double operation of elevated subject and miniaturized object of sight, actualizing the rhetorical conceit, the world seen panoramically *e specula*, of the text it prefaces.[65] Eventually *mappaemundi* enhanced the thematic purview of the *speculatio* by charting the evangelization of the *oikumene* within the economy of salvation; the series in copies of Beatus of Liebana's *Commentary on the Apocalypse* constitutes the earliest extant attempt.

Hugh of St Victor advanced a sophisticated, novel approach to the *mappa mundi*, which nevertheless remained rooted in the *speculatio* attendant upon the cosmic vision. He gave lessons on geography from a monumental wall map easily the size of the Ebstorf Map; fragments still existed at the Parisian abbey in the fourteenth century.[66] His purpose in

---

[64] My rapid précis synthesizes material from the studies cited in n. 6 above.

[65] Gautier Dalché, 'De la glose à la contemplation', 758–9.

[66] Patrick Gautier Dalché, *La 'descriptio mappe mundi' de Hugues de Saint-Victor: texte inédit avec introduction et commentaire* (Paris: Études augustiniennes, 1988), 192; Patrick Gautier Dalché, 'Nouvelles lumières sur la *Descriptio mappe mundi* de Hugues de Saint-Victor', in Gautier Dalché, *Géographie et culture*, 1–27, esp. 18–19; Patrick Gautier Dalché, '"Réalité" et "symbole" dans la géographie de Hugues de Saint-

taking his students through the world region by region was to facilitate their understanding of the literal, historical sense of Scripture, crucial for extended meditation on the deeper realities of divine revelation. Hugh ultimately worked, however, to shift his students' perspective from the inert materiality of physical space (correlate of the biblical *littera*) to a higher plane of consciousness whereby they might comprehend synoptically not only the design of all Creation but also the temporal ordering of God's plan for human redemption. Hence a *mappa mundi*, radically simplified to a generalized cartographic resume of the terrestrial arena, also appears as a component in his visualization of the *machina universitatis* embraced by Christ enthroned, his arms extended, head, feet and hands outside the system. The double aspect of the *mappa mundi* in Hugh's teaching, an up-close view from the vantage point of being in the world and a distant view from above, functioned to dramatize the Gregorian concept of the contemplative's interior transcendence and existential condition of detachment from earthly things. The cosmic implications of the Victorine programme remain fully active in images, discussed above, which show Christ *syndesmos* embracing only the *orbis terrarum* in the form of a *mappa mundi*, or, as in the case of the first of the Psalter maps, the torso of the Majesty astride it.[67]

The Ebstorf Map stages exactly the contrary motion. As Gautier Dalché points out, the Ebstorf Map 'attests to a reversal of priorities as to the place of the cartographic image in the economy of the meditative support'.[68] The *oikumene* takes over the pictorial composition, becoming its primary object, while highly attenuated, textual allusions to the *machina universitatis* (e.g., winds, heavens, six days of Creation) have a much-diminished impact given their relegation to the surrounding margins. The operative term that effects this reversal, I would add, is the inscription of the Christ's members *within* the world circle.

Taking a semiotic approach, Cornelia Herberichs has discussed how the map's dimensions factor into its modes of communication (*Medialität*).[69] The visual processing of such an enormous work compels interplay between multiple levels of reading and referentiality. Comprehension of

---

Victor', in *Ugo di San Vittore: Atti del XLVII Convegno storico internazionale, Todi, 10–12 ottobre 2010* (Spoleto: Centro italiano di Studi sul basso medioevo, 2011), 359–81.

[67] The model of the cosmic vision, codified in monastic circles, applies even to the Hereford Map, an argument I advance in a monograph now in preparation, *From Panoramic Survey to Mirror Reflection: Art and Optics in the Hereford Mappa Mundi*.

[68] Gautier Dalché, 'Pour une histoire des rapports entre contemplation et cartographie au moyen âge', 29–35, quote at 34–5.

[69] Herberichs, '. . . *quasi sub unius pagine visione coadunavit*', esp. 204–14.

a synoptic totality, the identification of God's body with the world, is possible only from a distance, while decipherment of individual elements (inscriptions, for instance) proceeds only at close range. Movement back and forth, searching out single objects (e.g. the tombs of apostles and martyrs) and 're-cognizing' their place within the overarching structure, triggers potentially limitless creative associations. Together the map's iconographic conception, physical format and reflexive exploitation of its status as material representation correlate with a particular visual and contemplative regimen that gives priority to speculating on God's presence in the world.

Promoting a true indexical relation to the real for which it substitutes, the Ebstorf Map turns readers into wayfarers. By dint of sheer size, the work beckons spectators to enter into the fullness of the sensible world and embrace corporeal sight as instrumental to the apprehension of things. The line of text just above the marginal T-diagram addresses readers in the second-person, '[there are things] you will find, if you take care to look in', or in/spect (*invenies, si procures inspicere*).[70] The design or composition (*forma*), by virtue of its exclusive focus on Christ's immanence and omnipresence within the *oikumene*, encourages meditation on the *vestigia dei* in nature and in history. And how else, during one's earthly sojourn, to *see* all the things of God's creation, the innumerable marvels, times and places beyond direct experience, if not through pictorial images of them made by human hands? The *res viarum* cartographically arrayed for visual delectation thus have the potential to lead, per Romans 1:20, to insight into the *invisibilia dei*.[71] The theory of speculation underlying the work transforms the second-order representation that is the *mappa mundi* into a medium of divine reflection in its own right.

By analogy with the Veronica (or with the full-body linen imprint of the legend behind the *Volto Santo*), the Ebstorf map thus likens the world to a cloth on which God impressed the *forma* of his Eucharistic body. This artistic conceit has its closest parallel in the theological conversion of the world into a mirror, an approach most cogently articulated by

---

[70] Kugler 1:40 no. 5/5.
[71] The speculative principle that the map 'offers ... to wayfarers ... the pleasure of the most pleasing sight of things along the way' (*prestat ... viantibus ... rerumque viarum gratissime speculationis dilectionem*) recalls Hugh of Saint-Victor's approach to his meditative figure of Noah's ark. In *De vanitate mundi*, the teacher invites the student on a walk through the ark, wherein all the works of creation and restoration are arrayed, and promises 'that manifold delight in such great things will meet us on our way'. *Spatiosa quidem, sed non fastidiosa erit ista deambulatio, ubi intrinsecus tantarum rerum varia oblectatio transeuntibus erit obvia*, PL 176: 720D, as translated in *Hugh of Saint-Victor: Selected Spiritual Writings*, Religious of C.S.M.V. (tr.) (New York: Harper & Row, 1962), 182.

Bonaventure on the basis of Victorine ideas. Perceiving the Creator's reflections therein initiates the soul's journey of return to God.[72] The Rhenish Dominican Henry Suso (c. 1295–1366) specifically advocated such a meditative practice in the context of his involvement with the *cura monialium*.[73] The Ebstorf Map overlaps to an uncanny extent the first part of the Franciscan's programme of contemplative ascent as outlined in his *Itinerarium Mentis in Deum*. Reading the former through the optic of the latter, however, does not aim to prove that the *Itinerarium* is the source text behind the cartographic project realized for the Saxon Bene-dictine nunnery. Rather the comparison is heuristic, intended to suggest how the Ebstorf Map might have assisted novices in the opening stages of spiritual formation. The paradigm of macro- and microcosm that governs both the map's overarching iconographic conception and its internal spatio-temporal order also structures Bonaventure's speculative regimen.

Just as the universe was created in six days, so the *minor mundus* ascends the ladder of illumination in six steps (1:5).[74] The map could have provided a substrate for visual exercises, corresponding to the first two steps (first two chapters of the *Itinerarium*), whereby the novice – well before she is able to contemplate God as the Alpha and Omega (1:5) – learns to contemplate God outside herself, i.e. through and in his vestiges in the sensible, material world (1:2). The lowest rung of the speculative ladder consists in having the aspirant put before herself the entirety of the sensible world as a mirror, through which she passes over to God on two historical models, the Hebrews' transit from Egypt to the Promised Land and Christ's passage from this world to the Father (1:9). The map at once supplies the equalizing plane of a reflective surface and prominently features the signposts marking out the exemplary Old and New Testa-ment paths (the Exodus, Christ's resurrection) that cut through appear-ances.[75] Sensory perception comes into play as she learns to appreciate

---

[72] Gautier Dalché, 'Pour une histoire des rapports entre contemplation et cartographie au moyen âge', 39.

[73] Hamburger, 'Speculations on speculation', 353, 360–82, and J. Hamburger, 'The use of images in the pastoral care of nuns: the case of Henry Suso and the Dominicans', in J. Hamburger, *The Visual and the Visionary: Art and Female Spirituality in Late-Medieval Germany* (New York: Zone Books, 1998), 197–232.

[74] References to the *Itinerarium* are from the Quaracchi Edition of the *Opera Omnia S. Bonaventurae*, 5 (1891), 295–316. For an English translation, see Bonaventure, *The Journey of the Mind to God*, Philotheus Boehner (tr.) and Stephen F. Brown (ed.) (Indianapolis: Hackett, 1993).

[75] The Israelites' passage through the Red Sea is indicated by the patch of dry land on which is inscribed *Sinus Arabicus* (another name for the Red Sea); Kugler 1:82–3, no. 27/12. Adjacent to the dry patch is *Mons Synai*, the largest single mountain on the Map, no. 27/14. Other relevant legends beneath the mountain to the north include: *Transitus*

the attributes of the Creator *through* the Creation. First, she considers things in themselves in terms of their externalities according to their weight, number and measure. Weight is that property which directs things to their particular location (1:11). Second, ruminating on how this world 'tends toward its origin, descent and end', she traces the successive temporal dispensations of nature, Scripture and grace (1:12). Third, she investigates the natures of things by applying reason (1:13). Training her analytic powers of discernment on visible things, she comes to understand how they reveal God's attributes: the book of creatures manifests the infinity of His power, the book of Scripture, the immensity of His wisdom, and the *body of the church* (*corpore ecclesiae*), among which range the divine sacraments, the immensity of His goodness (1:14). Studying the cartographic representation in a manner concordant with Bonaventure's recommendations for studying the world of things, she sees literally realized before her very eyes the scriptural verses inscribed at Christ's feet, that God's wisdom reaches mightily end to end and disposes all things sweetly.[76] The map's reader should 'see, hear, praise, love and worship, magnify and honour God in all creatures lest the whole *orbis terrarum* rise together against' her (1:15).

Moving next to the second rung, the novice contemplates the Creator *in* his vestiges, 'inasmuch as He is in them through His essence, power and presence' (2:1). This she achieves by means of the species, which, generated by all knowable things of the macrocosm, enter the microcosm through the five senses and impress themselves on the soul (2:2–6). In these species or likenesses, 'as in mirrors, can be seen the eternal generation of the Word, the Image and the Son' (2:7). All created things of the sensible world lead to God, because, of the 'first principle and eternal origin', there are shadows, echoes and pictures, vestiges, likenesses and spectacles divinely given as signs (2:11). The fullness of signification, however, rests in the sacramental order (2:12). Of course, as a visual compendium of graphic entities in the form of words and pictures, the Map itself generates species that enter into and impress themselves on the beholder; the work of human artistry participates in the chain of mirroring, mediating the things, places and ages in which the *invisibilia dei*, according to Romans 1:20, can be perceived and understood (2:13). The artist's handiwork, both in its execution and in the process of its

---

filiorum Israel, no. 27/16; *Manna hic pluit Dominus*, 1:80–1, no. 26/24; *Desertum Sin*, no. 26/25 and the names of several stations of the desert wanderings, *Raphidin*, *Helym*, *Marath*, nos. 26/23, 26, 27.

[76] Above Christ's right foot is inscribed *usque ad finem* [*fortiter*, implied], above the left; *suaviter disponensque* [*omnia*, implied], excerpted from Wis. 8:1; see Kugler, 1:321, 323 nos. 58/24 and 58/42, 2:144–5.

comprehension, emulates the divine aesthetics and semiotics that Bonaventure insists structures the Creation.

The practice of speculation as outlined by Bonaventure would demand
rigorous analysis of the Ebstorf Map's composition and content,
prompting mental recompilation of pictorial and verbal data with respect
to shifting categories and criteria. No simple didactic aid but rather a
'machine', to adopt Mary Carruthers's word, for meditation and cognition, the map would prepare the novice to 're-enter the mirror of mind'
now aglow with the light of divine things external to it (2:13).[77] The
spiritual journey at stake for viewers of the Ebstorf Map is not merely
imagined pilgrimage to earthly sites consecrated by sacred events and
relics, but rather the contemplative ascent itself.

Proper attention to the work can bring the novice metaphorically into
the atrium before the Tabernacle (3:1, 5:1), but there, at the threshold to
Bonaventure's third stage, the contemplation of God within the self, its
utility in a programme of spiritual ascent ends. Still, the map shows what
direction the novice must take going forward. According to Christine
Ungruh, the pair of images at the map's zenith, the earthly Paradise to the
north and the Holy Face on axis, appealed to the nuns' longing for the
heavenly paradise and anticipated beatific vision face to face.[78] No
doubt. But before the pictorial couplet indulges any optimism on that
score, the scene of the Fall and the Face *looking out* delivers the stern
message that, like the first parents, every sinner must submit to God's
penetrating gaze. In his sermon on conversion to the monastic life,
Bernard of Clairvaux develops the story of the Fall into an exemplum
of the soul's nakedness in the sight of God, precursor to repentance.[79]
The Holy Face of the Ebstorf Map – apart from its ecclesial reference and
Eucharistic resonance – puts aspiring contemplatives on notice that they
are not only beholders but also beheld. The staring effigy reminds the
novice to comport herself as if always under divine surveillance in
accordance with the founding document of western monasticism, the
Benedictine Rule.[80] Echoing this summons to unreserved introspection,
compunction and inner purification is the map's penitential orientation.
East, of course, is at top, but the map also points north, the cardinal

---

[77] Mary Carruthers, *The Craft of Thought: Meditation, Rhetoric, and the Making of Images,
400–1200* (Cambridge University Press, 1998), for her use of the term 'machine', esp.
198–203, 228–31, 254–7, 272–6.

[78] Ungruh, 'Paradies und *vera icon*'.

[79] Suzannah Biernoff, *Sight and Embodiment in the Middle Ages* (New York: Palgrave
Macmillan, 2002), 117–18.

[80] *Regula Sancti Benedicti* 7.13, in *Biblioteca Benedictina Intratext*, available at http://www.
intratext.com/X/LAT0011.HTM, accessed 24 July 2013.

direction allegorically associated since the time of Gregory the Great with dual movement of penance, aversion from sin and conversion to God.[81] Christ's right hand leads to the correctly aligned view of his resurrection.[82]

[81] Marcia Kupfer, *The Art of Healing: Painting for the Sick and the Sinner in a Medieval Town* (University Park, Pa.: Pennsylvania State University Press, 2003), 78–82.

[82] This chapter draws on research undertaken with the support of fellowships from the Center for Advanced Study in the Visual Arts at the National Gallery of Art, Washington DC, and the National Endowment for the Humanities. My heartfelt thanks to Paul Binski, Patrick Gautier Dalché, Jeffrey Hamburger, Margriet Hoogvliet, Herbert Kessler, Asa Mittman, Felipe Pereda, Christine Ungruh and Thomas de Wesselow for their critical readings of earlier versions. Where it shines, it reflects their insights; where it lacks lustre, my own shortcomings.

6 'After poyetes and astronomyers'
English geographical thought and early English print

Meg Roland

Although England had a vigorous history of map production during the high-medieval period, a long cartographic quiet in terms of English world-map production ensued.[1] As Lawrence Worms succinctly sums up English cartographic practice in the early print era, 'the history of map printing in the British Isles before [1555] is simply told: there was virtually none'.[2] By contrast, the production of Claudius Ptolemy's *Geography*, inclusive of maps based on his calculations, powerfully intersected with Continental print technology: the text was actively studied by humanist scholars, with translations from Greek manuscripts into Latin undertaken in the early fifteenth century and followed by cartographic printing in Italy and then northern Europe. The period from the mid 1470s to the mid 1550s was distinguished by the prestige and influence of Ptolemaic geography among European scholars, rulers and printers before the work of the second-century BCE Greek geographer and mathematician gave way, in turn, to geographic principles and representations based on the work of Gerardus Mercator.[3]

If a concurrent cultural process took place in England as on the Continent, however, it was largely undocumented by English-produced cartographic artefacts; that is, by English printed world maps. In light of this 'blank space' in the early print history of world map production in England, how can we assess the ways in which writers and readers in England synthesized and disseminated the emerging Ptolemaic paradigm and changing conceptions of the world? English geographical thought during this period cannot be traced along the surface of an English-made

[1] While the production of regional maps had currency in England, world maps do not seem to have been produced or, alternatively, to have survived. Sarah Tyack (ed.), *English Map-Making 1500–1650* (London: British Library, 1983); Catherine Delano Smith and Roger J. P. Kain, *English Maps: A History* (University of Toronto Press, 2000).

[2] Laurence Worms, 'Maps and atlases', in *The Cambridge History of the Book in Britain, vol. IV: 1557–1695*, J. Barnard and D. F. McKenzie (eds.) (Cambridge University Press, 2002), 228–45, at 229.

[3] See J. Simon, ch. 1; M. Small, ch. 7, this vol.

world map; contextualized within the emerging English book trade, it can be found, instead, amid the pages of literary and astrological early print books. Geographical knowledge in England during the late fifteenth and sixteenth centuries was disseminated via multiple and protean cultural forms – a dispersed and many-layered process of social construction that integrated classical and medieval geographic traditions, practices of a transnational book trade, the literary genres of romance and travel, and a broadly defined geographic discourse.[4] These forms were dynamically reinterpreted within a new medium – the public sphere of print.

Geographical trends in Tudor England

The influence of Ptolemaic geography and astronomy on European humanists was one of the significant intellectual forces of the sixteenth century. One of the primary markers of geographic sophistication in the fifteenth and early sixteenth centuries was the translation and production of Claudius Ptolemy's *Geography*, not exclusively as a scientific or geographic text, but as a text intimately tied to humanist discourse.[5] In addition, by the late fifteenth century, the *Geography*, often produced with coloured and illuminated maps, 'took pride of place in a surprisingly large number of great men's libraries'.[6] Ptolemaic maps presented a different visual representation of the world than the medieval *mappa mundi*. Maps based on Ptolemy's projection were typically, though not exclusively, oriented with north at the top; in addition, Jerusalem no longer held the position of symbolic sacred centre and coastlines began to represent more closely the outlines familiar to modern viewers. The *Geography* was first printed in Bologne in 1477, followed by editions produced in Rome, Ulm, Vienna, Strasbourg and, by 1541, in Lyons.[7] The text and maps of the

[4] Keith Lilley distinguishes the use of the term 'geographic knowledge' from the term 'geography' as a means to signal how geographical ideas, thoughts and representations have scope well beyond the discipline's conventional boundaries, either modern or historic. See K. D. Lilley, 'Geography's medieval history: a neglected enterprise?', *Dialogues in Human Geography* 1 (2011), 147–62.

[5] Patrick Gautier Dalché, 'The reception of Ptolemy's *Geography* (end of the fourteenth to beginning of the sixteenth century)', in *The History of Cartography*, vol. 3: *Cartography in the European Renaissance, Part 1*, David Woodward (ed.) (University of Chicago Press, 2007), 285–364.

[6] Lisa Jardine, *Worldly Goods: A New History of the Renaissance* (New York: Nan A. Talese, 1996).

[7] Gautier Dalché provides a comprehensive list of editions of the *Geography*, 'The reception of Ptolemy's *Geography*'; for a less comprehensive list, see also Henry N. Stevens, *Ptolemy's Geography: A Brief Account of all the Printed Editions down to 1730* (Amsterdam: Theatrum Orbis Terrarum, 1973), reprint of the 1908 edition, published by H. Stevens, Son and Stiles, London.

Geography provided a means by which to reimagine visually the physical space of the known world by virtue of the 28–32 maps which accompanied the text. Despite its impact among scholars and intellectuals, no edition was printed in England during the fifteenth or sixteenth centuries. Scholars have noted the absence of English cartographic production and have concluded that, up to the beginning of the sixteenth century, 'England was a geographical backwater, separated from the technological advances of the Continent' or similarly noted a perceived 'lag in English geography far behind the standard reached on the Continent'.[8]

By the end of the sixteenth century, however, English geographers and printers were re-engaged in cartographic output, producing the first national atlas as well as maps which were 'as detailed and accurate as any in the world'.[9] What underlay this long cartographic quiet in England? Were English readers, printers and intellectuals temporarily uninterested in or technologically unable to reproduce geographic advances in the early print era? Kathy Lavezzo has suggested that a geographic marginalism had long been cultivated in England as a form of cultural exceptionalism and that this mindset carried forward into the early print era.[10] The limitations of English print technology and fierce international competition may provide another part of the answer.[11] The absence of fifteenth- or sixteenth-century English world maps could possibly reflect a certain business acumen on the part of English printers: virtually every Continental printer who undertook the expensive production of an edition of the *Geography* went out of business shortly thereafter. But a lack of English-produced maps does not necessarily signify a corresponding lack of geographic curiosity or interest in mapping.[12] Looking beyond the evidence of map production, it becomes apparent that English readers and printers *did* participate in the changing geographic culture of Europe: missing from assessments of an isolated or indifferent English geographical imagination is the international context of the English book trade and the role of literary and popular texts in the development of a narrative English geographical

---

[8] D. K. Smith, *The Cartographic Imagination in Early Modern England: Re-writing the World in Marlowe, Spenser, Raleigh and Marvell* (Aldershot: Ashgate, 2008), 41; E. G. R. Taylor, *Tudor Geography: 1485–1583* (New York: Octagon Books, 1968), 13–14.

[9] Smith, *Cartographic Imagination*, 41.

[10] Kathy Lavezzo, *Angels on the Edge of the World: Geography, Literature, and English Community, 1000–1534* (Ithaca: Cornell University Press, 2006). Lavezzo argues that Englishmen during Cardinal Wolsey's time maintained a 'distinctly medieval perspective' (120).

[11] Worms, 'Maps and atlases', 229.

[12] See Nicholas Howe, *Writing the Map of Anglo-Saxon England: Essays in Cultural Geography* (New Haven: Yale University Press, 2008).

imagination; a narrative that drew upon the celestial, or astrological, strand of the classical geographic tradition.

The early print era in England – the transitional period between what is periodized as 'medieval' and 'Renaissance' – was marked by what Anne E. B. Coldiron has identified as 'a cross-cultural traffic in technology, ideas, and aesthetics'.[13] Although the map trade was transacted through diffuse channels not perfectly synonymous with the book trade, there was, nonetheless, a significant overlap, especially in terms of geographical books and books containing maps.[14] The book trade between England, Italy, France and the Low Countries was robust, as the library inventories of Oxford and Cambridge scholars from the time readily attest, and English printers during the period 1476 to 1557 had 'powerful connections to France and the rest of the Francophone continent',[15] as well as to Italian and northern European booksellers. These connections obviate, at least in part, a geographic isolationism posited for England during this time. At present, we lack a more detailed understanding of geographical thought in England in the early print era due, in part, to the paucity of documentation regarding the import and ownership of Continentally printed maps and Latin geographical texts; the trail is difficult to trace, as well, because of the fragile nature of maps themselves. World maps mentioned, for example, in an inventory of Cardinal Wolsey's goods do not specify if they are Ptolemaic-based maps, nor do the maps themselves survive.[16] Still, inventories of libraries at Cambridge University, for example, show that copies of Ptolemy's 'tabulai', printed in Venice and Ulm, were owned by university scholars along with unspecified *mappae* or *mappaemundi* as well as copies of Ptolemy's *Almagest,* his treatise on astronomy, and *Quadripartitum* (known as the *Tetrabiblos,* in Greek), his treatise of astrology. The scholars who owned these books valued the study of astronomy for its role in establishing church feast days, but also 'for its links with astrology and medicine'.[17]

---

[13] Anne E. B. Coldiron, *English Printing, Verse Translation, and the Battle of the Sexes, 1476–1557* (Aldershot: Ashgate, 2008), 2.

[14] Worms, 'Maps and atlases', 228–9.

[15] Coldiron, *English Printing*, 2. See also Henry Stanley Bennett, *English Books and Readers* (Cambridge University Press, 1965).

[16] See David Starkey (ed.), *The Inventory of King Henry VIII: Society of Antiquaries MS 129 and British Library MS Harley 1419* (London: Harvey Miller, 1998), 287–8. Kathy Lavezzo also discusses the two maps listed in the inventory and observes that their location at Hampton Court, Wolsey's residence, makes it likely that they were part of Wolsey's map collection: Lavezzo, *Angels on the Edge*, 116.

[17] Roger Lovatt, 'Introduction', in *The University and College Libraries of Cambridge, Corpus of British Library Catalogues 10*, Peter D. Clarke (ed.) (London: The British Library, 2002), i–xcii, at lxxviii.

Consideration, then, of book import and ownership practices in England challenges the tendency to equate map *production* with *use* and allows for the reassessment of English geographical thought in terms of cultural contact rather than nationalist isolation.[18] In addition, visual maps are 'inherently international . . . requiring little in the way of translation' and, as such, are highly mobile objects of trade.[19] Thus, a transnationalist book and cartographic milieu raises the possibility of a potentially livelier geographic culture within England than the map production record suggests. Making an analogous case about the role of translation as a vital part of early print literary production in England, Coldiron points out that 'we are better equipped to understand early modern English literature if we read it in its full, polyglot, international context'.[20] This same polyglot, international context offers a rich potential for reassessing English geographical culture in the early print or Ptolemaic revival period. To posit a strongly disinterested readership of such texts not only minimizes the international nature of the English book trade during this time but, at least for English humanists, scholars and printers, seems an improbable scenario.[21] Thus the book history record from the late fifteenth and early sixteenth centuries puts pressure on the geographical record and may offer possible answers to the puzzling absence of English geographic production during a period of dynamic Continental reformations of cartographic representation.

Negotiating printed geographies

'Geography' (by definition, 'earth-writing') was traditionally part of a threefold system of spatial studies, consisting of cosmography, geography and chorography.[22] One way to reconstruct English geographical thought in the fifteenth and sixteenth centuries is to recognize that the terrestrial strand of geography is one 'to which geographical science is

---

[18] For a discussion of the porous sense of national identity in fifteenth-century England, see Kenneth Hodges, 'Why Malory's Launcelot is not French: region, nation, and political identity', in *PMLA* 125 (2010), 556–71. Hodges challenges the idea of pre-modern proto-nationalism.

[19] Worms, 'Maps and atlases', 228.   [20] Coldiron, *English Printing*, 20.

[21] As just one example, Desiderius Erasmus translated an edition of the *Geography* from the Greek, published in Basle in 1533; although this was only two years before the death of Thomas More, it seems likely, given More's geographic fiction, *Utopia*, and his correspondence and friendship with Erasmus, that geographic topics were part of their intellectual exchange.

[22] Denis Cosgrove, *Geography and Vision: Seeing, Imagining and Representing the World* (London: I. B. Tauris, 2008), 17. On the distinction between these three branches, see also Simon, ch. 1, this vol.

now perhaps too completely confined'.[23] This perceptual 'confinement' no doubt contributes to modern scholars' identification of England as largely deficient in terms of geographical development during the late medieval and early modern period. Denis Cosgrove cautioned against a narrowly defined, modern conception of geography:

> So exclusively has the contemporary secular vision become limited to 'space-ship earth', that many students of geography scarcely appreciate the significance of planetary arrangements for patterns of climate on the earth's surface. Yet until very recently, geographical learning began with the aid of an orrery or celestial diagram ... whose geometry binds terrestrial and celestial spheres.[24]

The great Renaissance cosmographers such as Peter Apian, Oronce Fine, Sebastian Münster, and Martin Waldseemüller all drew upon Ptolemy's *Geography*, which describes the climates and formations of the earth, but also on *The Almagest*, Ptolemy's work of mathematical astronomy, which included material on constellations, fixed stars, eclipses and the movements of the celestial spheres. Signifying the lack of English-produced maps as a lack of geographic sophistication on the part of English readers arises, in part, from a conflation of the various strands of geography – the terrestrial, the celestial and the astrological – as they were understood in the fifteenth and sixteenth centuries, a tendency that defines the period through the lens of the post-Renaissance ascendancy of the visual.

In addition to the influence of the book trade and the reading of astrology as part of geographical studies, narrative geography was also a significant element of geographical reading in the late medieval and early modern period. As Small notes, narrative geography, rather than cartography, was far more influential throughout Europe than modern scholars have recognized, and included an emphasis on Ptolemy not primarily as a geographer but as an astronomer.[25] In particular, there seems to have been a greater interest on the part of English printers and writers (including Chaucer[26]) in the celestial and astrological components

[23] Cosgrove, *Geography and Vision*, 18–19.   [24] Cosgrove, *Geography and Vision*, 19.

[25] Small, ch. 7, this vol.; Gautier Dalché, 'The reception of Ptolemy'. See also Kim H. Veltman, 'Ptolemy and the origins of linear perspective', in *Atti del convegno internazionale di studi: la prospettiva rinascimentale, Milan 1977*, Marisa Dalai-Emiliani (ed.) (Florence: Centro Di, 1980), 403–7, who similarly argues that astronomy and geography were interconnected and that astronomy has been overlooked for its contribution to the development of linear perspective.

[26] See, for example, Chaucer's references to Ptolemy, including The Wife of Bath's warning against marriage by reference to portents that she, spuriously, 'attributes' to 'Ptolomee's *Almageste*': 'The same words writeth Ptolomee; Rede in his *Almageste* and take it thee' (ll. 182–3). Geoffrey Chaucer, 'The Wife of Bath's Prologue', *The Riverside Chaucer*, 3rd edn., Larry Benson (ed.) (Boston: Houghton Mifflin, 1987), 107.

of the work of 'Master Ptolemy'.[27] Perhaps print technology in England lacked the technical capabilities to print the *Geography* with maps, perhaps import met the relatively small demand for expensive editions of the *Geography*, but perhaps Ptolemy's *Almagest* and *Quadripartitum* spoke more powerfully to the English geographical imagination. So, while England was not an innovator in cartographic production during this period, neither was it completely disengaged from Continental geographic culture, nor lacking modes of geographic expression. In this regard, 'Ptolemaic revival' may be a useful term in conceptualizing this transitional period – one in which geographical thought and print culture productively co-developed, refashioning literary genres, geographic writing, and, eventually, cartography.[28]

Geographic writing and cartography thus developed contiguously across the fifteenth and sixteenth centuries, as earlier romance and travel narratives continued to have currency for Tudor readers, while at the same time medieval geographical works continued to be actively translated and printed.[29] In the absence of English-made maps, English geographical thought may be traced through these literary genres,

---

[27] For a discussion of the various editions of astronomical and astrological works printed in the early print era in England, see Henry Stanley Bennett, *English Books and Readers* (Cambridge University Press, 1965), 116–18. They include Caxton's *Mirrour of the World*, Trevisa's translation of Bartholomew's *De proprietatibus rerum*, *The Kalender of Shepherds*, *The Compost of Ptholomeus*, *The boke of demaundes of the science of Phylosophye and Astronomye Betwene kynge Boctus, and the Phylosopher Sydrake*, *De Astonomia*, Andrew Borde's *Pryncyples of astronomye*, William Salisbury's *The description of the sphere of frame of the worlde*, and, in 1552, Anthony Ascam's scholarly work *A lytel treatyse of astronomy*.

[28] It is well to note that the period in which Ptolemaic geography flourished in Europe defies the demarcation of the year 1500 as the boundary between the medieval and early modern periods, a boundary that has been increasingly contested. David Woodward, a historian of cartography, Keith Lilley, a cultural geographer, and William Kuskin, a literary scholar, each challenge the overdetermination in their respective fields of the transition from the medieval to the early modern as a moment of radical change or rupture. These scholars point to a building consensus in geographic and literary studies: the 'firewall of 1500' is a myth. See David Woodward,'Cartography and the Renaissance: continuity and change', in *The History of Cartography*, vol. 3, Woodward (ed.), 3–98; K. D. Lilley, '*Quid sit mundus?* Making space for "medieval geographies"', *Dialogues in Human Geography* 1 (2011), 191–7; idem, *City and Cosmos: The Medieval World in Urban Form* (London: Reaktion Books, 2009); William Kuskin, '"The loadstarre of the English language": Spenser's *Shepheardes Calendar* and the construction of modernity', *Textual Cultures: Texts, Contexts, Interpretation* 2 (2007), 9–31, at 11. See also William Kuskin (ed.), *Caxton's Trace: Studies in the History of English Printing* (University of Notre Dame Press, 2006), 1–31. Kuskin argues that the desire to 'find a singular moment' of transformation between the medieval and modern, typically located in the fifteenth century, is 'a desire for inhuman purity' (19).

[29] A special volume of *Arthuriana* 19.1 (2009) is dedicated to late medieval French Arthurian Romance and its continued popularity during the Renaissance. See, in particular, the introduction by Carol J. Chase and Juan Tasker Grimbert, 3–6, and Jane H. M. Taylor, '"Hungry shadows": Pierre Sala and his Yvain', 7–19.

including Thomas Malory's Arthurian romance *Le Morte Darthur* (completed by 1470 and first printed in 1485), William Caxton's translation of *Mirrour of the World* (first printed in 1481), and editions of *The Kalender of Shepherds* and *The Compost of Ptholomeus*, popular compendiums of astrological, spiritual and health advice (first printed in the late fifteenth century with subsequent editions throughout the sixteenth century). English readers in the early sixteenth century, as evidenced by the popularity of texts such as *The Kalender of Shepherds* and *The Compost of Ptholomeus*, developed a conceptual geography within the context of astronomy, astrology and prognostication – information that was self-consciously fashioned from the pseudo-authority of Ptolemy but attributed, more generally, to 'poyetes and astronomyers'.[30] While Henri Lefebvre discounted the mutual influence of humanism and science in the early modern period, Gautier Dalché instead makes a clear case that astrology 'should never be overlooked in a study of the reception of Ptolemy's geography'.[31] Such divisions, he argues, reflect a later disciplinary divide between poetics and science not pertinent to early humanist intellectual discourse. Literary and popular texts from this period exemplify the connections made by late medieval and early modern English readers between geography, astronomy, astrology and literature, a discourse not solely dependent on, but co-generative with, the emerging field of print cartography. For English readers and writers of this period, the 'reading' of geography was not so much a break with the classical and medieval past as a confirmation and expansion of it. Geography did not merely represent a world 'out there', but one that was intimately connected to the self through the practice of astrology, a practice framed by the interpretation of textual and celestial signs – the habitus of poets and astronomers.

Thomas Malory and Arthur's Roman War campaign

By the 1530s, Henry VIII and certain members of his court developed an understanding of maps as vital tools of administration and statecraft, an understanding that Peter Barber has identified as marking 'a watershed' in cartographic history.[32] Barber, map curator at the British Library, has identified Henry VIII as the pivotal figure in the development of what he

---

[30] *The Compost of Ptholomeus*, Robert Wyer, printer, 1530? [STC 20480], sig. O(2)r.

[31] Gautier Dalché, 'The reception of Ptolemy's *Geography*', 352.

[32] Peter Barber, 'England I: pageantry, defense, and government: maps at court to 1550', in *Monarchs, Ministers, and Maps: The Emergence of Cartography as a Tool of Government in Early Modern Europe*, David Buisseret (ed.) (University of Chicago Press, 1992), 27.

calls 'map consciousness'.[33] While the concept of a newly formed Renaissance 'map consciousness' is contested,[34] it seems clear that the use of maps as a spatial, geographic tool for governmental strategy gained currency during Henry's reign. But what precipitated this emerging Tudor understanding of geography and maps?

To answer this question, we may look, in part, to *Le Morte Darthur*, the last great Arthurian romance of the medieval period. Garrett Mattingly, a twentieth-century historian of the Tudor period, asserts that the young Henry VIII 'learnt his geography and politics mostly out of Froissart and Malory'.[35] Mattingly meant this as a deprecation of Henry's early geographical imagination, but just what would the geography of a reader of Malory look like? At first glance, the assumption – as it was for Mattingly – is that it would be a hazy romantic geography of undistinguished fields, forests and 'fayre champaynes' upon which adventures and quests are played out. However, Henry's early reading of Malory as well as Froissart's *Chronicles* may have been an integral part of his nascent cartographic awareness, a reflection of ideas about representational and symbolic space that was already in process in the genres of chronicle and romance. Medieval romances, with their common motif of expulsion-and-return, presented stories well-suited to redactions and revisions that accommodated the flux of geographical imagination and thought in the fifteenth and early sixteenth centuries.[36]

It is well known that Malory added specific English place names for English locales, as if to ground the vague landscape of Arthurian romance in the very soil and cityscape of fifteenth-century England. Joseph Parry has noted that Malory's narrative 'often unfolds simultaneously ... in two kinds of locations': one strand follows the fall of Lancelot through a recognizable, realistic geography, identifying town names and marking distances recognizable to Malory's contemporary readers; the second narrates, as Parry puts it, 'a nonlinear kind of map ... unable to connect the effects of human action with human causes'.[37] Examples of this dual geography include the simultaneous location of Arthur's burial in Glastonbury Abbey with his transmigration into the otherworldly Avalon, and the identification of the city of Winchester as the locale of the legendary Camelot. In addition, Malory suggests, for example, St Paul's Cathedral as the site of the sword-in-the-stone

---

[33] Ibid. 26.   [34] See Lilley, 'Geography's medieval history'.

[35] Garrett Mattingly, *Catherine of Aragon* (Boston: Little Brown, 1941), 114.

[36] Helaine Newstead notes this common formula in her overview of English romances in 'Romances, General', in *A Manual of the Writings in Middle English: 1050–1500*, fascicle 1 (New Haven: Connecticut Academy of Arts and Sciences, 1967), 24.

[37] Joseph Parry, 'Following Malory out of Arthur's world', *Modern Philology* 9 (1997), 148–9.

episode, tells us that Lancelot guided his horse across the Thames near Westminster Bridge to rescue Guinevere from Meleagant, and provides the names of the specific counties that side with Mordred in the final battle. Thus, Malory infuses the landscape of romance with a new geographic specificity not in his French or English sources.

Identifying local place names was an integral part of geographic practice in the mid- to late fifteenth century, one that parallels the practice of producing dual map sets in editions of Ptolemy's *Geography*: a set with the ancient Greek or Latin place names and one with contemporary place names. Malory, writing in the late 1460s, may not have seen a manuscript edition of the *Geography*, but he does enact a fifteenth-century impulse to layer new geographical knowledge over source texts, a blurring of two geographic discourses that had, in the earlier medieval period, remained somewhat separate: the theoretical-literary geographic tradition, based on the textual authority of classical sources, and the pragmatic geographical tradition of wayfinding.[38] Further, during the first half of the fifteenth century, as Gautier Dalché reminds us, Ptolemy's *Geography* 'was appreciated by Italian humanists for features other than those we see as constituting the originality of his work. First and foremost, the *Geography* was seen as a compendium of ancient place-names'.[39] In light of this practice, we can place what has typically been identified by literary scholars as an apparently individual, or even idiosyncratic, interest on the part of Malory into a larger cultural context: that of the development of a geographical imagination attendant to the double-life of place, a converging of the classical geographic tradition and emerging spatial practices. Malory's text, especially his account of Arthur's Roman War campaign, can be seen as an active agent in this cultural process; indeed, eventually serving as seminal reading for England's first 'cartographic king'.

Malory's tale of the birth, life and death of Arthur includes an account of the Roman War episode, Arthur's foray out of England against the Roman Emperor Lucius. It is the part of Malory's text subject to the most active editorial intervention in the fifteenth century, indicative of the potent geopolitical content of the narrative. In the Roman War account, the young King Arthur marches across the Continent to answer by force the Roman emperor's demand for taxes and fealty, vanquishing, along the way, a giant at Mont St Michel and a host

---

[38] Natalia Lozovsky, '*The Earth is Our Book': Geographical Knowledge in the Latin West ca. 400–1000* (Ann Arbor: University of Michigan Press, 2000).

[39] Gautier Dalché, 'The reception of Ptolemy's *Geography*', 359.

of 'Saresyns' – mercenaries in the imperial army of Rome.[40] Just as Froissart's late fourteenth-century *Chronicles* provided a verbal map of the Crécy and Agincourt campaigns in Normandy,[41] Malory's Roman War account also functioned as a kind of verbal wayfinder. While Malory's work has been characterized as geographically naïve, Malory's place names have been shown to mark, not geographical indifference but, more often, modern textual misconstructions of his intended place name. P. J. C. Field clarifies, for example, the confusion caused over Malory's use of the name 'Roone' to indicate the French provincial city, Rouen. Previously thought to refer to 'the Rhine', this modern geographic misidentification inexplicitly lurched the narrative of Arthur's campaign far from its more reasonable route through Burgundy.[42] While Malory's geography is not precise, it does exhibit a concern with geographic realism.

The route of Arthur's campaign to Rome has itself been the subject of much scholarly scrutiny and controversy. For the portion of the campaign traversing Normandy, Eugene Vinaver suggests that the circuitous route apparently indicated in the narrative by the place name 'Sessoyne' was the result of Malory's decision to 'make Arthur's journey across the Continent resemble Henry V's itinerary', thus turning Arthur's campaign into 'a tribute to the victor at Agincourt'.[43] But William Matthews more convincingly argues that 'Sesia' was the likely city to which Malory was referring, producing a route that supports a more geographically informed Malory.[44] In Malory's source text, the *Alliterative Morte Arthure*, Arthur vanquishes Lucius only to learn that he has been usurped by Mordred and that his legendary rule has been shattered. Malory, however, famously defers this disastrous outcome in his version of Arthur's Roman War campaign, instead narrating a victorious return to a coalescing 'nation' flush with imperial success and the prospect, however fleeting, of assuming the symbolic mantle of Rome. The Continental foray yielded a regnal success well beyond the boundaries of England.

During the reign of Henry VIII, as Lavezzo has argued, a tradition of English self-represented geographic isolation and exceptionalism, in tension with centralized Roman authority, helped to support the imaginative leap Henry needed to enact an ecclesiastical break with

---

[40] Eugène Vinaver and P. J. C. Field (eds.), *The Works of Sir Thomas Malory*, vol. 1, rev. edn. (Oxford: Clarendon Press, 1990), 194.

[41] John Bourchier, Lord Berners, at the request of Henry VIII translated Froissart's *Chronicles* into English; it was printed by Richard Pynson in 1523.

[42] P. J. C. Field, *Malory: Texts and Sources*, Arthurian Studies 40 (Cambridge: D. S. Brewer, 1998), 76.

[43] Vinaver and Field (eds.), *The Works of Sir Thomas Malory*, vol. 3, 1368.

[44] William Matthews, 'Where was Siesia-Sessoyne?', *Speculum* 49 (1974), 680–6.

Rome. Lavezzo observes that 'before Henry VIII asserted England's sovereign independence from the church he (outrageously) pondered sacking Rome'.[45] But perhaps Henry's speculative sacking is not quite so outrageous: Henry had read Malory. In so doing, he had before him the model of a successful break with Roman authority and a verbal map from Malory's romance of how such a military campaign might be plotted across the continent. Henry's reading of Malory's Roman War account provides an example of the impact of 'poetic geography'[46] or 'topographesis';[47] that is, texts that use location as a significant component to structure ideological or cultural scripts. With less scepticism about Malory's geographic awareness, we can see that reading Malory's romance would indeed be vital reading for the young Henry VIII, providing a mental map of a possible Continental campaign and suggesting a geopolitical possibility – a break with Roman authority – that would come to fruition under his reign.

William Caxton and *Mirrour of the World*

William Caxton, printer of *Le Morte Darthur* and the first to establish a print house in England, can be seen as a kind of nodal point in terms of late medieval and early modern geographic discourse: in London, Caxton printed an excerpt of John Trevisa's *Polychronicon* under the title *The Discription of Britayn* in 1480 and, in 1481, translated and printed the universal history and geography, *Mirrour of the World*. There is little doubt that William Caxton, as governor of the English nation in Bruges prior to setting up his print shop in London, was well acquainted with the holdings of the Ducal libraries of Burgundy and with Continental geographic printing. While earlier scholars tended to cast Caxton as a provincial figure, more recent scholarship has acknowledged the international aspect of Caxton's career and the way in which he 'forged new cultural links between the Yorkist and Burgundian courts'.[48] Indeed, it is not too great a claim that Caxton 'should be recognized as the most

---

[45] Kathy Lavezzo, *Angels on the Edge*, 26.

[46] John Gilles, *Shakespeare and the Geography of Difference* (Cambridge University Press, 1994), 5.

[47] Henry S. Turner, 'Literature and mapping in early modern England, 1520–1688', in *History of Cartography*, vol. 3, Woodward (ed.), 424. On the way in which literature and cartography are 'densely interwoven' see Tom Conley, 'Early modern literature and cartography: an overview', in *History of Cartography*, vol. 3, Woodward (ed.), 401–11.

[48] Alexandria Gillespie, 'Caxton and after', in *A Companion to Middle English Prose*, A. S. G. Edwards (ed.) (Rochester, NY: D. S. Brewer, 2004), 310.

cosmopolitan of English literary figures' in the late fifteenth century.[49] Much research remains to be done on the import of Continental cartographic materials and their dissemination in England during this period, but it seems likely that William Caxton played an integral role. Despite the lack of Ptolemaic-based cartographic production in England, Continentally produced world maps seem likely to have been in circulation in England, primarily as a result of the vigorous book trade between England, the Low Countries and Italy, in which Caxton was an active participant. Unfortunately, import records rarely record maps or individual book titles, instead noting the number of barrels of 'diverse histories'. The links between medieval English intellectuals and Italian humanists, more fully explored in terms of Chaucer's work, is less fully understood in terms of late medieval geographic discourse.[50] Nonetheless, we do know, for example, that Duke Humphrey (whose collection formed what would become the initial basis of the Bodleian Library) ordered a manuscript copy of the *Geography* from his Milanese book merchant in 1451,[51] and that by the mid sixteenth century, maps and globes were certainly in the libraries of English gentlemen.[52] Thomas Elyot's 1531 *The Boke Named the Gouernour*, for example, provides evidence of the presence of maps or globes in English homes at least by the early sixteenth century. Elyot famously remarks upon the pleasure he experiences when he 'beholde[s] the olde tables of Ptholomee where in all the worlde is paynted', articulating an understanding of geography that closely associates aesthetics, history and cartography.[53]

Caxton's source manuscript for his translation of *Mirrour of the World* was the French *Image du Monde*, written in 1250. The *Mirrour* is composed of three parts: a universal history, a medieval geography and an astronomy based on Ptolemy's *Almagest*. Although Caxton's edition is notable for containing the first printed maps in England, its reputation suffers as a negative exemplum of retrograde medieval geographical thought.

---

[49] Jennifer Goodman, 'Caxton's Continent', in *Caxton's Trace: Studies in the History of English Printing*, William Kuskin (ed.) (Notre Dame: University of Notre Dame Press, 2006), 101–23, at 101.

[50] John Tiptoft is one example of an English intellectual with strong ties both to Italian humanists and to Caxton. An active book buyer, Tiptoft also translated Cicero's *De Amicitia*, printed by William Caxton as *Of Old Age, Of Friendship, Of Nobility* in 1481. See Henry Lathrop, *Translations from the Classics into English, from Caxton to Chapman, 1477–1620* (New York: Octagon Books, 1967), 24.

[51] Gautier Dalché, 'The reception of Ptolemy's *Geography*', 319.

[52] See, for example, John Evans, 'Extracts from the private account book of Sir William More of Loseley, in Surrey, in the time of Queen Mary and Queen Elizabeth', *Archaeologia* 36 (1855), 284–310. More's 1556 inventory includes world maps, a globe and a perpetual calendar.

[53] Thomas Elyot, *The Boke Named the Gouernour*, 1531. Facsimile. No. 246 (Menston: England: Scholar Press, 1970), f. 37.

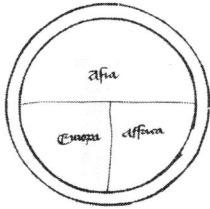

Figure 6.1 The tripartite world, *Myrrour of the World*, printed by
William Caxton, 1481.
Source: STC 24762, British Library.

Caxton's *Mirrour* contains very simple T–O woodcuts, with hand lettering,
startlingly medieval when compared with the Ptolemaic-based maps
being printed on the Continent at this time (Fig. 6.1). The technical demand
of printing copper-etched maps and the fact that the *Geography* was in
Latin (a language from which Caxton did not typically translate) both
offer an explanation why Caxton never undertook to translate and publish
Ptolemy's work himself. But the influence of new representations of the
world can be detected in Caxton's prose, signalling a negotiation on his
part with the alternative geographic frameworks circulating in the late
fifteenth century. For example, Caxton inserts the statement into his edition
of the *Mirrour* that, 'And how be it that the Auctour of this book saye that
thise contrees ben in Affryke, yet, as I vnderstonde alle thise ben within the
lymytes and boundes of Europe'.[54] Caxton is acutely aware that geographic
information is in flux and asks readers 'to correcte and amende' any errors
and that 'they repute not the blame on me, but on my copie'.[55] In this,
Caxton continues the practice of geographical writing which had long
been considered 'dynamic' and 'subject to revision'.[56]

The simple woodcut maps and illustrations of the cosmos in Caxton's
Mirrour are carried over from his manuscript source, but Caxton's text
also includes a second group of original woodcuts. Of particular interest
is the image that accompanies the beginning of the text proper – that of
a scholar engaged in study of a globe. The text surrounding this image
includes the axiom that 'God sees all, hears all, knows all', and 'holdeth
all thyngs in his honde'.[57] Here, the scholar reproduces the divine act
of holding the world in his hand, now in the form of a globe, an image
that precedes the English production of terrestrial globes by Molyneux by

---

[54] Oliver Prior (ed.), *Caxton's Mirrour of the World* (London: Early English Text Society by
Oxford University Press, 1966), 94 [STC 24672].
[55] Prior (ed.), *Caxton's Mirrour*, 7.   [56] Simon, ch. 1, this vol.
[57] Prior (ed.), *Caxton's Mirrour*, 9.

100 years. Caxton's text also includes an original image of a Mediterranean scholar, perhaps a representation of Ptolemy although not expressly identified (Fig. 6.2). Caxton made subtle but significant changes to both the text and images of his source, suggesting that the *Mirrour* is not merely medieval geography's last gasp in England, but a complex and early shift toward an amalgam of classical theoretical geography, Christian universal history and eyewitness geographical narrative. As such, the *Mirrour* stakes out a framework for English geographical culture as Ptolemy's works are assimilated over the next fifty years.

There is evidence that Caxton's *Mirrour* met a growing interest in geographic and astronomical knowledge in England: it was reprinted by Caxton in 1490 and again by Andrew Lawrence in 1527. In Lawrence's edition, however, Caxton's initial T–O map images become blank circular spaces on the page, empty signifiers of a medieval geographic past. Lawrence's text is another juncture for English geographical narrative: while his title and verbal text are from Caxton's *Mirrour of the World*, Lawrence's initial image – that of a scholar in his study – is taken from Guy Marchant's 1493 edition of *Le Compost et Calendrier de Bergiers*, a text which was subsequently translated into English and published as *The Kalender of Shepherds* and as *The Compost of Ptholomeus*, two texts that would exert considerable influence in the development of popular geographical thought in England throughout the sixteenth century.

The Kalender of Shepherds and *The Compost of Ptholomeus*

The Kalender of Shepherds (not to be confused with Spenser's late sixteenth-century poem, *The Shepheardes Calendar*[58]) forms what might be called a 'multi-text',[59] a loosely affiliated body of early printed books which share a title, image or content but that contain wide variation between editions. First translated into English from the French in 1503,[60] *The Kalender of Shepherds* was reproduced in numerous English versions

---

[58] Robert Lane discusses Spenser's appropriation of the popular form of the calendar in *Shepheards Device: Edmund Spenser's 'Shepheardes Calendar' and the Institutions of Elizabethan Society* (Athens, GA and London: University of Georgia Press, 1993).

[59] For an explication of the term 'multi-text', see Iain Macleod Higgins, *Writing East: The 'Travels' of Sir John Mandeville* (Philadelphia: University of Pennsylvania Press, 1987). Higgins uses this term to refer to the body of editions, translations and redactions that make up the *Travels* of John Mandeville.

[60] For a discussion of the various editions of the *Kalender of Shepherds* see Martha Driver, 'When is a Miscellany not miscellaneous? Making sense of the "Kalender of Shepherds"', *The Yearbook of English Studies*, vol. 33, *Medieval and Early Modern Miscellanies and Anthologies* (2003), 199–214.

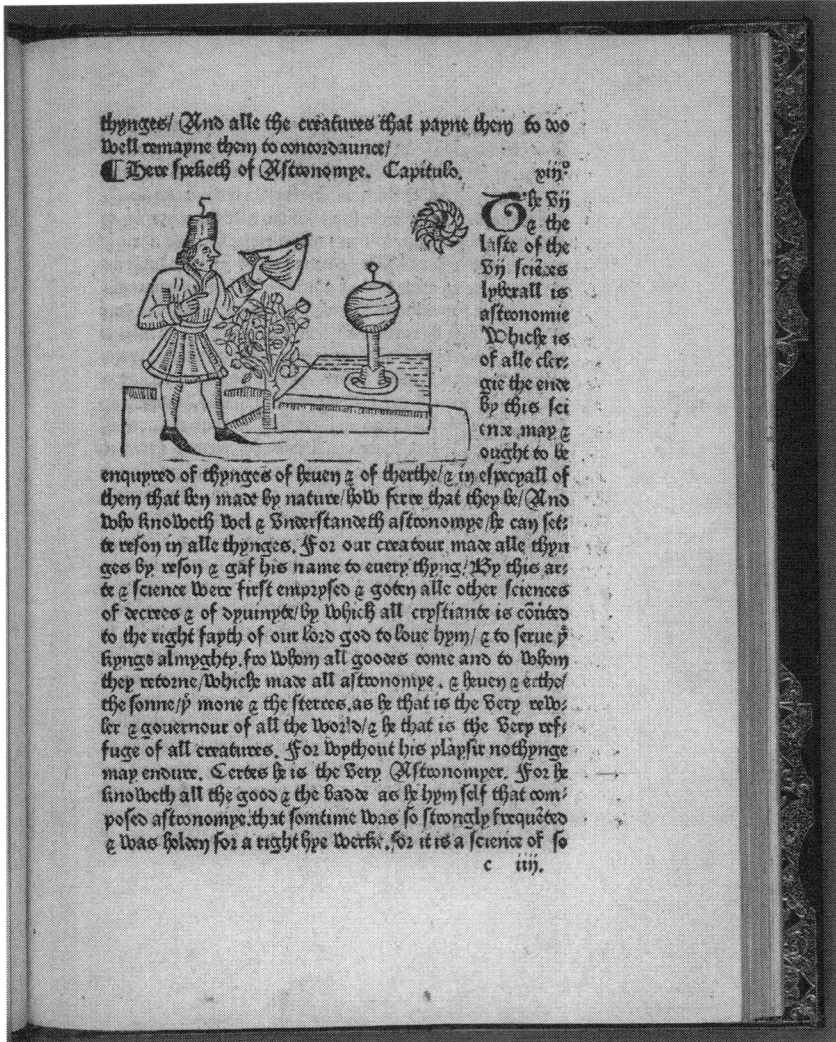

Figure 6.2 Astronomer with his instruments, *Myrrour of the World*, printed by William Caxton, 1481.
Source: STC 24163, Huntington Library.

throughout the sixteenth century, but in two strands: one maintaining the title of *The Kalender of Shepherds* and the other changing the title to *The Compost of Ptholomeus*. Although they are often described as containing 'nothing from Ptolemy', other than the 'falsification' of authorial attribution,[61] these texts do, in fact, have a general articulation of some of the astrological matters set forth in Ptolemy's *Quadripartitum*; for example, planetary and zodiacal influences, the relationship of zonal geography and ethnography, and the reckoning of a healthy man's life expectancy. Whereas Ptolemy, for example, provides a complex and esoteric analysis of life expectancy, *The Kalender* picks up the same topic but simplifies the probability to a straightforward seventy-two years. While the *Quadripartitum* is primarily a treatise on astrology, it also contains some geographical information,[62] indicating the ways in which these subjects were less rigorously distinct in classical and medieval geographical writing.

Initially, editions of the *Kalender* did not include terrestrial geography, *per se*: they functioned primarily as a *computes* or calendar of Catholic feast days as well as a compilation of astrological prognostication, Christian prayers, proverbial morality, health advice, and sententious or comic verse (such as 'The Assault on the Snail'). But when the definition of geographical thought is contextualized as inclusive of astrology and astronomy, these small volumes can be reassessed as contributing to the popular dissemination of new geographical referents. And, eventually, terrestrial geography becomes specifically integrated into this protean cultural form.

In the *Kalender of Shepherds*, the putative source of the astrological and health information is initially an unnamed, ancient shepherd. In the Prologue to Richard Pynson's 1506 and Julian Notary's 1518 editions, a classical shepherd is invoked as the text's source:

As here before tyme there was a Sheparde kepynge shepe in the feldes/ whiche was no clerke ne had no vnderstandinge of þᵉ letterall sence/ nor of no maner of scripture nor wrytynge/ but of his naturall wytte & vnderstandinge sayd. How be it þᵗ lyuynge & dyenge be all at þᵉ pleasure of almyghty god.[63]

The authentication for the information in the text was a 'natural' and pastoral figure of wisdom, devoid of book learning. In the Prologue, it is also stated that 'this boke was made for them that be no Clerkes to brynge them to great vnderstandynge',[64] thus identifying itself as a text for a non-elite readership yet at the same time offering access to the very tradition of classical learning it obscures, and intimating a connection

[61] Lathrop, *Translations from the Classics*, 330.
[62] O. A. W. Dilke, with additional material supplied by the editors, 'The culmination of Greek cartography in Ptolemy', in *The History of Cartography Volume 1: Cartography in Prehistoric, Ancient, and Medieval Europe and the Mediterranean*, J. Brian Harley and David Woodward (eds.) (University of Chicago Press, 1987), 177–200, at 180.
[63] *Kalender of Shepardes*, Julian Notary, printer, 1518 [STC 22410], sig. A(5)v.
[64] Ibid., sig. A(6)r.

between occult knowledge and the act of reading. The frontispieces in both the 1506 and 1518 editions depict a rustic shepherd, complete with bagpipe and staff – an image which proliferates through subsequent editions. But despite this 'natural' source of authentication in text and imagery, these editions also include an image of a scholar seated in his study, a version of which subsequently appeared in Lawrence's edition of the *Mirrour of the World*. This image of the scholar both augments and stands in tension with the primary authenticating image of the text – that of the shepherd. Via this image of scholarship, then, readers of *The Kalender of Shepherds* participated in an aspirational identification with the role of a learned scholar, suggesting that the reader, too, is engaged in the powerful symbolic action of contemplation, reading and learning. In Notary's 1506 edition, Ptolemy is merely cited in the table of contents in relation to the twelve signs of the zodiac but not mentioned in the text. In Pynson's 1518 edition, Ptolemy is referenced both textually and visually, again in relation to the zodiac, but as a very minor reference in the text. Subsequent editions of the *Kalender of Shepherds* were brought out, in various forms, by a number of printers during the sixteenth century, among them Colwell, deWorde and Copland, and, in these editions, Ptolemy remains a minor referent.

Beginning in the 1530s, a strand of the multi-text breaks off: the text is condensed, new images are added, others are eliminated, and the title is changed to *The Compost of Ptolomeus/Prince of Astronomeye, Translated out of Frenche in to Englysshe. For them/ that wolde haue knowlege of the Compost*' (Fig 6.3). These editions, initially published by Robert Wyer, make a significant modification: the name of Ptolemy is increasingly inserted into the verbal text, shifting the authentication from the ancient shepherd to Ptolemy. The introduction is similar to that of the *Kalender of Shepherds*, but now reads: 'So as we vnderstnde, *Ptholomeus sayeth*, that lyuynge and dyenge is all at the pleasure of almyghtye God'.[65] The Catholic feast day calendar is eliminated, along with much of the Christian moralizing and, generally, a narrower focus on the astrological components. Neither the woodblock image of the shepherd nor that of the scholar carries over once the text is renamed *The Compost of Ptholomeus*; instead, the symbolic function previously vested in the figure of the scholar or shepherd is now conflated into the single figure of Claudius Ptolemy, 'Prince of Astronomeye'. Interestingly, the woodblock image of Ptolemy in the 1552 edition of *The Compost* first appeared in Marchant's 1499 edition as an illustration of Prester John, a richly

---

[65] *The Compost of Ptholomeus*, Robert Wyer, printer, 1530? [STC 20480a], sig. A(6)r (emphasis added).

Figure 6.3 Title page, *The Compost of Ptholomeus/Prince of Astronomye*, printed by Robert Wyer, 1530.
Source: STC 20480, Huntington Library.

suggestive previous use.[66] In addition to the images of Ptolemy using astronomical instruments and textual attributions to Ptolemy in editions of *The Compost*, both strands of this multi-text exhibit nascent spatial practices in several ways: there are images and verbal text that 'map' the zodiac, diagrams which map the veins and body, and the adoption of Caxton's innovation of rubricating or mapping the text with a table of contents. Like their medieval antecedents, these spatial practices were grounded in the negotiation of the self in relation to the cosmos and form a kind of 'hidden' geographic practice in English reading culture. Over time, specifically terrestrial geographical elements were added to editions of *The Compost*: a chart of linear distances between cities and, eventually, a simple world map.

In his editions of the *Compost*, Wyer not only strengthened the association of the verbal and visual text with Ptolemy, but also incorporated specifically geographical information: Wyer appends a 'Rutter', a navigational chart of the distances between various port cities, consequently increasing the function of the text as a source of geographic information and signalling that even *before* the addition of 'The Rutter', *The Kalender* and *The Compost* were understood to be part of a broader geographical discourse, part of a continuum that initially mapped the sky and body, but that now also 'mapped' the earth. The coalescing factor is the link between astrology and geography, an association which locates narratives within the popular reception of Ptolemy's *Almagest*, *Tetrabiblos* and *Geography*.

For English readers in the early print era the images of and attribution to Ptolemy thus narrate and mediate an encounter with emerging geographical thought. The textual and visual attribution to Ptolemy created a kind of aura for the text that mystified the diffuse authorship of the work, and that subsumed a fascination with the occult and Catholic ritual into a pseudo-scientific discourse. In addition, the association with Ptolemy served as symbol of humanist learning, now popularized as part of an emerging social identity of non-clerical literacy and vernacular-based learning. Through the appropriation of Ptolemy, a figure both of mystification and symbolic capital, the print history of the *Calendar* and *Compost* suggests an interest in geographical thought in England during the late fifteenth and early sixteenth centuries belied by the map production record, an interest that was redirected to popular astrological

---

[66] *Le co[m]post et kalendrier des bergeres*, Guy Marchant, printer, 1499, sig. A.(1)r. For a discussion of the use of this image in various texts and the influence of the *Kalender of Shepherds* on English illustration, see Martha Driver, *The Image in Print: Book Illustration in Late Medieval England and its Sources* (London: British Library, 2004), 178–80.

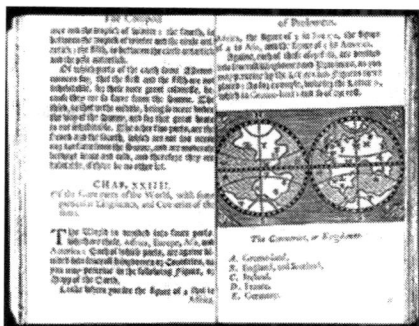

Figure 6.4 'The world is devided into foure parts', *The Compost of Ptolomeus, Prince of Astronomie*, printed by Henry Gosson, 1600(?) *Source:* STC 20482, British Library

texts. For English readers, these texts signal an engagement with emerging geographic discourse, but one that encompassed the celestial and astrological as well as terrestrial aspects.

Sometime between 1600 and 1638 (the dating is uncertain), M. Parsons produced an edition of *The Compost of Ptolomeus* for Henry Gosson, a printer of, among other things, the geographically exotic *Pericles, Prince of Tyre*, partially attributed to Shakespeare. The Gosson edition of the *Compost* contains a woodblock image of a world map (Fig. 6.4), complete with a legend listing continents and nations, and, of particular interest, instructions to the reader on the symbolic process of decoding a map:

> The world is duided into foure parts, which are these, Affrica, Europe, Asia, and America; Each of which parts, are againe diuided into seuerall Kingdomes or Countries, as you may perceiue in the following figure, or Mapp of the Earth.
>
> Looke where you see the figure of 2 that is Affrica, the figure of 3 is Europe, the figure of 4 is Asia, and the figure of 5 is America.
>
> Again, each of these aforesaid, are deuided into seuerall Kingdomes and Prouines, as you may perceiue by the Letters and figures there placed: As for example, looke for the Letter A, which is Green-land: and so of the rest.[67]

What is significant here is not so much the map itself – it is a simple woodcut far less sophisticated than earlier Continental map production or contemporaneous English regional or city map production. But, like

---

[67] *The Compost of Ptolomeus*, printed by M. Parsons for Henry Gosson, 1638? [STC 20482], sig. D(8)r.

the addition of the Rutter, the addition of this map into the verbal text indicates that *The Compost* had already been considered part of English geographic discourse by its readers and printers.

The appropriation of Ptolemy into these editions invites the reconsideration of these texts as part of the popular reception of Ptolemy's work in England in the late fifteenth and early sixteenth centuries. Just as terms such as 'chaos theory' or 'relativity' are circulated in popular culture today, the connection to specific Ptolemaic scientific principles in these popular texts was tenuous, yet generated a certain cultural currency. A Continental example of the scholarly reception of Ptolemy's work can be found in Sebastian Münster's 'Kollegienbuch' (1515–18), a lecture notebook which includes his transcription of sections of the *Geography*, forty-four maps, a calendar with astronomical illustrations, calculations of distances, and mathematical, astronomical and geographical excerpts. Gautier Dalché's argument in regard to the 'Kollegienbuch' is applicable, albeit in a popular culture context, to editions of the *Mirrour, Kalender* and *Compost*: the production of these texts forms part of the 'continuing relation between the Ptolemaic text and maps, theoretical astronomy, astrology, medicine, and the construction of astronomical instruments'.[68] Although the term 'compost' in its sixteenth-century use signified a form of calendar or computation of astronomical information, the term 'compost' in its modern sense is also a useful one in terms of the geographic framework of the early sixteenth century – a slowly heated mass of old and new knowledge continually redigested into contingent representations of the world. Though often seen as spurious or an embarrassment to the history of English geographical development, editions of the *Kalender* and the *Compost* thus provide a means by which to trace the dissemination of geographical thought in early print other than the more academic mid-century texts, *The Castle of Knowledge* (1556) and *The Cosmographical Glasse* (1559).[69]

'Belatedness', reconsidered

The entry of England into Renaissance world mapmaking includes Emery Molyneux's 1592 globes, Richard Wright's 1599 *Chart of the World* and John Speed's *Prospect of the Most Famous Parts of the World*,

---

[68] Gautier Dalché, 'The reception of Ptolemy's *Geography*', 352.

[69] Although the *Kalender* and *Compost* are often described as popular, low-culture texts, Martha Driver has shown that editions of the *Kalendrier* were owned by both aristocratic and middle-class readers in France; Richard Pynson, who published editions of the *Kalender* in England, was also the official printer to Henry VIII and, earlier, Caxton's *Mirrour*, was published for a well-to-do readership in London. See Driver, 'When is a miscellany not miscellaneous?', 201; Bennett, *English Books*, 16.

produced in 1627. Accompanying Speed's map is a new authenticating figure for the text – that of Speed himself, an English mathematician and cartographer, and a lessening of the central role of astrology in geographic discourse in England. R. J. Mayhew identifies this period of the 1590s as the point after which the content of English geographical writing began to cohere with the geographical discoveries of expedition voyages and, as he argues, to 'transform' into a form replete with the systematic 'methodization' of modern geographical practice. Citing a long scholarly 'perplexity' concerning the means by which the 'generic conventions of cartography and of prose geography were transformed by the nexus of Renaissance learning, discovery, and print',[70] Mayhew's study raises important questions regarding the too-easy linking of the Renaissance with a synchronous shift in geographic representation, an association born of the presumed rupture between the medieval and the modern. Considering geography, instead, as a 'textual tradition',[71] Mayhew identifies the complex transition from medieval to early modern geographical thought and the role of prose geography, yet persists in the identification of a 'belated transformation' of geographical thought in England compared to the more synchronous development of geographical thought, cartography and exploration on the Continent.[72] E. R. Taylor's definitive summary of Tudor geography strikes a similar note: according to Taylor, topics of geographical debate 'found no echo in London for two generations' and the relatively few examples of English geographical writing prior to 1550 'merely establishes the rule of neglect – the exception of interest'.[73] However, in addition to the book trade, intellectual exchanges between Continental and English humanists, and diplomatic missions between England and Italy, the early sixteenth century is also marked by periods of religious exile, a phenomenon Taylor acknowledges as potentially contributing to the exposure of certain English readers to Continental geographic thinking.[74] But with his focus on geography books printed in England or printed in English abroad, Taylor's study is primarily an assessment of English *originality* in geographical/textual production rather than inclusive of the transnational,

---

[70] R. J. Mayhew, 'Geography, print culture and the Renaissance: "The road less travelled by"', *The History of European Ideas* 27 (2001), 352.

[71] R. J. Mayhew, 'The character of English geography c. 1660–1880: a textual approach', *The Journal of Historical Geography* 24 (1998), 385.

[72] Mayhew, 'Geography', 355.

[73] E. G. R. Taylor, *Tudor Geography: 1485–1583* (New York: Octagon Books, 1968), 1.

[74] While Taylor, (ibid. 25), questions the facility of obtaining foreign geographical books for England's book buyers, he acknowledges the role that students and religious exiles played in bringing Continental books back to England.

readerly and popular geographical culture that has formed the basis of the present study. Further, in his summary of native English geographical works prior to 1550, Taylor gives short shrift to Caxton (none at all to Lawrence), and does not consider *The Kalendar of Shepherds* as part of the geographical record, dryly commenting that it will be 'convenient to mention and dismiss' the editions of *The Compost of Ptolemy* printed by Robert Wyer.[75] These 'little text-books of judicial astronomy', Taylor asserts, 'have no bearing on geographical science'.[76] While Taylor is right to distinguish between the métier of prognostication and what would develop as modern geographical science, the link between these books and the development of English geographical thought has been significantly undervalued.

Both Mayhew and Taylor fundamentally frame English geographical thought in the era of the Ptolemaic revival as suffering from a form of belatedness. Mayhew's approach is significant for the way that it investigates the reading habits and generic expectations of English readers rather than merely posing an intellectual blank. Still, the assumption has stubbornly remained that English readers and intellectuals were critically disengaged from Continental geographical thought; that is, that the English geographical imagination was primarily one of latency or lack. Mayhew singles out Caxton's edition of *The Description of Britain*, for instance, as an acute example of the lack of Renaissance systematization: the versified section on Wales, he contends, results in a work characterized by a lack of methodology and in which, he laments, 'for about one tenth of that text, prose succumbed to verse'.[77] It is my hope that the current study queries this assumption of belatedness and suggests, instead, some initial insights into the particular character of late medieval and early modern English geographical thought rather than an evaluation of its vacuity.

***

In looking to Malory's late medieval romance, to Caxton's geographic publications, and to the tradition of *The Kalendar of Shepherds* and *The Compost of Ptolemy*, we can shift the question – from why the English suffered from a lack of exploration-based geographical interest – to a study of how geographical thought *was* constructed during this period, evaluating it for its own aesthetic, and, in so doing, upsetting the assumptions of inevitability in terms of modern geographic paradigms. Such a recuperation of English geographical imagination in the age of the Ptolemaic revival relies on the inclusion of texts and reading culture in ways that seem decidedly non-geographic to modern sensibilities. The current study brings to light some possible avenues by which to

---

[75] Ibid. 14   [76] Ibid.   [77] Mayhew, 'Geography', 357.

consider the role of astrology, literature and book import, and the way that pseudo-Ptolemaic texts spoke to a particular form of geographical thinking. The belatedness perceived in the wake of Renaissance methodization of geography has obscured to some extent a countervailing English model of a poetical and astrological geography, incrementally infused with terrestrial geographical ideas.

Thus, the absence of English-produced world maps or editions of Ptolemy's *Geography* did not forestall an engagement with new models of geographical thought in literary and popular texts before Wright's and Speed's cartographic publications. The various editions of the *Mirrour*, *Kalender* and *Compost* can be seen as vital agents in the circulation and articulation of a narrative-based geography, one that sought to map the heavens, the body, the book, and, eventually, the earth as part of an integration of the geographical authority of the classical past and the increasingly visual and spatial representations of the present. With their description of astrological influences, schematics of the body, rutter and map, these texts represent a way of thinking about spatial relations that continued well into the early modern period – the connection of the soul, the body, the earth and the cosmos. Indeed, editions of *The Compost* reinforce this habitus of mind, concluding: 'And therefore is man called the little world'.[78]

The development of geographical knowledge and imagination in England in the fifteenth and sixteenth centuries, both Ptolemaic and pseudo-Ptolemaic, can be productively traced within a broad cultural context – one which includes early print romances, geographic narratives and astrological texts. Rather than employing either a strictly map-based definition of 'geography', or map production as the primary basis for assessing English geographical knowledge, we may look to geographic discourse in literary and popular works not only as antecedents to the 'blossoming' of Renaissance cartography in England, but as forming a geographical imaginary unique to this period of cartographic quiet in England. The role of import, literary texts, narrative geography and the influence of the astrological strand of Ptolemy's work are key attributes of a mutually constitutive discourse of 'poyetes and astronomyers'.[79]

---

[78] *The Compost of Ptolomeus*, printed by M. Parsons for Henry Gosson, 1638? [STC 20482], sig. I(8)v.

[79] I would like to thank the Huntington Library for the Farfel Fellowship and the opportunity to work with the Huntington's collection of geographical books from the early print era. The National Endowment for Humanities seminar 'Reformations of the Book', held in Antwerp and Oxford and organized by Dr John King, provided key research support. Drs Sarah Tyack and Catherine Delano Smith, workshop leaders of the History of Maps and Mapping seminar at the University of London Rare Book School, provided me with suggestions for further research. My thanks, too, to my colleague Dr Perrin Kerns.

7   Displacing Ptolemy?

The textual geographies of Ramusio's
Navigazioni e viaggi

Margaret Small

The sixteenth century in Europe is noted for its cartographic innovation and experimentation.[1] The introduction in full of Ptolemy's *Geography* into the Latin West, between 1406 and 1409, reinforced an increasing interest in (geo)graphic depictions, in coordinate systems and in cartographic projections, which as Woodward, Gautier Dalché and Power among others have shown were developing from the thirteenth century onward.[2] Together with the development of Ortelius's great atlas of the world, and Mercator's projection, a significant increase in maps of all types, from portolan charts to regional maps, to the great world maps, occurred during the following two centuries. However, cartographic depictions represent only half the story of sixteenth-century mapmaking.

Scholars of sixteenth-century geographical thought have been keenly aware that mapping the world was not simply about visual geographies – maps – and that a narrative, textual tradition of geography continued from the medieval period.[3] They have not, however, always acknowledged that this narrative tradition itself constituted a way of mapping the world. Some of the leading geographers of that time, or at any rate purveyors of geographical knowledge, preferred to map the world through words rather than images. The work of Giovanni Battista Ramusio, one of the foremost geographers of the sixteenth century, reveals a tension between these narrative and cartographic geographies, by almost completely rejecting cartographic depiction in his work. This was not because of a lack of interest in maps – indeed Ramusio claimed that he was led to

---

[1] D. Woodward (ed.), *Cartography in the European Renaissance: The History of Cartography 3* (Chicago University Press, 2007).

[2] D. Woodward, 'Roger Bacon's terrestrial coordinate system', *Annals of the Association of American Geographers* 80 (1990); P. Gautier Dalché, *La Géographie de Ptolémée en Occident* (Turnhout: Brepols, 2009), 140. Power, ch. 4, this vol.

[3] D. Woodward, 'Cartography and the European Renaissance: continuity and change', in *Cartography in the European Renaissance: The History of Cartography*, ed. D. Woodward, vol. 3.1 (Chicago University Press, 2007), 3–24, at 7.

begin his work after looking at unsatisfactory Ptolemaic maps – but rather because he found that visual maps were unsuited to presenting the changing world.[4] Maps caught a moment in time whereas text, in the form that he used it at least, allowed him to keep pace with a world whose outlines seemed to be changing almost daily, as news of new lands came back to Europe from explorers of new worlds. As a result, Ramusio ultimately resorted to a textual mapping of the world. He was not the only sixteenth-century geographical writer to choose a textual over a cartographic portrayal, but given his close connections with Venetian cartographers, it was clearly a deliberate choice.[5] As the key exponent of a new discipline of geography which was founded on mapping the world through the words of the eyewitness traveller rather than through images, Ramusio is the focus of this chapter.

Geographical traditions in Ramusio's Europe

As Denis Cosgrove notes, the

distinction between cartographic and narrative forms of geographical descriptions evidenced by Ptolemy and Strabo informed partially distinct streams in early modern geography. The discoveries of the sixteenth and seventeenth centuries that gave geographical science such prominence in the early modern world were interpreted and disseminated by cartographers such as Martin Waldseemüller, Giacomo Gastaldi and Gerard Mercator, and narratively by writers such as Sebastian Münster, Gianbattista Ramusio and the Hakluyts.[6]

The reason, according to Cosgrove, was not that the cartographers did not have texts, or that the narrative geographers did not use maps, but that the focus of the cartographical and textual traditions *diverged*. Certainly there was a degree of separation between cartographic and

---

[4] G. B. Ramusio, 'All'eccellentiss. M. Ieronimo Fracastor. Gio. Battista Ramusio', in *Giovanni Battista Ramusio, Navigazioni e viaggi*, ed. M. Milanesi, vol. 1 (Turin, 1978), 4–5 (hereafter *NV*).

[5] Giacomo Gastaldi, the leading Venetian cartographer, knew Ramusio so well that he was even entrusted with educating Ramusio's son Paolo. G. B. Parks, 'Ramusio's literary history', *Studies in Philology* 52 (1955), 134.

[6] D. Cosgrove, *Geography and Vision: Seeing, Imagining and Representing the World* (London: I. B. Tauris, 2008), 7. Cosgrove's inclusion of both Mercator and Münster in this list is somewhat misleading and shows how modern perceptions have affected our view of these divergent traditions. Sebastian Münster included so many maps in his *Cosmographiae Universalis Lib VI* that had he not written any text, he would be famous for his cartography; Mercator, though justly known for his maps, expounded his philosophy of geography in the texts which accompanied his atlas: S. Münster, *Cosmographiae Universalis Lib VI* (Basle, 1550), Mercator, *Atlas sive Cosmographicae Meditationes de fabrica Mundi et Fabricati Figura*, 3rd part completed by his son Rumold Mercator (Amsterdam, 1595).

textual geography in the period, but this separation has been amplified by recent scholarship. Two distinct traditions of scholarship have developed. One strand has focused on cartographical techniques, developments, ideas and presentations of the sixteenth century, often to the exclusion of any examination of textual geography.[7] A second less studied one has tended to examine textual geographies, often ignoring maps.[8] More recently this has begun to change as the interaction between maps and text has been discussed more critically, while at the same time there has been a renewed emphasis on the question of continuity with the geographical ideas of the preceding centuries.[9]

Classical ideas played a significant role in shaping the sixteenth-century ideas of the nature and structure of the world, transmitted both directly and through the mediation of medieval writers such as Isidore, Bede, Orosius and, later, Roger Bacon and Robert Grosseteste.[10] Such authors had not only kept alive many of the ideas circulating in the classical world, as Lozovsky and Merrills have shown elsewhere in this volume,[11] but also had grafted them onto their own geographical theories, which helped to influence the later sixteenth-century writers.

---

[7] Some of these studies have also included studies of textual geography. The third volume of J. B. Harley and D. Woodward's *History of Cartography* contains a series of articles by Tom Conley, Henry S. Turner, Nancy Bouzrara, Franz Reitinger, Theodore J. Cachey Jr and Neil Safier and Ilda Mendes dos Santos on cartography and literature in various western European countries. Even these scholars, however, have a tendency to separate maps from their accompanying texts. Franks Lestringant and Monique Pelletier are among the few scholars who examine narrative geographies as maps in any real degree. F. Lestringant and M. Pelletier, 'Maps and descriptions of the world in sixteenth-century France', in *Cartography in the European Renaissance: The History of Cartography*, ed. D. Woodward, vol. 3 (Chicago University Press, 2007), 1463–79. There is as yet no corresponding work on 'textual geography' to match Woodward's volumes on cartography.

[8] See for example, F. Lestringant, *Mapping the Renaissance World: The Geographical Imagination in the Age of Discovery* (Berkeley: University of California Press, 1994), whose work on mapping is almost entirely concerned with textual description. See also M. McLean, *The Cosmographia of Sebastian Münster: Describing the World in the Reformation*, St Andrews Studies in Reformation History (Aldershot: Ashgate, 2007), who again focuses in large part on the text rather than the maps of the *Cosmgographiae Universalis Lib VI*.

[9] See Lilley, 'Introduction', this vol.

[10] See for example, W. Tillinghast, 'The geographical knowledge of the ancients considered in relation to the Discovery of America', in *Narrative and Critical History of America*, ed. J. Winsor, vol. 1 (Boston, 1884, reprinted New York, 1967), 1–58; A. Grafton, A. Shelford and N. Siraisi, *New Worlds, Ancient Texts: The Power of Tradition and the Shock of Discovery* (Cambridge, Mass.: Harvard University Press, 1992); W. G. L. Randles, 'Classical models of world geography and their transformation following the Discovery of America', in *The Classical Tradition and the Americas: European Images of the Americas and the Classical Tradition*, ed. W. Haase and M. Reinhold, vol. 1 (Berlin and New York: Walter de Gruyter, 1994), 56–61.

[11] Merrills, ch. 2; Lozovsky, ch. 3, this vol.

This continuity of geographical thought was spurred on by the age of exploration, as writers such as Fernandez Armesto, Wey Gómez and Christine Johnson have demonstrated.[12] Fernández-Armesto was among the first to draw attention to the importance of medieval ideas in the age of discovery, and others have begun to follow his lead. Wey Gómez, for instance, looks particularly at the influence of the ideas to be found in the works of Aristotle, Albertus Magnus and D'Ailly, arguing that by the time of Columbus, the idea of an unnavigable torrid zone between the tropics was disputed, and that the tropics were seen as a potentially profitable area of exploration. Christine Johnson's recent book on Germany and German traders argues that aspects of sixteenth-century exploration were easily absorbed into the way in which German scholars and merchants viewed the world, which included a classical as well as a medieval inheritance.[13] These writers have brought to light aspects of the continuity with the medieval period that have sometimes been overlooked in studies of early modern geography. They have all tended to focus on the continuity of geographical ideas between the sixteenth century and the preceding era, but most of them have tended to focus on explorers' texts rather than on the works of stay-at-home geographers. The influence of both medieval and classical thought on the way sixteenth-century geographers visually and verbally mapped the world is, however, evident. The classical writers and the medieval authors were held in different degrees of reverence, however. While virtually all classical writers, regardless of their ability, were held in some esteem, even by an author such as Ramusio who was trying to map the world through travellers' words, for the most part it was the medieval travellers and chroniclers rather than theoreticians who were given much credit by the sixteenth-century writers. There is, for instance, no direct mention of Isidore, Orosius or Bacon in any of the texts Ramusio edited.

Although sixteenth-century writers thought that Classical Antiquity and their own age surpassed all others, this did not, however, mean that they also disregarded geographical knowledge inherited from the intervening period. Thus Ramusio, for example, while he certainly rated classical theoretical geographers more highly than medieval ones, did also include various works by medieval travellers, including an edition of Marco Polo's

---

[12] F. Fernández-Armesto, *Before Columbus: Exploration and Colonisation from the Mediterranean to the Atlantic 1229–1492* (Basingstoke: Palgrave, 1987); N. Wey Gómez, *The Tropics of Empire: Why Columbus Sailed South to the Indies* (Cambridge, Mass.: MIT Press, 2008).

[13] C. R. Johnson, *The German Discovery of the World: Renaissance Encounters with the Strange and Marvelous* (Charlottesville, Va.: University of Virginia Press, 2008).

travels.[14] Ramusio was not himself interested in the date at which a text was written, but rather in whether it could be considered to have authority. He valued autoptic material no matter when it was created, and classical, medieval and sixteenth-century texts were of equal importance to him in providing what he thought was the most relevant information for creating a verbal map of the far-flung reaches of worlds old and new.

Ramusio's worlds and his work

Ramusio is one of the less-studied sixteenth-century geographers, yet he was also among the most influential.[15] Between 1550 and 1559 he published a massive compilation of geographical travel narratives entitled *Navigazioni e viaggi*. This three-volume compendium consists of more than seventy carefully edited narratives, but it contains only thirteen maps, several of which did not appear in the first edition. Unlike the majority of sixteenth-century geographers Ramusio wrote very little in his own words. He introduced some of the texts which he published with his own discussions or *Discorsi*, but for the most part he let the words of the traveller stand so that the reader could see the regions and peoples described through the eyes of the beholder. Yet the texts were not chosen and published at random: they were assembled so that each text complemented the next in constructing a picture of the newly discovered or explored regions of the world. The arrangement of the text allowed a reader to create a mental map of the world in an age of reconnaissance. Maps in the form of graphic images were not always as suitable for the task of mapping the expanding world. Ramusio thus faced similar problems to Roger Bacon, who, writing in the thirteenth century in the aftermath of increased contact with the Mongol empire, likewise 'was conscious of the fact that geographical knowledge was in dramatic flux from both high theoretical and practical perspectives'.[16] In Ramusio's day, too, the Ottoman threat on Europe's eastern borders had furthered an already powerful interest in the Levant; the Portuguese voyages to

[14] G. B. Ramusio, 'I viaggi di Marco Polo Gentil uomo veneziano', *NV*, vol. 3, 7–297. See also Lilley, 'Introduction', this vol.

[15] In 2007 Jerome Barnes completed a doctoral thesis on Ramusio's Images: 'Giovanni Battista Ramusio and the history of discoveries: an analysis of Ramusio's commentary, cartography and imagery in *Delle navigationi et viaggi*' (University of Texas, 2007), and in the same year Fabio Romanini: *Se fussero piu ordinate e meglio scritte; Giovanni Battista Ramusio correttore ed editore delle Navigationi et vaiggi* (Rome: Viella, 2007). Aside from these two works, the most recent substantive work on Ramusio was written by Marica Milanesi in 1984: *Tolomeo sostituito: studi di storia delle conoscenze geografiche nel XVI secolo* (Milan: Unicopli, 1984).

[16] Power, ch. 4, this vol.

southern and eastern Africa and as far afield as Japan had brought unimagined semi-mythical areas into the compass of the knowable; the Spanish westward voyages had disclosed a whole new world. The difficulty for the geographer lay in assessing, assembling, attempting to verify and processing the flood of information about these previously unknown regions. For Ramusio this meant that he himself needed to develop a whole new geographical format.

Until Ramusio's work the majority of textual geographies belonged to one of three categories. There were encyclopaedic works such as that of Isidore, discussed earlier in this volume, small-scale regional works, or, which were increasingly popular in the sixteenth century, large-scale single-authored cosmographies such as those of Antonio Nebrija and Sebastian Münster.[17] There were also a few collections of travel narratives such as Montalboddo's *Paesi Novamente Retrovati* and Grynaeus's *Novus Orbis*, but these were more or less random collections.[18] Ramusio by contrast followed a clear arrangement over the course of his three volumes. The structure of the *Navigazioni e viaggi* made it comparatively easy for the reader to create a mental image of the world without reliance on maps. The first volume focused primarily on Africa, the second on Asia and the third on the New World. Ramusio intended to add a fourth volume on Antarctica and South America but he died before this was accomplished.[19] Marica Milanesi, the leading authority on Ramusio, argues that he was not dividing the world strictly according to continents but more according to human usage, so that the first book was given to areas covered by the spice trade, the second to the interior of Asia and the third to the newly discovered regions.[20] But although Ramusio did pay attention to human geography in selecting his material, considerations of physical geographical play a larger part in the layout of the books than Milanesi implies: the regions covered by the volumes were the parts of the world which were not properly described in the ancient geographies, and which he felt Ptolemy had inadequately mapped. As he himself explained, he was spurred on to make the compilation when he saw how deficient Ptolemy's maps of Asia and Africa seemed in the light of more recent information. He thought that if he could gather together eyewitness

---

[17] S. Münster, *Cosmographia*, first published 1503; A. Nebrija, 'Introductorium Cosmographiae', ed. V. B. Sanchez (Seville: Lebrija, 2000).

[18] F. Montalboddo, *Paesi novamente retrovati* (Vicenza, 1507); Grynaeus, *Novus Orbis Regionum ac insularum veteribus incognitarum una cum tabula cosmographica, et aliquot alii consimilis argumenti libellis, quorum omnium catalogus sequenti patebit pagina* (Basle, 1532).

[19] Milanesi, *Tolemeo sostituito*, 41.

[20] M. Milanesi, 'Giovanni Battista Ramusio e le *Navigazioni e viaggi* (1550–1559)', in *L'epopea delle scoperte*, ed. R. Zorzi (Florence: Olschki, 1994), 96–7.

narratives about the regions which the ancients had not known, these narratives, in conjunction with portalan maps, could provide the basis for accurate, longitudinally and latitudinally correct maps 'which would give the greatest satisfaction to those who delight in such knowledge'.[21]

It is hard to over-emphasize the importance of Ptolemy's geography in the early stages of this age of 'reconnaissance'. There is virtually no modern book on geography in the sixteenth century which does not give some space to Ptolemy's role in the formation of geographical ideas in the period.[22] This is in part because the Renaissance writers themselves (and Ramusio was no exception) cited Ptolemy as the most important of all geographers, and yet by the mid-sixteenth century, although Ptolemy still held pride of place for almost all geographers, his text was no longer suitable for their purposes. While his theories about how to project a sphere onto a plane surface and divide the world into a longitudinal and latitudinal grid system that covered the whole earth were still relevant, his data were frequently flawed even when they were concerned with parts of the world from which he had eyewitness reports; and when it came to dealing with the southern reaches of Africa, with southern and eastern Asia and with the Americas, all that he had to offer was his comprehensive grid system.

The kind of information which Ramusio intended his narratives to supply could indeed have been used by a cartographer to create maps to substitute for those of Ptolemy, but Ramusio in fact provided something more radical, but also more traditional, than a simple substitution of Ptolemy. It was radical in that he presented the geography of the whole world through the words of a multiplicity of individual travellers; it was traditional in that he reverted to the long-established practice of describing the world in words rather than images. He took on what he himself called the wearisome task of combining the narratives together to form a verbal picture of the world which would supplement the descriptions of the ancients, and which would, in effect, displace Ptolemy's mathematical and cartographic style of geography, replacing it with an almost entirely textual one from which maps were largely excluded.[23]

Maps and texts: diverging traditions?

The few maps which Ramusio included in his *Navigazioni* are restricted to the first and third volumes. His publisher, Tommaso Giunti, mentions that some had been created for the second volume, but the entire volume

[21] Ramusio, 'All' Ieronimo', 4–5.   [22] See Simon, ch. 1; Roland, ch. 6, this vol.
[23] Ramusio, 'All' Ieronimo', 4.

was destroyed in a fire at Giunti's publishing house in 1557.[24] Ramusio died the same year, and though Giunti reset and reissued the text in 1559, significantly he included no images.[25] All our knowledge of Ramusio's use of cartographical material is therefore to be gleaned from the first and the third volume. Nonetheless this limited source base provides a clear indication of Ramusio's attitude towards maps. In the first edition of the first volume – the work which arguably most clearly indicates what Ramusio thought of the book before he had the opportunity to react to critical reception – he included only one map. By the second edition of the first volume published in 1554 he had added three small-scale maps of Africa, India and Southeast Asia, and in the first edition of the third volume published in 1557 he bound nine maps in with the text, making a total of thirteen maps in the two volumes. Perhaps expense was a factor in Ramusio's initial reluctance to include maps, but woodcutting, which was the technique he used in the *Navigazioni*, was relatively cheap, costing only about a twelfth of a copper plate engraving.[26] It seems more likely that it was inclination rather than cost which deterred Ramusio from including more maps.

There were certainly maps available to him to publish. He himself owned a Ptolemaic atlas, although, given his statement about the inadequacies of Ptolemaic maps for the areas he was covering, it is hardly surprising that he did not publish any of these.[27] He also, however, had access to the portolan charts which he mentioned in his introduction, and he had even published a small-scale map of the New World in 1534, but none of these made it into his work.[28] I would argue that the reason that Ramusio found little place for maps among his narratives was that they worked directly contrarily to the function of the text. Even the few maps which did find their way into the *Navigazioni* demonstrate the problems that a static image posed for mapping a world in a time of rapidly expanding geographical knowledge. Ramusio was trying to construct a format for his book which was sufficiently flexible in a time when geographical knowledge was constantly altering and the shape of the known world was changing daily. The whole format of a compilation was to

---

[24] T. Giunti, 'Tommaso Giunti ai lettori', in *NV*, vol. 3, 3.   [25] Ibid. 4.

[26] D. Woodward, *Maps as Prints in the Italian Renaissance: Makers, Distributors and Consumers* (London: British Library, 1996), 32.

[27] R. A. Skelton, 'Introduction', in *Delle Navigatione et Viaggi Raccolte da M. Gio. Battista Ramusio*, 3 vols. (Amsterdam: Theatrum Orbis Terrarum, 1967–70), vol. 1, pp. i–xvi, at xii.

[28] This map was possibly drawn by the cartographer Andrea di Vavassore. A. Holzheimer and D. Buisseret, *The Ramusio Map of 1534: A Facsimile Edition*, The Hermon Dunlop Smith Center for the History of Cartography Occasional Publications (Chicago University Press, 1992), 7.

provide a structure that permitted easy supplementation as knowledge about the world expanded, so that the world or portion of the world described was not frozen at the time of a certain discovery.

Ramusio used the flexibility of this format to the full. Subsequent editions could be, and were, embellished by the simple addition of new travellers' reports. For instance, Ramusio in his own lifetime added Barbosa's and Barros's narratives to the second edition of his first volume in 1554; others continued to expand the work after his death, adding, for example, reports of travel to Sarmatia to the 1583 edition of his second volume.[29] On more than one occasion these reports contained contradictions, but if Ramusio' concern was with attempting to present an authoritative text for the reader, he left it to his readers to assess and distil the information they wanted. He was an editor, not a creator. He provided the tools, not the answers.

Despite working first as Secretary to the Venetian State and then as Secretary to the Council of Ten, Ramusio retained strict control over the texts of the *Navigazioni,* and all 2 million words of the three volumes were edited by him.[30] He clearly felt that even without travelling he had the skills and knowledge to edit the geographical texts. He was a schooled humanist, and although he is now known chiefly for his contributions to geography, he was as much a classicist as a geographer, and his first editorial activities were all in Greek and Latin.[31] He used many of the same skills which he applied to classical authorities in his efforts to create reliable editions of travellers' reports. His skills were certainly better suited to dealing with texts than images, however, and he was content to leave the drawing of the few maps which embellish the volumes to his friend Giacomo Gastaldi, one of the leading Venetian cartographers of the period.[32] The editorial control which Ramusio held over the texts was not so strictly applied to the maps.

---

[29] G. B. Parks, 'The Contents and Sources of Ramusio's Navigationi', in *Terza edizione delle navigationi e viaggi raccolta già da M. Gio. Battista Ramusio vol. 1 (Venice, 1563, 1583, 1606), facsimile reprint with introduction by R. A. Skelton and analysis by G. B. Parks* (Amsterdam: Theatrum Orbis Terrarum, 1967–70), 2–27.

[30] To get some impression of how demanding Ramusio's role as Secretary to the Venetian State was, one need only look at Marino Sanuto's *Diarii.* Ramusio and his work for the State is referred to in all but a handful of the fifty-eight volumes: *I diarii di Marino Sanuto* (1969). On his membership in the Council of Ten, see Giunti, 'Tommaso Giunti ai lettori', 5.

[31] His first edited work was an edition of Quintilian which Ramusio completed for Aldus Manutius after the death of its first editor Andrea Navagero. G. B. Parks, 'Ramusio's literary history', *Studies in Philology* 52 (1955), 132.

[32] Piero, 'Della vita e degli studi di Giovanni Battista Ramusio', *Nuovo Archivo Veneto, Nuova Serie* 4 (1902), 105.

It is not known when he began to start collecting geographical material. His interest in geography would have been easily fostered in sixteenth-century Venice, for though Venice's commercial and political power was beginning to decline, its interest in geography was expanding. It was a centre both of trade and of publication, and contemporary reports show that there was a widespread market for classical and other geographical material.[33] Throughout the 1520s letters from some of the leading Venetian figures including the diplomat Andrea Navagero and Cardinal Pietro Bembo refer to geographical works which they had sent to Ramusio. He had also, through his close friend, the Paduan scholar and physician Hieronimo Fracastoro, begun to correspond with Gonzalo Fernandez de Oviedo, a Spanish historian of the New World who had spent much of his life in the Americas.[34] By this time it is clear that Ramusio was well on his way to becoming one of the most geographically knowledgeable men in Venice with an extensive library of geographical and explorers' texts and maps. As Fracastoro, the possible inspiration for making the compilation, once jokingly remarked, he had correspondence from the New World, from Greenland and the Arctic, from the Equator and even from the Antarctic.[35] He seems to have had a gift for making friends and an ability to inveigle geographical material out of almost anyone, although the bulk of Spanish and Portuguese geographical material remained beyond even his reach.[36]

Mapping 'new worlds'

The single map which Ramusio initially included in the first volume of the *Navigazioni* was a map of the Nile basin (Fig 7.1). It illustrates a specific section of the *Navigazioni* – namely two discourses written by

[33] D. Cosgrove, 'Mapping new worlds: culture and cartography in sixteenth-century Venice', *Imago Mundi* 44 (1992), 65–89.

[34] G. F. Oviedo, *Historia general y natural de las Indias, Islas y Tierra-firme del Mar Oceìano, por el Capitan G. Fernandes de Oviedo y Valdeìs ... Publiìcala la Real Academia de la Historia, cotejada con el coìdice original, enriquecida con las enmiendas y adiciones del autor, eì ilustrada con la vida y el juicio de las obras del mismo, por D. J. Amador de los Rios*, vol. 2 (Madrid: Academia del Historia, 1851), Book 33, ch. 53, 543; Book 38, prohemio, 63; H. Fracastoro, A. Fumano, N. Arco and G. B. Ramusio, *Carminum editio II: Mirum in modum locupletior, ornatior, & in II. tomos distributa ... In hoc Italicae Fracastorii epistolae adjectae, nunc primum summo studio quaesitae, & congestae; inter quas eminent longiores illae amaebaeae, seu potius libelli, Jo. Baptistae Rhamnusii & Fracastorii De Nili incremento ... (Appendix)* (Padua, 1739), 68.

[35] H. Fracastoro et al., *Carminum*, 68–9.

[36] The Iberian governments instituted a strong censorship policy, which made it difficult for other nations to obtain their material. J. B. Harley, 'Silences and secrecy: the hidden agenda of cartography in early modern Europe', *Imago Mundi* 40 (1988), 57–76.

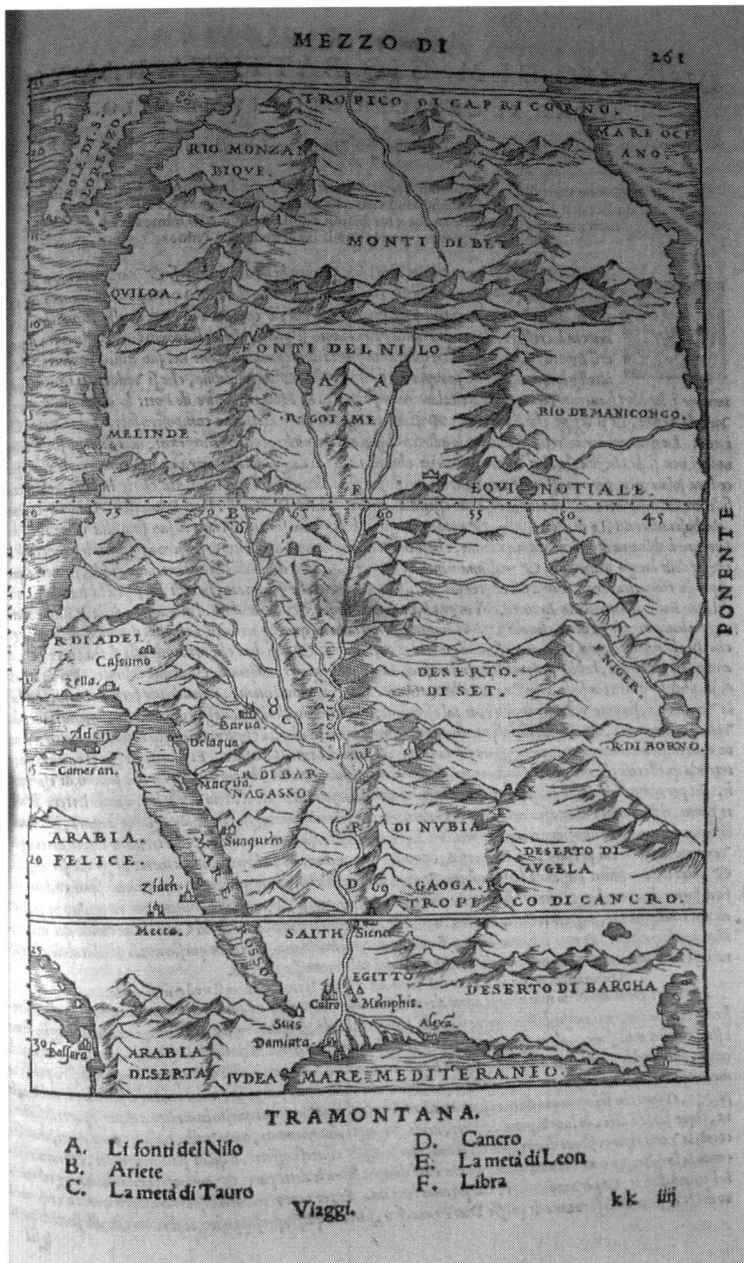

Figure 7.1 Map of the Nile river basin.
Source: Delle navigationi e viaggi, vol. 1, p. 261; reissued from the 1550 edition, reproduced by permission of Cadbury Research Library, Special Collections, University of Birmingham.

Ramusio and Gieronimo Fracastoro on the reasons for the flooding of the Nile – and is set immediately before these *Discorsi* in the text. Like all the maps in the *Navigazioni*, it is a woodcut probably carved by Matteo Pagano, with the north at the foot of the map.[37] The map shows the sources of the Nile as two lakes. These lakes were the two lakes that featured as sources of the Nile on Ptolemaic maps and were shown on most maps of the Nile in the first half of the sixteenth century.[38] Otherwise the map has remarkably few specific details. Both Ramusio himself and his cartographer Gastaldi relied for information about the geography of the Nile on the Ethiopian travel narratives of Brother Francesco Alvarez. It must have been from Alvarez's description that Gastaldi included the first cartographical depiction of Ethiopia extending south of the equator showing the kingdom of Goiam at about six degrees south.[39] The mountain kingdoms of Amara and Bogamidri, which, according to Alvarez, lie above the Nile and are riddled with fields, valleys and fountains, are also depicted. They are, however, unlabelled on the map. Although Ramusio published Alvarez's description elsewhere in his first volume, the map is not intended as an illustration of his narrative but of Ramusio's own *Discorso*.[40]

The map has a detailed coordinate system which enables the viewer to locate the few places and features that are recorded, but these are incidental; rather, the function of the map is to chart the flooding of the Nile. In his Discourse about the Nile floods Ramusio discusses the state of the Nile floods at different points along its course, which he designates A, B, C and D, and relates these to the position of the sun in certain signs of the zodiac.[41] In other words, he uses an astrological calendar to time the flooding. Gastaldi then inscribed both letters and zodiac signs on the map to give a visual representation of Ramusio's argument. The map charts the progression of the floods through time and space down to the coast. The sources of the Nile are at the two points marked A. At point B under the sign of Aries the river floods just north of the equator; by point C, at approximately ten degrees north, the river is flooding halfway through Taurus, and so on. In a sense it did not matter if further

---

[37] Woodward, *Maps as Prints*, 34.

[38] F. Relano, 'Against Ptolemy: the significance of the Lopes-Pigafetta Map of Africa', *Imago Mundi* 47 (1995), 49–66.

[39] R. A. Skelton, 'Notes on Maps', in *Prester John of the Indies*, tr. and ed. C. F. Beckingham and G. W. Huntingford, 2 vols. (Cambridge: The Hakluyt Society, 1961), vol. 2, pp. 562–8, at 567.

[40] F. Alvarez, 'Viaggio nella Etiopia al Prete Ianni fatto per don Francesco Alvarez portughese', in *NV*, vol. 2, 198.

[41] G. B. Ramusio, 'Discorso di messer Gio. Battista Ramusio sopra il crescer del fiume Nilo, allo eccellenitssimo messer Ieronimo Fracastoro', ibid. 234–5.

discoveries were made about the course of the Nile so long as the latitudinal degrees still coincided with the zodiac signs.

When compared with the great map of Africa which Ramusio included in the 1554 edition of the first volume and in all subsequent editions, it seems dated and lacking information. Many of the features are also quite different. This is because in the meantime Ramusio had come across some new material. Between 1550 and 1554 he obtained de Barros's description of Africa, which he then published in the second edition of the *Navigazioni*. Barros's *Historia* included a statement about a vast lake from which flowed all the great rivers of Africa, including the Congo, the Zambesi and the Nile.[42] Gastaldi shifted his map of Africa about to accommodate this new interior, although he continued to represent the Ptolemaic twinned lakes, merely enlarging the western one; and he redrew the hydrography of southern Africa to make it agree with Barros's description. He also used Leon of Africa's description of North Africa, which Ramusio had already published to great effect, taking sites and latitudes from Leon's work. Even so, Gastaldi's map does not entirely complement Ramusio's texts. His mapping of the region around Morocco is extremely scanty compared to Leon's text, which describes kingdom after kingdom, their general topography, and the rivers which separate them, remarkably few of which appear on the map. Ramusio did not suppress the first map since it still functioned as an illustration to his text on the Nile flood, but the rapidity with which its geographical details became out of date is in itself a good justification for Ramusio's initial reluctance to include maps with his texts.

A third map of Africa appeared in volume 3 of the *Navigazioni* (Fig 7.2). But this map, although it was probably adapted by Gastaldi for Ramusio's publication, was one of four originally drawn to accompany the description of four voyages to the New World, Madagascar, Africa and Sumatra, recorded by an unknown French pilot, all of which were included by Ramusio along with the narratives.[43] Its main function was to support the detailed description of the coastline, complete with longitudinal, latitudinal and league references, given in the text, though,

[42] G. Barros, 'Dall "Asia" di Giovan di Barros', ibid. 1077.
[43] S. Grande, *Le carte d'America di Giacomo Gastaldi, Contributo alla storia della cartographia del secolo XVI* (Turin: Carlo Clausen, 1905), 74–6. The modern translator Bernard Hoffman states that these maps were made entirely by Gastaldi and were not original to the text, but they do not fit with Gastaldi's own contemporary cartographical developments, nor does Ramusio's statement support this idea. They were adapted but not designed by Gastaldi. B. Hoffman, 'Account of a voyage conducted in 1529 to the New World, Africa, Madagascar, and Sumatra, translated from the Italian, with notes and comments', *Ethnohistory* 10 (1963), 10.

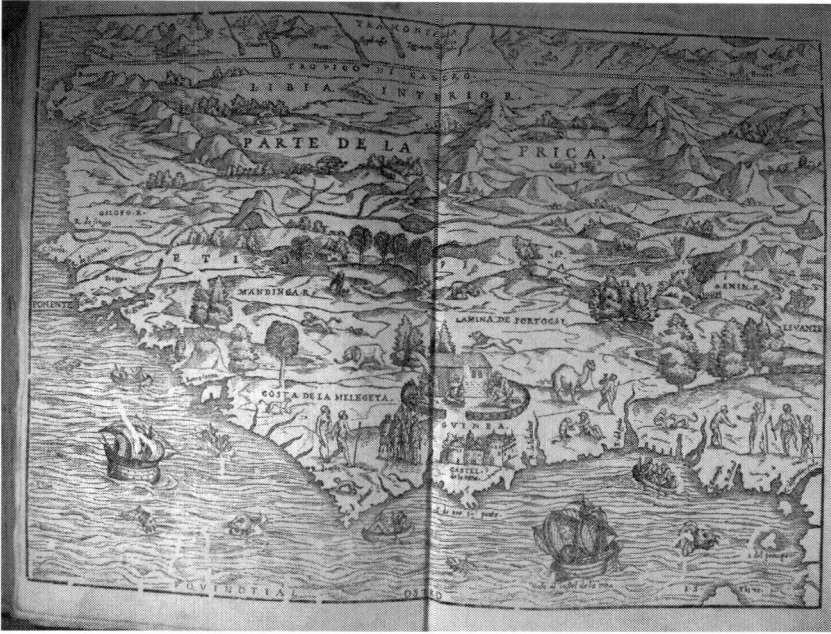

Figure 7.2 'Parte del Africa'.
Source: Delle navigationi e viaggi, vol. 3, pp. 370-1, from 1606 edition; reissued from 1563, woodcut original 1557 edition, reproduced by permission of Cadbury Research Library, Special Collections, University of Birmingham.

since it lacks a scale or a grid system, it does not do full justice to the description and cannot stand separate from the text for the text gives it scale. The treatment of the interior is largely pictorial, with little emphasis on geographic detail. It depicts encounters between Africans and Portuguese, and shows a Portuguese fort, and a stylized and clearly somewhat fantastic depiction of a 'native' dwelling and other animals and features commonly associated with Africa. The hydrography is similar to that of all Gastaldi's maps of Africa in that the river Niger flows westward across the map.[44] It cuts across a great plain and is surrounded to the north by mountains, as Leon of Africa describes. The close relationship of the map to the text has been respected by Ramusio even in some details which

[44] R. Biasutti, 'La carta dell'Africa di G. Gastaldi (1545–1564) e lo sviluppo della cartografia africana nei sec. XVI e XVII', *Bollettino della Società Geografica Italiana* series V, 9 (1920), 411.

must, by the time he published it, have been fairly certainly wrong, notably
the position of Ethiopia, which, in accordance with the text, is shown
directly south of Barbary rather than on the east coast where it actually
belonged.[45] In some ways it serves a diametrically opposite purpose to the
idea behind the compilation. It shows the state of knowledge about an area
at a particular time, and according to one particular narrative.

The inclusion of the four maps accompanying the description of the
voyages by an unknown captain was not entirely Ramusio's choice.[46]
They came from Fracastoro, and were inserted in large part to please
him.[47] Indeed, Fracastoro seems to be behind the inclusion of the
majority of maps in Ramusio text. Ramusio himself explains that

> Since your Excellency [Fracastoro] has exhorted me several times in your letters
> that I should wish to have four or five plates made of the part of this world which
> has recently been rediscovered, in imitation of Ptolemy, showing how much was
> known of it up to the present time, what shores were included in the navigation
> maps made by the Spanish pilots and captains ... I have not wanted to fail in
> obedience to your commandments, and I have got Mr Giacomo de' Gastaldi, a
> Piedmontese and an excellent cosmographer, to reduce one universal map of it
> into small compass, and then produce that drawing divided into four plates, with
> the greatest care and diligence of which he is capable, so that studious readers
> may see how much information is available about it.[48]

He then continues with the statement mentioned above that he has included
the four maps which the Parisian gentleman had sent to Fracastoro.

Jerome Barnes argues that 'the inspiration for the maps was credited to
a suggestion given Ramusio by Fracastoro but it is hard to imagine the
idea did not originate with Ramusio considering his obvious interest in
the subject'.[49] It seems more likely, however, that they were included
largely out of deference to Fracastoro than out of any real interest on
Ramusio's part. Fracastoro, according to Ramusio, wished the world to
know how much had been discovered by the time of the publication of
the *Navigazioni*, the shores that Spanish captains had investigated, the
ongoing state of discovery in Brazil and Peru. Fracastoro was not

---

[45] 'Discorso d'un gran capitano di mare francese del luogo di Dieppa sopra le navigasioni
fatte alla Terra Nuova dell'Indie Occidentali chaimata la Nuova Francia, da gradi 40 fino
a gradi 47 sotto il polo artico, e sopra la terra del Brais, Gunea, isola di San Lorenzo e
quella di Sumatra, fino alle quali hanno navigato le caravelle e navi francese', in *NV*,
vol. 6, 1919–20.

[46] Ramusio never names the captain, and so I have left him unnamed. It is generally though
not universally believed that they were written by a French Captain called Parmentier.
See Skelton, 'Introduction', 12, and Grande, *carte d' America*, 75–6.

[47] G. B. Ramusio, 'Discorso di messer Gio. Battista Ramusio sopra il terzo volume delle
Navigazioni e Viaggi', in *NV*, vol. 5, 16.

[48] Ibid., 17.   [49] Barnes, 'Ramusio and the history of discoveries', 117.

Ramusio's patron, but he did provide Ramusio with geographical material and he also seems to have been the person with whom Ramusio debated geographical matters the most. Ramusio clearly respected his judgement even if he did not always agree with it. In terms of maps, he and Ramusio had quite different outlooks. Fracastoro was content with a time-specific map – indeed he wanted one. Ramusio, by contrast, was looking for complete knowledge rather than time-specific information.

In his selection of texts Ramusio paid no attention to political change over time. A travel narrative was still deemed to have veracity even if entire empires had fallen and governments changed between the date of travel and the date of Ramusio's publication. The fifth-century BCE narrative of the Carthaginian captain Hanno's voyage down the west coast of Africa sits next to a 1535 Portuguese *routier* of a journey from Lisbon to St Thomas Island; he set the early sixteenth-century travels of Ludovic of Varthema to East Asia immediately before a narrative of the voyages of Iambolus taken from the first-century BCE historian Diodorus Siculus.[50] Ramusio was building a geography of the world, piece by piece. If he thought that a narrative contained relevant, accurate information, then he included it. He was, however, clearly aware that the outlines of the world were seemingly changing almost in front of his eyes. New narratives were included between the first and second editions; Jacques Cartier's voyages to the east coast of Canada showed the inadequacies of the unnamed captain's earlier descriptions. The captain himself used the explorations of Verazzano as a starting point for his own explorations.[51] Maps can show charted and uncharted territory, but they still freeze an image. Had Barbosa been used as the basis of the maps of Africa, not Barros and Alvarez, Mozambique and Quiloa would have been shown as islands.[52] Maps necessarily highlighted the state of knowledge at a particular time, not the expanding pool of knowledge. They were therefore peripheral to Ramusio's purposes at best, and at worst could undermine them. Ramusio did not even follow through on his promise to have Gastaldi draw up subsections of the map of the Americas. All he included was the spherical projection map (Fig 7.3).

---

[50] 'Navigazione da Lisbona all'isola di San Tomé', in *Giovanni Battista Ramusio, Navigazioni e viaggi*, ed. M. Milanesi, vol. 1 (Turin: Carlo Clausen, 1978), 567–88. Hanno, 'La navigazione di annone, capitano de' Cartaginesi nelli parti dell'Africa fuori delle colonne d'Ercole', in *NV*, vol. 1, 547–50. D. Siculus, 'La navigazione di Iambolo mercatane dai libri di Diodoro Siculo, tradotta di lingua greca nella toscana', in *NV*, vol. 1, 897–901. L. Varthema, 'Itinerario di Lodovico Barthema', in *NV*, vol. 1, 763–892.

[51] Indeed Grande argues that Girolamo da Verazzano may have made the map. Grande, *Carte d' America*, 77.

[52] O. Barbosa, 'Libro di Odoardo Barbosa, portoghese', in *NV*, vol. 2, 548–9.

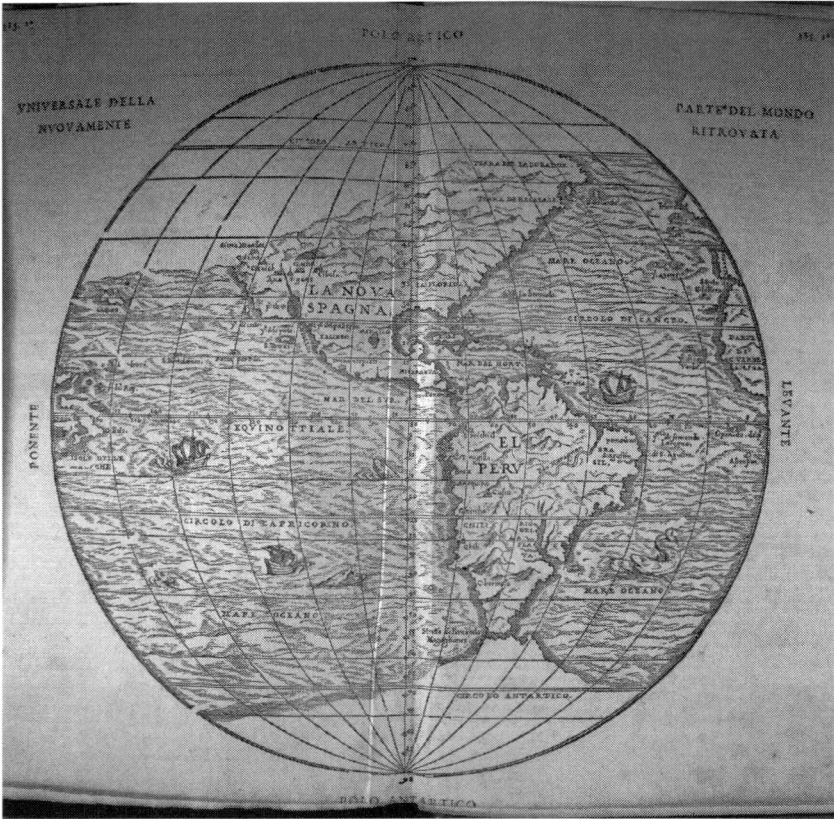

Figure 7.3 'Universale della parte del mondo nuovamente ritrovata'.
Source: Delle navigationi e viaggi, p. 385, from 1606 edition; reissued
from 1563, woodcut original 1557, reproduced by permission of
Cadbury Research Library, Special Collections, University of
Birmingham.

This map certainly corresponds to Fracastoro's desires to show the state
of progress of discovery – some of the interior of the Americas is fancifully
filled with mountains, but for the most part, Gastaldi did not hypothesize
about the unexplored regions. He did not even label the region north of
Nova Spagna *terra incognita* since he did not speculate on whether it was
land or sea, and was not at this stage sure whether Asia and America were
connected or not. The west coast is surprisingly well-mapped with the
locations of cities discovered by the Coronado expedition recorded in
1540 depicted, but the east coast is inadequate, once again demonstrating
no evidence of a familiarity with Jacques Cartier's voyage up the

St Lawrence recorded in the same volume of the *Navigazioni*. Despite containing some very up-to-date information, the map was out of date even by the time it was issued. The pace of mapping could not keep up with the pace of acquiring new geographical information.

The only three maps in the last volume for which we can attribute the choice to incorporate them solely to Ramusio are the three plans of cities in the New World. Their purpose was not so much geographical as rhetorical. They are none of them realistic portrayals of the cities. The map of Tenochtitlan is a variation of a well-known map of the city which had circulated since the first edition of Cortes's letters was published in 1524.[53] It is a distinctly idealized and Europeanized view of the city which used Cortes's narratives as a starting point, but combined it with European ideas of a townscape. In the centre of the city is shown a supposedly Aztec temple which bears more resemblance to European towers than to Aztec pyramids. Ramusio and his friends saw Tenochtitlan and Venice as New and Old World counterparts of one another and this is reflected in the map.

The map of Cuzco is, as far as we know, an original made to accompany Pedro Sancho's *Account of the Conquest of Peru*.[54] There is an image of an Inca being carried in a sedan chair up to the great temple, but the map is clearly more a Europeanized idea of what the great Inca city was like, than an actual representation. Both the city and the landscape are simply products of the European imagination. The city is portrayed as a square, but this is a misinterpretation of Spanish sources which were describing the grid pattern within the city.[55] The city itself was not square, but rather puma-shaped with the great Plaza Huacapayta at its heart.[56] This is far from its portrayal in the Ramusio compilation. The third of these native cities to be shown, Hochelaga, is again idealized. Cartier, having visited Hochelaga (where Montreal is now situated), had left a record of the city, saying it was round, surrounded by three timber stockades each within the other. There is one gate and within that were fifty houses each about fifty yards long, and in the middle was a great courtyard with a large fire in it.[57] The map bears a resemblance to the description, but is certainly not an accurate depiction.[58]

[53] R. Kagan, *Urban Images of the Hispanic World: 1493–1793* (New Haven: Yale University Press, 2000), 89.

[54] Ibid. 69.

[55] L. Nuti, 'The perspective plan in the sixteenth century: the invention of a representational language', *Art Bulletin*, 76 (1994), 122.

[56] G. Magli, 'Mathematics, astronomy and sacred landscape in the Inka Heartland', *Nexus Network Journal*, 7 (2005), 1.

[57] J. Cartier, 'Le Navigazioni di Jacques Carthier', in *NV*, vol. 6, 978–80.

[58] Cosgrove, 'Mapping new worlds', 83.

These urban plans were clearly imaginative rather than accurate portrayals of the New World cities. To some extent they complemented the textual descriptions, but the style of plan and architecture were Europeanized. Their purpose was to show both the variety and excellence of native construction in the New World, and also to demonstrate how it reflected a European world. For this reason, for example, the artist of the Cuzco map adopted the bird's-eye perspective that was popular in Europe, and portrayed a design of a city which embraced the Renaissance enthusiasm for symmetry and proportionality. By 1556, when Ramusio published this volume, both Tenochtitlan and Cuzco had been largely destroyed by the Spanish. Hochelaga, although Ramusio did not know it, had also vanished; but by including them Ramusio had visually demonstrated that the New World was capable of artistic and architectural achievements as great as anything in the Old.

Navigating Ramusio's geographies

Most of the maps included in the *Navigazioni* prove how difficult it was to complement the texts and create an image which was able to accommodate the rapidly changing geography of the 'Age of Reconnaissance'. The three maps which were added to the second edition of Ramusio's first volume – the maps of Southeast Asia, Africa and India – did achieve some of this purpose: they presented new and seemingly comprehensive images of the regions as they had become known, but it must be remembered that these were later additions. All the other maps corresponded to certain narratives but not to the overall portrayal of the wider world which Ramusio was providing. Even Gastaldi's innovative map of the New World showed locations at a fixed point in time and in an incomplete state of knowledge. They were 'static images', in a book which was intentionally formatted to allow the inclusion of new geographical material as it became available. What Ramusio set out to do was to provide an expandable, reliable basis of source material to document the geography of a world in a rapidly expanding stage of discovery. His governing principle was that eyewitness accounts were reliable evidence so that an explorer's report, provided it met his criteria of accuracy and credibility, was valuable no matter when it was created. He was not interested in change over time, but rather with an accurate portrayal of the world.

In creating such a geography, Ramusio used a variety of traditions. He relied upon the travellers, if not the theorists, of the previous two millennia. He was writing in an era when Ptolemy had been the driving force of the new geography, a force which had done much to promote global expansion and had put ideas of scale and cartographic representation at the centre of European ideas of geography. However, Ramusio set aside Ptolemaic geography. Instead he built up a geography that had more connection with Isidore, and his encyclopaedic coverage and careful systematic ordering, than with the post-Ptolemaic cosmographies of his own era. Ramusio built on classical, medieval and contemporary works to create something new akin to the chorography Jesse Simon describes at the outset of this volume: where both chorography and geography 'could exist as either a textual or a pictorial artefact'. However it was not the scale of the depiction which separated the two, but rather the amount of detail contained within. Thus, if one had enough space one could realistically create a chorography of the entire world.[59] While Ramusio certainly had neither the space nor the inclination to create a visual depiction of what Simon has described as a chorography, he did amass texts which provided fine details on the new discoveries, and furnished a verbal one. He built up an image of the world, place by place.

His texts were never intended to be the final point, however. He created a verbal displacement of Ptolemy, showing all the inaccuracies in the Alexandrian's scales, coastlines and descriptions. He also provided the tools for others to make a graphic displacement. The information he recorded was intentionally timeless, intended to show the physical and human geography of the world, but not the political geography. In his own time, the world was changing too persistently for a fixed visual image to be viable, but Ramusio had never denied the usefulness of cartographic expression. After all, his geographical undertakings had begun with an examination of Ptolemy's maps. As the rush of geographical exploration slowed down, more and better maps became possible. His texts then provided an invaluable starting point for cartographers who were trying to piece together the outlines of the world and create a new cartography. Gastaldi himself continued to use Ramusio's narratives and improve on the maps which he had made for the *Navigazioni*, creating, for example, influential maps of Asia and Africa that stood separate from Ramusio's text.[60] As late as 1608, more than fifty years after Ramusio published his compilation, documents from the *Navigazioni* were being added to the Ortelius bibliography as key sources for his map of

---

[59] Simon, ch. 1, this vol.

[60] W. G. L. Randles, 'South-East Africa as shown on selected printed maps of the sixteenth century', *Imago Mundi* 13 (1956), 78–84; T. Suarez, *Early Mapping of Southeast Asia* (Singapore: Periplus Editions, 1999), 131.

Southeast Asia in the edition of the *Theatris Orbis Terrarum*.[61] The *Navigazioni* itself had provided a verbal map of the world, but had also clearly achieved Ramusio's purpose of providing a source of information that, in combination with other material, allowed the creation of a map which supplanted any Ptolemaic one.

---

[61] http://www.orteliusmaps.com/book/ort_text166.html (accessed 3 December 2011).

Part II

Geographical imaginations

8   Gaul undivided

Cartography, geography and identity in France at the time of the Hundred Years War

Camille Serchuk

Gallia est omnis divisa in partes tres. 'All Gaul is divided into three parts.' The opening passages of Julius Caesar's *Commentaries on the Gallic Wars* define the target of Roman conquest in geographic terms. Caesar goes on to describe the boundaries of Gaul, its inhabitants, and their character. He does so – at least in part – to enhance the value and importance of his military triumph. The heirs to the vanquished Gauls – French medieval readers and translators of *the Gallic Wars* – consumed, copied and transformed Caesar's text, preserving some features and emphasizing them, while erasing others. Although the text and its description of the divisions of Gaul enjoyed considerable popularity and influence in France throughout the Middle Ages, at the time of the Hundred Years War, when France was politically divided, French texts often replaced the tripartite form of Caesar's Gaul with other models of geographical description. This essay will consider the changing character of geographical thinking about France by the French in the later Middle Ages – expressed in languages both visual and verbal. It will show how geography became entwined with contemporary French identity, particularly with reference to the ways that authors, artists and mapmakers received, transmitted and ignored the tradition of Caesar's divided Gaul.

Fought with the English, primarily on French soil, the war occasioned monumental division in France. Instigated by a dispute about the inheritance of the French crown, as well as over feudal rights owed the French king by his English rival, the Hundred Years War waxed and waned from 1337 to 1453.[1] The calamities of the war were visited on France both physically, in terms of lost and plundered territory, and politically, in terms of the huge rifts that the war wrought in the fabric of French

---

[1] Some of the best-known general sources regarding the Hundred Years War include: Christopher Allmand, *The Hundred Years War: England and France at War, c. 1300–c. 1450* (Cambridge University Press, 1988); Anne Curry, *The Hundred Years War*, 2nd edn. (London: Macmillan, 2003); Jean Favier, *La Guerre de Cent Ans* (Paris: Fayard, 1980); Jonathan Sumption, *The Hundred Years War*, 3 vols. (Philadelphia: University of Pennsylvania Press, 1999–2009).

society. Not only did the French struggle against the English, but also, because of bickering, rivalry and ultimately murder among the peers of the realm, the French were also divided against one another, and thus the Hundred Years War also became a civil war. The Armagnacs (French supporters of the French king Charles VII), the Burgundians (French but for some critical years supporting the English) and the English divided France into three fractious parts that joined into fragile alliances. As divisions, they were not as coherent or distinct as Caesar's provinces, but they divided France painfully if temporarily.

A small group of texts and images made by and for supporters of Charles VII that describe French geography at the time of the Hundred Years War testifies to the ways in which the context of the war shaped perceptions and descriptions of French territory. In their efforts to show the French united to resist the threat of English rule, these texts and images celebrated a unified France in which the divisions of Gaul/France, be they geographical or political, past or present, were blurred or erased. The extent of their descriptions of France, the placement of geographical excurses in the larger works to which they belong, their components, and their elaboration with non-geographical material all reveal efforts to further this cause and to bind the history and identity of the French to their land and to emphasize unity over division.

De bello Gallico and its geographical legacy in late medieval France

Caesar's *Commentaries on the Gallic Wars* were read in Latin by those who could do so, but the text was best known in a French compilation of the early thirteenth century called *Li Fet des Romains*, which drew not only on Caesar's commentaries but also on the works of Lucan, Sallust and Suetonius.[2] As Mireille Schmidt-Chazan and Gabrielle Spiegel have shown, the editor of *Li Fet des Romains* transformed the text from a Roman to a French one, reframing the narrative from one designed to celebrate Caesar's triumph to one that celebrated the French, using French place names rather than the Latin ones.[3] This can be seen right

---

[2] L. F. Flutre and K. Snyders de Vogel, *Li Fet des Romains compilé ensemble de Saluste et de Suetoine et de Lucan*, 2 vols. (Paris and Groningen: E. Droz and J.-B. Wolters, 1938). For a discussion of *The Gallic Wars* in late medieval France, see Colette Beaune, *Naissance de la nation France* (Paris: Gallimard, 1985), 40–6.

[3] Mireille Schmidt-Chazan, 'Les traductions de la "Guerre des Gaules" et le sentiment national au Moyen Age', *Annales de Bretagne et des Pays de l'Ouest* 87 (1980), 387–407; Gabrielle M. Spiegel, *Romancing the Past: The Rise of Vernacular Prose Historiography in Thirteenth-Century France* (Berkeley: University of California Press, 1993), 129–51.

from the outset: whereas Caesar opens his account by describing the divisions of Gaul, *Li Fet des Romains* begins instead with a statement about the size of Gaul: 'France estoit molt granz au tens Juilles Cesar. Ele estoit devisee en.iij. parties' (France was very large at the time of Julius Caesar; it was divided in three parts).[4] This reworking of Caesar's opening confirmed to the French reader the glory of Gaul, and set its divisions squarely in the past (tense). *Li Fet des Romains* was a popular and influential text, widely copied and excerpted throughout the thirteenth and fourteenth centuries. Although its dim view of the English and the Germans may have resonated with contemporary sensibilities, by the time of the Hundred Years War it had fallen out of fashion and was considered outdated.[5] After the war, humanist appetite for ancient texts at the end of the century renewed interest in *De bello Gallico*, and two new translations were commissioned in 1472 and 1485.[6] In other words, although Julius Caesar's account of the conquest of Gaul was known throughout the Middle Ages, its popularity appears to have waned precisely at the moment when France appeared to be irreparably divided.

The shadowy legacy of Caesar's divided Gaul on French geographical thinking at the end of the Middle Ages is nonetheless manifest in a unique visual representation of France that was clearly shaped by the context of the war. In the pages that follow, I will examine this map and its unusual approach to the geographical space of France. By comparing it to a selection of textual and pictorial examples both contemporary to it and before and after – although by no means a comprehensive survey, which would be beyond the scope of this article – I will show how descriptions of French geography composed during and immediately following the war often departed considerably from ancient and high medieval precedent in order to circumvent the convention of Caesar's tripartite Gaul. These variations suggest that conceptions and representations of geography were fluid and dynamic and that they were shaped by the growing sense of French identity that was sparked by the war.

There are several *mappaemundi*, but only one *French* stand-alone map of France that survives from the period of the Hundred Years War. Preserved today in Paris as Bibliothèque Nationale MS Fr 4991 f. 5v, it belongs to a manuscript of the French genealogy and chronicle called

[4] Flutre and Snyders, *Li Fet des Romains*, 79. Cf. Spiegel, *Romancing the Past*, 130.
[5] Beaune, *Naissance*, 42.
[6] Schmidt-Chazan, 'Les traductions de la "Guerre des Gaules"', 387–8. Robert Bossuat also observes that French translations of the *Commentaries* were produced in the last third of the fifteenth century, after the war had ended and coinciding with renewed humanist interest in Roman history. R. Bossuat, 'Traductions françaises des *Commentaires* de César à la fin du XVe siècle', *Bibliothèque d'Humanisme et Renaissance* 3 (1943), 253–411.

A tous nobles, and it dates from the middle of the fifteenth century (Fig 8.1).[7] It shows France surrounded by water, criss-crossed with rivers and densely populated with towns, of which the most prominent is Paris. The manuscript in which it appears is primarily illustrated with portraits of the French kings (appropriately, for a genealogy), and some churches, cities and historical scenes (primarily battles) also appear, united by the device of a vine that signifies the genealogical tree running across each page. The map is located near the beginning of the text, in place of the portrait of the mythical King Marcomir, who purportedly ruled in the fourth century. The placement of the map at this spot in the text was certainly deliberate, and may borrow from the more common practice of placing a world map at the beginning of a universal chronicle.[8] This pictorial tradition may also have roots in the tradition of the *Historiarum adversum Paganos Libri VII* (*The Seven Books against the Pagans*) of Orosius, which famously begins with a description of the geography of the world in order to provide a setting for its historical narrative that follows it.[9] And indeed, it is tempting to see the legacy of Orosius in the setting of the geography of France at the beginning of the text of *A tous nobles,* since in the prologue the author names Orosius as one of his sources. Yet the map, which would have appeared on folio 2 verso,[10] is not quite at the opening of the volume; rather, it appears shortly thereafter, at the moment in the text when France begins to assume a distinct identity. Instead of providing a setting for the account of how refugees from the Trojan War find their way down the Seine to settle Lutetia, it marks the place in the narrative when one of the Trojans' descendants, Marcomir, changes the name of Gaul

---

[7] Paris, Bibliothèque nationale de France, MS. Fr. 4991, f. 5v. For a discussion of the dating of the map, see Camille Serchuk, 'Picturing France in the fifteenth century: the map in BNF Ms Fr 4991', *Imago Mundi* 58 (2006), 133–49, and C. Serchuk, '"Ceste figure contient tout le royaulme de France": cartography and national identity at the time of the Hundred Years War', *Journal of Medieval History* 33 (2007), 320–38. For a discussion of *A tous nobles,* see Marigold Anne Norbye, 'The king's blood, royal genealogies, dynastic rivalries and historical culture in the Hundred Years War: a case study of *A tous nobles qui aiment beaux faits et bonnes histoires*' (PhD dissertation, University College London, 2004).

[8] Anna-Dorothee von den Brinken, 'Mappa mundi und Chronographia: Studien zur imago mundi des abendländischen Mittelalters', *Deutsches Archiv für Erforschung des Mittelalters* 24 (1968), 118–86; David Woodward, 'Reality, symbolism, time and space in the medieval world map', *Annals of the Association of American Geographers* 75 (1985), 510–21.

[9] For a longer discussion of Orosius, the form of his work and the role of geography in it, see Lozovsky, ch. 3, this vol.

[10] An error in the binding of the manuscript in the seventeenth century places the map today on f. 5v.

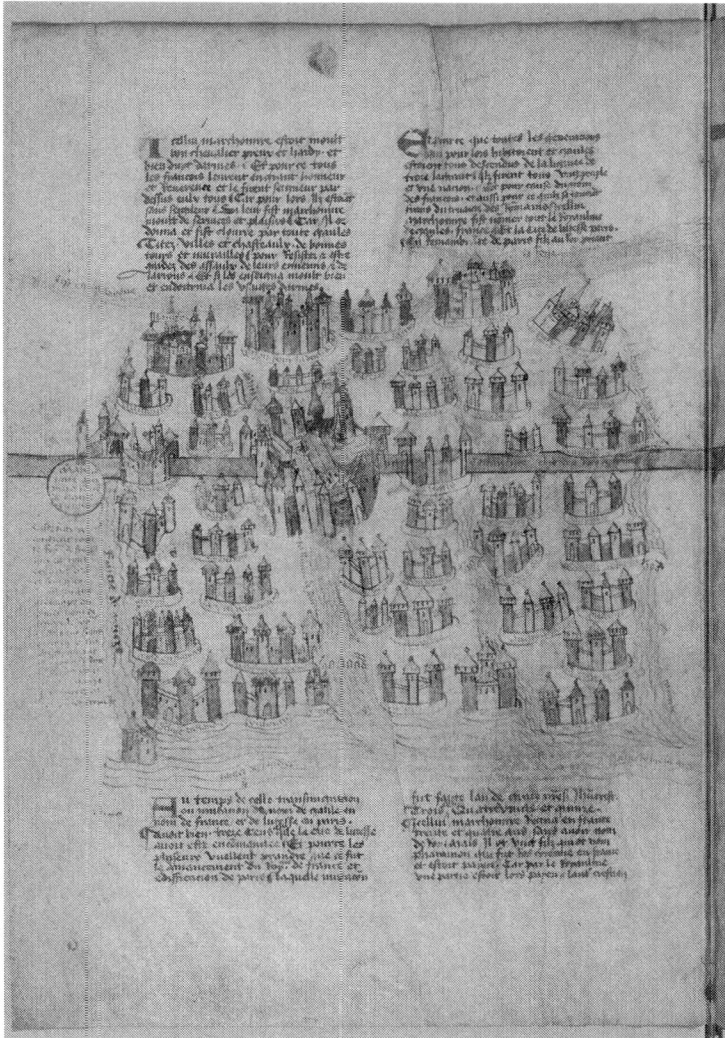

Figure 8.1 Map of France, from *A tous nobles*, *c.* 1458.
Source: Paris, Bibliothèque nationale de France, MS Fr. 4991, f. 5v.

to France, when France first comes into being. The origins of France are further emphasized by the rubric on the vine that runs across the page, which explains 'ycy commence l'arbre des rois de france' or 'here begins the [genealogical] tree of the kings of France'. The rubric ties the beginning of the (temporal) lineage of the French monarchs to the origins of the (spatial) territory they rule.[11]

The relationship of French rulers to the territory they ruled is still not fully understood by modern scholars.[12] How did the kings define or understand France? Did they conceive of the territory as fixed or as having changed since the days when it was known as Gaul? The text of *Li Fet des Romains* seems to suggest that the territory was very large, but does not explain how it might have changed over time. The text of *A tous nobles* itself offers no clarification of this question; it does not describe either the content or the limits of French territory, either historically or at the time of the composition of the text. The text and the little roundel to the left of the map (*Marcomir qui mua le nom de gaule en france* or Marcomir who changed the name of Gaul to France) both imply continuity between Gaul and France, without specifically accounting for the boundaries of the territory. Yet a rubric next to the map states that France has grown and been reduced over time, although it does not say how.[13]

Although the text and its rubrics ignore any definition of the boundaries of France, the map, by contrast, describes France as the territory bounded by the Saone and the Rhône, the Meuse, the Flemish and Breton seas (that is, the English Channel), and the Gironde. Although other sources call the Channel the *Mer d'Angleterre* or the English Sea, this map uses the more ambiguous term *Mer de Bretaigne*. These water boundaries are a formal device, since the map does not include sites that lie beyond them. At the time the map was made, the rivers no longer defined the boundaries of France; instead they demonstrate the mapmaker's acquaintance with respected ancient authorities. Julius

---

[11] Cf. Daniel Birkholz's consideration of territory and rule on the Gough Map in his *The King's Two Maps: Cartography and Culture in Thirteenth Century England* (New York: Routledge: 2004), 122.

[12] Robert Fawtier, 'Comment le roi de France, au début du XIVe siècle, pouvait-il se représenter son royaume?', *Mélanges offerts à Paul-E. Martin par ses amis, ses collègues, ses élèves: Mémoires et Documents publiés par la Société d'Histoire et d'Archéologie de Genève* 40 (1961), 65–77; Bernard Guénée, 'La géographie administrative de la France à la fin du Moyen Age: élections et bailliages', *Le Moyen Age* 67 (1961), 293–323; Gérard Sivery, 'La description du Royaume de France par les conseillers de Philippe Auguste et par leur successeurs', *Le Moyen Age* 90 (1984), 65–85. More recently, Patrick Gautier Dalché has argued persuasively that revenue lists would have given the king a sense of the realm, and that a map was not necessary to understanding space. Patrick Gautier Dalché, 'De la liste à la carte: limite et frontière dans la géographie et la cartographie de l'occident medieval', in *Castrum 4. Frontière et peuplement dans le monde Mediterranean au Moyen âge*, Jean-Michel Poisson (ed.) (Rome and Madrid: Casa Velázquez and École française de Rome, 1992), 19–31, at 29.

[13] For a discussion of the rubric, see Serchuk, 'Picturing France', 143.

Caesar in the *Gallic Wars*, Pliny in the *Natural History*, Orosius in the *Seven Books against the Pagans*, Solinus in the *Polyhistor* and Isidore in the *Etymologies* all mention such boundaries.[14] In these works, the Seine, the Loire and the Garonne mark the internal divisions of Gaul, and the Rhône and the Rhine define external boundaries, along with the Atlantic ocean, the Pyrenees and the Mediterranean. River boundaries are also a defining feature of *Francia occidentalis* as described by the Treaty of Verdun in 843, the division of the Empire of Charlemagne that reportedly named the Rhône, the Saône, the Meuse and the Escaut [Scheldt] as the fluvial limits of the territory. Despite the significance of the treaty that marked the independence of the territory of France from that of the Empire, its details were certainly murky to at least one fifteenth-century author, who said: 'Charles the Bald had the western kingdom, which one presently calls the kingdom of France. [It extends] from the Meuse river until the kingdom of Aragon and Navarre, between the sea and the Rhône river'.[15] Thus river boundaries were retailed – and reconfigured – in conjunction with both their historical importance and also contemporary knowledge of or desire for territory.[16]

The water boundaries served several purposes for the mapmaker. They were part of his conception of Gaul/France because he was reliant on ancient authors or historical documents, and because he wanted to enhance the authority of his own description by drawing on them. The fluvial frame also created the illusion that France had fixed boundaries on all sides and was a discrete territory. Sources contemporary to the map suggest that the French envied the English for the security and distinct identity that were the benefits of having a long coastline and being 'surrounded by the sea'.[17] The desire to be independent and autonomous

[14] For more on Isidore's geographical thought, see Merrills, ch. 2, this vol.

[15] 'Charles le Chauve et le royaume d'Occident que l'on dit a present le royaume de France, depuis la riviere de Meuse jusques au royaume d'Arragon et de Navarre, entre la mer et la riviere de Rosne': Bibliothèque nationale de France, Ms. Fr. 5059 ff. 42r–v. Cited by Nicole Pons, 'La Guerre de Cent Ans vue par quelques polémistes français du XVe siecle', in *Guerre et société en France en Angleterre et en Bourgogne XIVe–XVe siècle*, P. Contamine, C. Giry-Deloison and M. Keen (eds.) (Lille: Centre d'Histoire de la Région du Nord et de l'Europe du Nord-Ouest, 1991), 143–69, at 155.

[16] Indeed, the river boundaries varied; according to some ancient sources, the Rhine, rather than the Saône was the eastern limit. These boundaries were politically sensitive from the Middle Ages through to the twentieth century. Peter Sahlins, 'Natural frontiers revisited: France's boundaries since the seventeenth century', *The American Historical Review* 95 (1990), 1423–51. See also Norman D. Schlesser, 'Frontiers in medieval French history', *The International History Review* 6 (1984), 159–73, and also Norman J. G. Pounds, 'France and "Les Limites Naturelles" from the seventeenth to the twentieth centuries', *Annals of the Association of American Geographers* 44 (1954), 51–62.

[17] *Le Débat des hérauts d'armes de France et d'Angleterre, suivi de The Debate between the heralds of England and France by John Coke*, Léopold Pannier and M. Paul Meyer (eds.) (Paris:

may also have been a motivation for the renaming of the *Mer d'Angleterre* on the map; this erasure also eradicated any explicit evidence of the English from the map.

The map of France in Fr. 4991 is an admixture of awareness of earlier authorities and contemporary circumstances.[18] The water boundaries are not the only evidence of this; further liberties are taken with tradition regarding the internal divisions of the territory, which is not exactly divided into three distinct parts. Although the Seine and the Loire seem to mark three areas at the bottom of the map, the divisions become blurred at the top. This may be partly a function of the boundaries as the sources describe them, but it remains a possibility that by this arrangement of the rivers, the mapmaker wanted to offer some resistance to the notion of a divided France, immortalized in three parts since the time of Julius Caesar. At the time the map was made, after all, the unity of France was the cause for considerable rejoicing. The manuscript in which the map appears can be dated between 1456 and 1461, immediately following the war.[19] Only Calais remained in English control.[20] All of France was again entirely the dominion of the French king, and the mapmaker took pains to celebrate this cohesion, by means of his use of a consistent colour scheme, the encircling fluvial boundaries, the repetitive forms of the towns and their orderly arrangement. All of these elements added unity to the map.

Discomfort with internal divisions within Gaul/France is also evinced in the text in which the map appears. The traditional divisions of Gaul were not only geographical but also ethnographical: Caesar for example describes the territory of Gaul and also many tribes that inhabited it. This approach in particular is rejected on this page, though in the text rather than on the map: 'Et pour ce que toutes les generacions qui pour lors hibitoient et [sic] Gaules estoient tous descendus de la ligniee de Troie la grant, ilz firent tous ung peuple et une nacion' ('And because all of the inhabitants of Gaul at that time were all descended from the lineage of Troy the Great, they were all one people and one nation').

Firmin Didot, 1877). For an English perspective, see Kathy Lavezzo, *Angels on the Edge of the World: Geography, Literature, and English Community, 1000–1534* (Ithaca: Cornell University Press, 2006).

[18] A contemporary English map likewise balances present and past, see Peter Barber, 'The Evesham world map: a late medieval English view of God and the world', *Imago Mundi* 47 (1995), 13–33.

[19] The date of the manuscript with the map is discussed in Serchuk, 'Ceste figure', 327.

[20] For the geography of France in this period, see Auguste Longnon, *Les Limites de la France et l'étendue de la domination anglaise à l'époque de la mission de Jeanne d'Arc* (Paris: Palme, 1875).

The author's emphasis on the unity of the peoples of Gaul challenges Caesar's account of their variety; the tension with the ancient source originates with the divisions among the French at the time the text was composed, in the early years of the fifteenth century.[21]

The interest in a unified France – in terms of people and territory – in this map is also typical of other texts and documents produced in France around the middle of the fifteenth century. In some cases written before the end of the war, and in others after, these works share a vision of a unified France, under the rule of a legitimate French king, and in their description of the territory consistently downplay the traditional division of Gaul. Some of these authors and artists knew about this tradition or these traditions and tacitly rejected them. Instead, they describe the realm by means of its boundaries or geographical features; even then they do not follow the earlier sources closely. Fifteenth-century authors and mapmakers do not wholly reject the traditions of ancient and early medieval geography; instead they consistently prefer an approach that can be characterized as ambivalence about the formulae of the ancient texts rather than as an aversion to them. Because these works are quite different in their focus and indeed in their definition of territorial boundaries, it cannot be said that they were all relying on a common source. A rare pictorial expression of this ambivalence, the map in Fr. 4991 exemplifies the challenges faced by authors and artists of this period who sought both to respect and to undermine the construct of France presented by the ancient authorities. Its emphasis on unity rather than division and its deft interweaving of past and present demonstrate graphically what texts of the period attempt verbally. The visual language of the map facilitates the integration of geography and identity, a process achieved with varying degrees of subtlety in the other text and images examples to be considered here.

Textual geographies

Because the map in Fr. 4991 is unique – there is no other independent map of France that survives from this period – the closest comparable evidence for contemporary geographical thinking is textual. The ancient and early medieval written sources on Gaul – Julius Caesar, Pliny, Solinus, Orosius and Isidore – describe Gaul by means of its internal divisions: Gallia Belgica, Gallia Lugdunensis or sometimes Celtic Gaul, and Gallia Aquitania, and sometimes a fourth: Gallia Narbonensis. A few

---

[21] Norbye believes that the work belongs to the second decade of the fifteenth century.

ancient sources also describe the natural frontiers of Gaul, which included the Rhine, the Pyrenees, and the Ocean; sometimes the Rhône and the Mediterranean are mentioned, and even occasionally the Britannic Sea. The most thorough French compilation of these sources appears in a work known as the *Grandes Chroniques de France*, a compendious royal chronicle written in the thirteenth century at the Abbey of Saint Denis, outside Paris, and then subsequently redacted and expanded by other editors.[22] First composed in Latin and later in French, the *Grandes Chroniques* was heavily dependent on Latin sources and models, and its geographical excursus, in chapter five, makes specific reference to both Caesar and Pliny as well as '*autres philosophes*'.[23]

Since we have here mentioned two provinces of Gaul, which is now called France, it is proper at this point to describe all Gaul as Julius Caesar described it, who conquered it in ten years; Pliny and many other philosophers are in agreement with him. All Gaul is divided into three provinces: the first is Celtic, which amounts to the territory of Lyons, the second that of the Belgae, and the third that of Aquitaine. The province of Lyons, which begins at the Rhône and ends at the Gironde, contains many noble cities, whose names we have placed here, to make describing them easier ...

The fifth chapter of the *Grandes Chroniques* describes the reign of Clodio, who is the grandson of Marcomir, the king associated with the map in the *A tous nobles* manuscript. There is little in the text to signal why the

---

[22] See Gabrielle M. Spiegel, *The Chronicle Tradition of Saint-Denis: A Survey* (Brookline: Classical Folia, 1978). For a summary of the medieval textual traditions describing France, see the chapter entitled 'Le sens de l'espace', in Bernard Guénée, *Histoire et culture historique dans l'Occident medieval* (Paris: Aubier Montaigne, 1980), 166–78. Guénée considers descriptions of France by non-French authors like Brunetto Latini and Giraldus Cambrensis; I have focused here only on a small group of French authors.

[23] 'Mais pour ce que nous avons ci fait mencion de deus provinces de Gaule, qui ore est appelée France, avenante chose est que soit mise ici la distinction de toute Gaule en la manière que la descrit Jules César, qui en dis ans la conquist. A lui s'accorde Plinius et mains autres philosophes. En trois provinces principaus est toute Gaule devisée. La première si est Celte qui vaut autant à dire comme celle de Lyon; la seconde celle de Belge; et la tierce celle d'Aquitaine. La province de Lyon, qui commence au Rosne et fenist à Gironde, contient maintes nobles cités, desquelles nous avons ci mis les noms; car par les noms des cités sera plus légièrement la description entendue ...
Quant les François eurent conquis toutes ces provinces, ils les devisèrent en deus parties tant seulement. La partie devers Septentrion, qui est enclose entre Meuse et le Rin, apelèrent Austrie; celle qui est entre Meuse et Loire, apelèrent Neustrie, et par ce nom fu jadis Normendie apelée, avant que-Normans la prissent. La partie devers Lyons que les Bourgoignons pristrent, retint le nom de eus; pour ce fu-elle apelée Bourgoigne. Ci avons descrit le siège de toute Gaule au mieus que nous povons, selon les livres des anciens aucteurs' (*Grandes Chroniques de France*, Paulin Paris (ed.), (Paris: Techener, 1837), 14–16).

description should appear here: Clodio wanted to expand the boundaries of his realm, but he was hardly unique in this desire. Perhaps Clodio's name – for he was Clodio the Hairy (*Clodio le chevelu*) – offers the best explanation, for in his reign must have been the logical place for a discussion of *Gallia Comata*, or Hairy Gaul, as it was known to both Caesar and Pliny, both of whom are cited repeatedly.

Although copied numerous times in the fifteenth century, the *Grandes Chroniques* is a thirteenth-century text, and it is not, therefore, an effective tool by which to evaluate the status of French geographical thinking in texts from the period of the Hundred Years War. But a digest of the *Grandes Chroniques* was compiled in the first decade of the fifteenth century (indeed, it is in this text that the map discussed above appears). This digest, known as *A tous nobles*, is a genealogy of the French kings and an abbreviated chronicle of French history. According to its prologue, it draws on Orosius, and on the chronicle of popes and emperors of Martin von Troppau, known in French as the *Chroniques martiniennes*, and also the *Grandes Chroniques*. The audience for this text was primarily the French nobility, although a few English versions have survived.[24] One of these English versions is the only example of the surviving sixty-five copies of *A tous nobles* to include a geographical description of France.[25] Stylistic evidence suggests the manuscript, which is in roll format, was produced in England around 1415, and annotated in the late 1420s.[26] The geographical description retains many of the features of its source in chapter five of the *Grandes Chroniques*. It begins '*Gaulle est divisee en III provinces principaulx lions belgue et acquitaine lyons commence au rosne et fenist a Gironde*' ('Gaul is divided into three principle provinces, Lyon, Belgica and Aquitaine. Lyon begins at the Rhône and ends at the Gironde').[27] It condenses the text drastically, leaving only one or two qualities to describe each province rather than the long paragraphs of Pliny or even of the *Grandes Chroniques*. It mentions both ancient and medieval divisions: first it describes the provinces of Gaul, and then the two divisions of the Frankish Kingdom: Austrasia and Neustria. Unlike the French maker of the map in Fr 4991, the English editor of this text does not appear to have viewed divisions as painful scars that had to be smoothed over. Although this work was composed before the worst periods of division during the

[24] Norbye, 'The King's blood'.   [25] Manchester, John Rylands Library, MS. Fr. 54.
[26] Norbye, 'The King's blood', Appendix 9, note 35. Raluca Radulescu is exploring the English character of this roll in a forthcoming article about its textual connections to the Anglo-Norman *Brut*. I am grateful to Patricia Stirnemann and Claudia Rabel of the Institut de Recherche et d'Histoire des Textes for their insights into the English origins of the roll.
[27] I thank Dr Norbye for sharing her transcription of this passage with me.

Hundred Years War, such division posed no problem for the English editor, who closely followed his source, the *Grandes Chroniques*. The description of France in *A tous nobles* is the only example of a fifteenth-century text I have found that retains a reference to geographic divisions in Gaul; its uniqueness is certainly due to its English origin.

Closer in allegiance and context to the map in Fr 4991 is Bernard de Rosier's *Miranda de laudibus Francie et de ipsius regimine regni* of 1450, which also nonetheless describes a France divided, although in this case by two languages, French and Occitan. He hastens to add that the speakers of these languages are united in their faith and by the rule of their most Christian king. His description of the territory, furthermore, describes no division: 'France, also called by another name, long-haired Gaul, extends from the Rhône and the Alps and the Pyrenees mountains to the British Sea, and reaches the shores of the Gascon Sea'.[28] Bernard de Rosier's description of the territory of France is brief, but it relies on a form that was popular in the fifteenth century: emphasis on frontiers rather than internal features. A similar approach is found in Gilles le Bouvier's *Livre de la description des pays*, a vernacular geographic text written sometime between 1451 and 1457. Gilles le Bouvier, called the Herald of Berry, was in the employ of Charles VII, and produced an Armorial and a chronicle for him, as well as texts concerning the recovery of Normandy and the Guyenne.[29] He travelled widely in the service of the king, visiting both Rome and Trebizond, and in the preface to his text he gives the reader to believe that his entire text was drawn on his first-hand observations of 'the countries, the people, and other strange things' ('des païs, des hommes, et des aultres choses estranges').[30]

Gilles le Bouvier's text is contemporary to the map in Fr. 4991 and of all the texts considered here, comes closest to describing the territory of France as it appears on the map. His description of France opens the work, and as one might expect, is the longest and most detailed. He initially describes the climate and then describes the size of the kingdom by means of the days necessary to journey along a north-east to south-west axis, from Sluis to St Jean Pied-de-Port, and another on a north-west to south-east axis, from St Matthieu in Brittany to Lyon.[31]

---

[28] 'Francia alio nomine comata Gallia dicta est; que a Rodano et Alpibus Pireneisque montibus usque ad mare Britanicum tendit, equoris Vasconum littora plectit': Patrick Arabeyre, 'La France et son gouvernement au milieu du XVe siècle d'après Bernard de Rosier', *Bibliotheque de l'école des chartes* 150 (1992), 245–85, 268. My thanks to Joe Solodow and Marigold Anne Norbye for their help with this passage.

[29] Gilles le Bouvier, dit Héraut de Berry, *Le Livre de la description des pays*, E.-T. Hamy (ed.) (Paris: E. Leroux, 1908), 4–11.

[30] Ibid. 29.

[31] Ibid. 29–30. 'Et Premièrement du Royaulme de France, pour ce que c'est le plus bel, le plus plaisant, le plus gracieux et le mieulx pourporcioné de tous les aultres, car il a six

He shares this means of measurement with Bernard de Rosier, whose boundaries – the Rhone, the Pyrenees, the British sea – are more generalised.[32] Gilles's diagonal measurements may relate to his assertion later in the work that France is lozenge-shaped, neither long (rectangular) nor square.[33] Solinus describes the extent of Gaul from the Rhine to the Pyrenees and the Ocean to the mountains of the Cevennes,[34] which is a similar principle, and which may have informed Bernard's description, but I am not persuaded that Gilles closely relied on Solinus or on any other ancient source for his account.

Although Gilles describes the realm in great detail, two aspects of his geography are particularly distinctive. The first of these is his account of the boundaries of France, which is quite detailed. He describes a circuit that begins with Sluis (L'Ecluse) in Flanders, like the map, and says that from there to the kingdom of Navarre, the kingdom is bounded by the sea. From Navarre to Narbonne are the Pyrenees, and from Narbonne to Aigues Mortes is the Mediterranean. At Aigues Mortes, the boundary becomes the Rhône until it meets the Saône; and then to the Meuse, and from there, the Scheldt (l'Escaut). His boundaries echo those of the Treaty of Verdun, but he is also clearly indebted to a variety of ancient sources including Pliny and Caesar, although they are never specifically acknowledged. Gilles goes on to enumerate the navigable rivers and the many regions of France. He describes the features, agriculture and assets of every region, using the same language from section to section, only occasionally breaking the pattern to note, for example, that in Artois no wine is produced and people drink beer, or that in Poitou the land is good for growing grain and vines. He identifies hot springs and saltpans and sandy terrain, and in particular observes that the Bretons speak an incomprehensible language and, equally strangely, only drink water. Almost every region is described as a 'bon païs', although his praise is not uniform: every now and then he mentions that the people of a region, Burgundy, for example, are rude and unrefined, like people who live in the mountains.

---

moys d'esté et six moys d'iver, ce que n'a nul aultre royaulme. C'est assavoir esté y commence en avril et dure jusques en octobre que blez et vins sont recueillis. Et l'hiver dure d'octobre jusques en avril et n'est en ce dit royaulme l'esté trop chault, ne l'hiver trop froit selon ce qu'il est chault et froit en aultres païs. Ce dit royaulme a de long XXII journées; c'est assavoir depuis l'Ecluse en Flandres jusques a Sainct-Jehan-de pié de Porc [sic] qui est l'antrée du royaulme de Navarre et a de large XVI journées: c'est assavoir depuis Saint Mathieu de fine poterne en Bretaigne jusques a Lyon sur le Rosne.'

[32] Arabeyre, 'La France et son gouvernement', 268.

[33] *Le Livre de la description des pais*, 38.

[34] 'Galliae inter Rhenum et Pyrenaeum, item inter Oceanum et montes Cebennam ac Juram porriguntur': Solinus, *Polyhistor*, ch. 22. C. Julius Solin, *Polyhistor*, A. Agnant (ed. and tr.) (Paris: C. L. F. Panckoucke, 1847), 178–9.

Despite some criticism (notably of the Burgundians, who sided with the English), Gilles' work is eventually laudatory. Like Caesar and Isidore, he identifies the many different peoples of the realm, and their territories and habits; his description nonetheless seems more integrated than divided: these elements seem like facets, in other words, rather than divisions.

Although we know from his other works that Gilles was a partisan for the French cause,[35] only two passages in the *Livre de la description des pais* indicate his political allegiances. The first is in the section on what we now call the Île-de-France: after praising the people of this region and how hard they work, he observes that this region suffered terribly during the war ('ces païs ont esté moult foulléz de la guerre').[36] At the end of the description he speaks more generally of the kingdom, describing all the people of France as one. After praising the nobility, he says, 'the people of this kingdom are simple people and are not at all warlike people like other people, because their lords never send them to war if they can avoid it'.[37] He is referring here, of course, to the English, who were blamed by the French for starting and sustaining the war, and for ravaging the pristine beauty of France. So his discussion of the individual qualities of the peoples of France concludes with an observation about the people as a whole, and he reinforces his assertion of their collective qualities by setting them in opposition to the English. It is worth noting here that Caesar's account of the Gauls praises their bravery and military prowess; *Li Fet des Romains* enhances this further.[38] But in the context of the Hundred Years War, Gilles casts the French as the victims of the English and downplays their bellicose nature.[39]

Another member of Charles VII's inner circle also used a description of France and its unity as a means to accuse the English of waging war. It appears in a work called the *Mirouer historial abregié de France*, or the abridged historical mirror of France. This work is a hybrid French chronicle and universal history, with a bit of encyclopaedia thrown in. The title clearly links it to the *Speculum historiale* of Vincent de Beauvais, but it includes a wide variety of information, particularly focused on France, that is not part of the *Speculum*. Kathleen Daly, who has edited the text, believes that the author was Noël de Fribois, who was a notary

---

[35] For a discussion of the life and work of Gilles le Bouvier, see the preface of *Armorial de France, Angleterre, Ecosse, Allemagne, Italie et autre puissances, composé vers 1450 par Gilles le Bouvier, dit Berry*, Vallet de Viriville (ed.) (Paris: Bachelin-Deflorenne, 1866), 1–36.

[36] *Le Livre de la description des pais*, 50.

[37] Ibid. 52.   [38] Spiegel, *Romancing the Past*, 132–6.

[39] To some extent, this is a conventional means of reviling an enemy; cf. other examples cited in Philippe Contamine, 'C'est un très périlleux héritage que guerre', *Vingtième Siècle: Revue d'histoire* 3 (1984), 5–15.

and secretary to Charles VII. Royal accounts document a payment to Fribois for his work on the 'croniques de France'; many people believe him therefore to be a late editor of the *Grandes Chroniques*.[40] In any case, his pedigree as a passionate supporter of Charles VII and the French cause cannot be doubted; not only did he claim, in another work, to have been exiled by the English from his home in Normandy, but the tone and content of the work attributed to him is unambiguously pro-French. Only six copies of the *Mirouer* survive; the most extensive of these was produced for Charles of Anjou, Count of Maine, brother-in-law and cousin to the king. It can be dated to 1451–2.[41]

The description of France is a brief addition to a discussion, purportedly drawn from Vincent of Beauvais, but sufficiently confused in its details to make it impossible to identify securely. In the added text, Fribois defines the boundaries of the kingdom from the Rhine to the sea on this side, and from the Alps, which are what one calls the mountains at the entry to Italy, until the Mediterranean Sea that encircles Provence, and until the Pyrenees or Navarre. He goes on to say that these limits have long been the true body of this most Christian kingdom, and the valiant and worthy King Clovis and his successors since have held and peacefully delighted in all these lands, which are the former body of Gaul, as is established as the truth by authentic and approved sources.[42]

Fribois packs this brief passage quite densely, and jabs in every line at the English claim to French territory. His boundaries again seem to recall Solinus, although he adds a boundary: Solinus uses four points of reference and Fribois adds a fifth: like Gilles and unlike the map, he includes the Mediterranean. His insistence that Provence is encircled by the Mediterranean is a strange echo of Orosius's account of Gallia Aquitania, in which the province is described as encircled by the Loire.[43] Fribois can't have confused the two, and the parallel is curious. He never

---

[40] Noël de Fribois, *Abregé des Croniques de France*, Kathleen Daly (ed.) (Paris: Société de l'histoire de France, 2006), 27.

[41] *Abregé des Croniques de France*, 30–4, 37–40.

[42] Ibid. 239: 'Combien que depuis le fleuve du Rin jusques a la mer par deça, et depuis les Alpes, que on appelle les mons de l'entrée d'Ytalie, jusques a la Mer mediterraine qui enclot Prouvence, et jusques aux mons Pirenés ou de Navarre, tout fut anciennement du vray corps de ce tres crestien royaume; et que le tres vaillant et preux roy Clovis et depuis ses successeurs ayent tenu et joÿ paisiblement de tous iceulx pays, qui sont du corps ancien des Gaulles, appert clairement par la verité des plus autentiques et approuvees histoires.' The source is Oxford, Bodleian MS 968, f. 191.

[43] *Aquitanica Provincia oblique cursu Ligeris fluminis, que ex plurima parte terminus eius est, in orbem agitur* ('The province of [Gallia] Aquitania is shaped into a circle by the slanting course of the river Loire, which bounds the province for almost all of its length'): Paulus Orosius, *Historiarum adversum Paganos Libri VII*, I, 2. Karl Zangemeister (ed.) (Leipzig: B. G. Teubner, 1889), 11.

mentions Orosius, nor does he mention the provinces of Gaul described by him. Indeed, although the four provinces of Gaul named by Orosius might have been less problematic for Fribois than Caesar's three, since they did not lend themselves so readily to comparison with the divisions in France in Fribois's own day, Fribois avoids the question entirely and considers the whole of the territory rather than its parts. The continuity with Gaul, however, is integral to his discussion; like the makers of the map, he makes this historical dimension explicit. And like them, Fribois also connects the territory to its long line of kings. His remarks assert an ancient and enduring bond between the French and their land: they imply that control of the territory cannot therefore simply be usurped by the English. And by describing France as a 'the true body of this most Christian kingdom', Fribois denounces the English actions as a desecration of something sacred, comparable to the violation of the Host. Finally, like Gilles, he claims that Clovis and his successors were peaceable; this conforms to another widely held belief among the French at the time: that the bellicose nature of the English led them to wage war on everybody.[44] As it did for Gilles, the description of France provided Fribois with an opportunity to decry the harm done to the kingdom by the English.

The same polemical character, albeit without the geographical description, is shared with a text entitled *Débat des Hérauts de France et d'Angleterre*.[45] The *Débat* is a fictional debate, reported from a French perspective, between heralds of England and France; it was written between 1453 and 1461.[46] The heralds dispute the merits of their kingdoms, citing historical events, military skill, the plentitude of game, naval talent, architectural traditions, and the honour of their men and the beauty of their women.[47]

Although the *Débat* does not include a separate section that defines geographical boundaries or provinces, the heralds do speak at length about agricultural diversity and bounty and about some of the geographical features of their kingdoms, notably forests, ports (the French envy those of England) and rivers. The French herald proudly states that

---

[44] Such accusations were commonplace in wartime rhetoric. For example, in his discussion of the year 1436, the Bourgeois of Paris complains: *Les Angloys par leur droicte nature veullent touzjours guerrer leurs voisins sans cause*, in *Journal d'un Bourgeois de Paris, 1405–1449*, A. Tuetey (ed.) (Paris: Société de l'Histoire de Paris, 1881), 320.

[45] *Le Débat des hérauts d'armes de France et d'Angleterre, suivi de The Debate between the heralds of England and France by John Coke*, Léopold Pannier and M. Paul Meyer (eds.) (Paris: Firmin Didot, 1877).

[46] Charles VII was king when the text was written; he died in 1461. Internal evidence in the text suggests that it was composed after 1453.

[47] An English version dates to 1549. *The Debate betwene the Heraldes of England and Fraunce* appears in Pannier and Meyer's edition, 55–125.

one of the benefits of the rivers is that they allow for travel to Spain, Lombardy and Germany without a sea journey;[48] this is a distant echo of Solinus, who says that from Gaul it is easy to get to Spain and Italy.[49] Although it treats the geographical features separately, as points to be scored in the debate, the text is rich in geographical detail, and generally echoes the description of Gilles le Bouvier, including the conception of a continuous coastline from L'Ecluse to Bayonne. But while there is no explicit geographical description here, the French herald in the *Débat* does address the question of division of the realm, and with vehemence. There is division, he says, and he identifies its origin with the princes of the blood. But he quickly shifts the blame to the English, accusing them of exploiting the division to claim the crown for themselves, and sowing war and discord in every imaginable manner.[50] Elsewhere in the *Débat* the French herald condemns the offences of the English in war, which are innumerable his eyes: their tactics are despicable and their practices lack honour. But no passage of the *Débat* sputters with quite the same fury as the section on the role played by the English in the current war: there's no ambivalence here about division. The bitterness of the herald on this subject may explain some of the evasive manoeuvres around the geographical divisions of Gaul in other works.

The approaches of these authors to the description of France differ, as does the role of geography in each text. All of the works, however, manipulate and modify the ancient and early medieval models, in keeping, I contend, with their desire to portray a unified rather than divided France at the time of the Hundred Years War. These works differ markedly from the geographical descriptions in pre-war texts like the *Descriptio Galliarum* of Bernard Gui and the one in the *Imago Mundi* of Pierre d'Ailly, as well as in Jean de Saint-Victor's *Tractatus de Divisione Regnorum*, a geographical excursus that was intended to serve as the introduction to a universal chronicle. The *Tractatus*, composed in several campaigns between 1302 and 1326, closely follows Isidore, meticulously describing all of the divisions of Gaul.[51] Instead, Gilles le Bouvier, Noël de Fribois and the anonymous author of the *Débat* focus on the boundaries of France – on, in other words, its amplitude – rather than on internal divisions, administrative or regional.

---

[48] *Débat*, 47.

[49] *Ex isto sinu quoquo orbis velis, exeas in Hispanias, et in Italiam terra marique.*('And one can go from there to any place in the world, to Spain and to Italy by land and by sea'): C. Julius Solin, *Polyhistor* (ch. 22), 178–9.

[50] *Débat*, 23.

[51] Jean de Saint-Victor, *Traité de la division des royaumes: Introduction à une histoire universelle*, I Guyot-Bachy and Dominique Poirel (eds.) (Turnhout: Brepols, 2002), 238–47.

Visual geographies

This brief and selective survey of some of the evidence for geographical thinking about France in textual form at the time of the Hundred Years War invites consideration of the comparable surviving visual evidence in this period. Yet the interest in the distinct and unified character of France, so forcefully articulated by Noël de Fribois and Gilles le Bouvier finds no real parallel in contemporary maps. Few maps made in France survive from the Middle Ages; those that do belong to contexts other than that of the war.[52]

Earlier medieval maps that show France, however, can help to bring the geographical thinking – and ideology – of the map in Fr. 4991 into focus. First among these is the map of Europe from Lambert of Saint-Omer's *Liber Floridus* of *c.* 1120 (Fig 8.2).[53] Gallia is compressed between two ranges of mountains, the Alps and the Pyrenees. Although odd in contour and deprived of the names of its cities, Gallia is clearly structured by the authorities consulted by Lambert: labelled are Burgundia, Aquitania, Neustria, Narbona and Flandria and also the tribe of the Morini are all identified in Gallia, and the territory is carefully divided by the regular squiggles of rivers: the Seine, the Loire, the Rhône and Rhine. Informed by both ancient sources and the legacy of the Treaty of Verdun, Lambert slices up Gallia into wedges. The unsystematic nature of Lambert's text and accompanying maps and images has frustrated scholars who have tried to find or impose order on it, but Lambert's vision of Gaul is, if multivalent, simply, I think, an effort to express the conventions of his sources pictorially. He shows the fluvial boundaries and the important mountains, and labels the regions in the order they are described. The only element that could be considered idiosyncratic is the addition of the Morini, the tribe that settled Flanders at the time of the Roman Empire.[54] But Flanders was Lambert's own region, and, he may have simply have added more information to the map where he had it.

---

[52] Cf. Monique Pelletier and Henriette Ozanne, *Portraits de la France: les cartes, témoins de l'histoire* (Paris: Hachette, 1995); Numa Broc, 'Visions médiévales de la France', *Imago Mundi*, 36 (1984), 32–47.

[53] Ghent, University Library, MS. 92, f. 241v. Albert Derolez, *Lamberti S. Avdomari canonici Liber floridvs. Codex avtographvs Bibliothecae Vniversitatis Gandavensis Avspiciis eivsdem vniversitatis in commemorationem diei natalis* (Ghent: Story-Scientia, 1968); A. Derolez, *The Autograph Manuscript of the Liber Floridus: A Key to the Encyclopaedia of Lambert of Saint-Omer* (Turnhout: Brepols, 1998); Danielle LeCoq, 'La mappemonde du Liber Floridus ou la vision de Lambert de Saint-Omer', *Imago Mundi* 39 (1987), 9–49, esp. 28ff.

[54] Evelyn Edson, *Mapping Time and Space: How Medieval Mapmakers Viewed Their World* (London: British Library, 1997), 110.

Figure 8.2 Lambert of Saint-Omer, *Liber Floridus, c.* 1120.
Source: Ghent, Rijksuniversiteitsbibliotheek, MS 92, f. 241v.

By comparison to Lambert's map of Gallia, the map of France in Fr 4991 looks modern, ruggedly fortified and quite united. The red line that Lambert uses to distinguish Gallia from its neighbours does reinforce the definition of its external boundaries, which are otherwise muted, because the conventions of icons and rivers and other content are consistent throughout the map. The emphasis on regions, rather than cities, combined with the unlabelled house-shaped city signs on the map, serves to render Europe more unified than any of its component states. Lambert does not, therefore, afford Gallia special attention, nor does his map expose or explore contemporary allegiances or alliances.

Another, and maybe more valuable, comparison can be made with the depiction of France on a world map that appears on the final folio of the presentation copy of the *Grandes Chroniques de France*, made around 1274 for Philip the Bold (Fig 8.3).[55] Although this map has a number of sites from ancient history and crusader destinations, Europe is condensed to four signs: Constantinople, Athens, Rome and Paris, which is the smallest of them.[56] France is labelled, just above Paris, and another rubric appears below, *langobardia*, which, if nothing else, would seem to offer compelling evidence that the mapmaker borrowed heavily from a source that he didn't understand very well. Indeed, the placement of *langobardia* and the rubric identifying *Lacio* (below Rome) suggest that the mapmaker was working from a source with a great deal more detail than the scale of his own map would allow: he compressed France and Spain to such a great extent that they appear where Lombardy would have been on the original. Anne D. Hedeman and others have noted a connection between this map and the *mappaemundi* of the Sallust type, and, indeed, some of the Sallust maps share some of the compression of western Europe that is seen in this map.[57] However, although it is a French map in a French chronicle, the world map in the *Grandes Chroniques* evinces none of the geographical awareness of France that

---

[55] Bibliothèque Ste. Geneviève MS. 782, f. 374. Anne D. Hedeman, *The Royal Image: Illustrations of the Grandes Chroniques de France 1274–1422* (Berkeley: University of California Press, 1991), 11–29, 257–58, consulted at http://www.liberfloridus.cines.fr/textes/biblio_fr.html. I would like to thank Anne D. Hedeman for sharing her most recent thoughts about this manuscript, now published in *Imagining the Past in France: History in Manuscript Painting 1250–1500* (Los Angeles: J. Paul Getty Museum, 2010), 108–11.

[56] Danielle LeCoq, 'Elements pour une lecture d'une mappemonde médiévale', *Mappemonde* 88(1) (1988), 13–17, consulted at http://www.mgm.fr/PUB/Mappemonde/M188/p13-17.pdf

[57] I am not aware of any discussion of the Sallust maps that treats this compression as a salient feature. See Edson, *Mapping Time and Space*, 18–21. I am grateful to Chet Van Duzer for his insights about the Sallustian prototypes of the world map in MS 782.

Figure 8.3 *Mappamundi*, from the *Grandes Chroniques de France*, c. 1274.
Source: Paris, Bibliothèque Sainte-Geneviève, MS 782, f. 374v) (c)
Bibliothèque Sainte-Geneviève, Paris.

characterize the other examples considered here. Instead, it claims for
Paris the same status as the great cities of history. Danielle LeCoq has
suggested that the map emphasizes the place of Paris in the *translatio
studii*, as heir to the knowledge of the ancients, exemplified by Athens

and Rome, rather than any geographical feature.[58] As the sole French site pictured on the map, Paris serves as a synecdoche for all of France; it is unified, but in keeping with the conventions of the map, it is also extremely simplified. This image affirms the place of France on the map, as it were; it cannot properly be considered a map of France.

Two additional maps made outside of France also highlight the distinctive features of the map in Fr. 4991. The first is a map of Gaul that appears today in a fourteenth-century manuscript of Solinus in the Vatican.[59] Solinus does not say much about Gaul,[60] so this map is clearly not based directly on his description, but instead was meant to enhance and expand upon it. How this map, conceived independently, came to be associated with the manuscript is not yet clear. It includes names of tribes like the Pictavi, but also cities like Tolosa (Toulouse) and Narbona (Narbonne), and bigger regions like Hispania citerior (the Roman province of the same name). Nathalie Bouloux suggests that the inclusion of the Lotharingians might indicate that the map is a copy of a Carolingian map or of a Carolingian redaction of an ancient map.[61] The words are arrayed on the page to reflect their location in the territory, but only the rivers and the St Bernard Pass (here Montes Jovis) are described pictorially. Compared to the map in Fr. 4991, the territory seems smaller, and the parts are less connected to one another. It also seems – like Lambert's map, although more coherently – a visual explication or clarification of a written source or sources, whereas the map in Fr 4991 seems, because of its deft manipulation of visual rhetoric, less based in a text and more a product of ideology attached to an ostensibly geographical framework. The ideology that shapes Fr 4991 is one of French national identity; the map's conventions allow France to be compared implicitly to England. The war with England did not just foment this rivalry; it provoked both the English and the French to define national identity in relation to the other. English counterexamples can therefore illuminate the pointedness of a particular French approach. So just as the geographical excursus in the English copy of *A tous nobles*, with its reliance on the Caesarean divisions, provides a useful counterpoint to the resistance to division in contemporary French exempla, an English map serves to highlight the integration of ideology in the context of geography in the map in Fr. 4991.

The Evesham world map was made around 1390 for the prior of Evesham Abbey. It was revised about twenty years later, but by 1450

---

[58] Lecoq, 'Elements pour une lecture', 15.

[59] Bib. Apos. Vaticana, Ross 228, f. 35v. Nathalie Bouloux, *Culture et savoirs géographiques en Italie au XIVe siècle* (Turnhout: Brepols, 2002), 182–4.

[60] See above, n. 35.   [61] Bouloux, *Culture et savoirs géographiques en Italie*, 184.

was considered obsolete and the parchment was reused for another purpose.[62] As Peter Barber has noted, the representation of France on this map is remarkable for the clarity of its English perspective. France on the map is dominated by two signs, those of Calais, which is considerably larger than Rome, and of St Denis. Paris, on the other hand, is practically insignificant. Drawing on earlier English precedents, which themselves rely on ancient sources, this map is also criss-crossed with a network of lines that delineate the boundaries of regions and provinces, some modern, some ancient. Like the author of geographical excursus in *A tous nobles*, the maker of this map was not troubled by division; indeed, he appears to have taken pleasure from it. His vision of an English France is in every way the antithesis of the unified and inviolate kingdom presented in the map in Fr 4991.

The divisions of Caesar's Gaul are not clearly represented on any of these examples; as with the textual accounts, only ghostly evidence of their structure lingers. The divisions are first clearly marked on a map of Gaul in a late fifteenth-century luxury copy of Ptolemy's *Geography*, which was probably produced in the workshop of the Florentine Pietro del Massaio.[63] This map, however, clearly belongs to an Italian humanist context; it is not contemporary to the war (like the Solinus manuscript discussed above), nor can it be connected to French sources. It's not an image of Gaul born of French pride in a classical heritage; it is instead evidence of humanist reverence for the authority of classical texts.

To come full circle now back to the map of France in Fr 4991, its unusual structure and content cannot be easily explained by cartographic precedents; instead, it was probably created as a graphic means to express the nascent national sentiment in the text in which it appears.[64] According to the adjacent text of *A tous nobles*, which tells of the change of name, the fortification and the facility with weapons that were all inaugurated in the reign of Marcomir, the territory seems unified and well-protected. These narrative elements in the reign of Marcomir may have led the mapmaker to insert the map in that spot, rather than

[62] Barber, 'Evesham world map'. The map also appears in Richard Marks and Paul Williamson (eds.), *Gothic: Art for England 1400–1547* (London: V&A Publications, 2003), 146.

[63] Paris, BN Ms Lat 4802. Cf. Pelletier and Ozanne, 255. See also L. Duval-Arnould, 'Les manuscrits de la Géographie de Ptolémée issus de l'atelier de Piero del Massaio (Florence, 1469-vers 1478)', in *Humanisme et culture géographique à l'époque du concile de Constance autour de Guillaume Fillastre*, D. Marcotte (ed.) (Turnhout: Brepols, 2002), 227–44.

[64] Serchuk, 'Picturing France'.

in the reign of Clodio, which would have been more in keeping with the textual tradition of the *Grandes Chroniques*. The mapmaker wanted the map to illustrate the continuity from Gaul to France and from the past to present, as well as its unity and strength. The bold green line of the genealogy that runs through the map might at first be misconstrued as a dividing element, but instead it serves to tie the map of France to the rest of the pictorial campaign of the manuscript. This map of France, with its circuit of rivers and carefully drawn bridges, seems to resist division. The rivers that rush so boldly from the seas peter out before they can slash the territory in three. Instead, the map proffers an image of France undivided and indivisible.

Conclusions

The legacy of Caesar's *Commentaries* on late medieval geographical thinking considered here provides a means to examine how the context of the war shaped perceptions and descriptions of French territory. As the map and the textual examples indicate, the period of the war coincides with evidence of some resistance to the model of a divided territory. Writers in the circle of Charles VII, who reigned through the period of the most acute division, most vehemently rejected the formula of the tripartite division. Noël de Fribois, Bernard de Rosier and Gilles le Bouvier instead described the extent of the realm from its extremities, using intersecting axes from south-west to north-east and south-east to north-west. This paradigm had the virtue of also sidestepping the English presence in Aquitaine, which was located to the west between the termini of the two axes. By measuring on the diagonal, in the fashion of a surveyor, these authors relied on an ostensibly scientific authority to make France seem even bigger than it was, and also less divided. Each of these authors found a way to integrate a description of France into his work, and if there is consistency among their geographical thinking, there is diversity of both form and purpose. Each was tailored to serve the particular needs and contexts for which they were prepared.

Not every author rejected Caesar, of course. Caesar's text had enjoyed considerable authority before the war, and humanist attention to ancient authors subsequently led to a revival of interest in his description at the end of the fifteenth century. In fact, so complete was his rehabilitation that the description of Gaul is excerpted and copied as a stand-alone text.[65] It is also important to mention that although descriptions of

---

[65] For example, BN Fr 833, f. 92ff.

France composed during the war tended to avoid the division of Gaul into three parts, the *Grandes Chroniques*, which perpetuated the model, was still copied repeatedly in the fifteenth century. So the erasure of division was certainly not complete. Furthermore, in texts like those of Alain Chartier, a cleric and member of the court of Charles VII, who wrote a variety of political and poetic works in the 1420s, division becomes a rallying cry. Although Chartier does not focus on territorial matters, his repeated denunciation of division and his multiple calls for the unification of the peoples of France show how important this idea was in the context of the war. Similarly, not all dimensions of French geographical thinking were variable over the course of the fifteenth century. The integrity of boundaries is a concern that remains relatively unchanged. The relation of the centre to the periphery, a political metaphor dependent on a geographical conception of space, is also fairly consistent, and downplays the potential for division among the peripheral regions by assigning power to the centre.[66] The capital at Paris (or ostensibly at Paris, since in the fifteenth century the king himself was not always there) provided a central administration from which power emanated to the provinces. The importance of Paris as the centre is essential to all of the images and texts considered here.

Although not ostensibly geographical, the other political metaphor of the time that also emphasized coherence and completeness over division was that of the body politic. That the state might be described by means of this corporeal metaphor is well known; but in this period French political theorists explicitly expanded this metaphor beyond the structures of the state to the territory itself: Paris was seen as the head of the body of France.[67] That the members of this body were not to be separated from it through violent acts of war is implicit in these texts as well. And certainly distinctions between body and territory are blurred in some maps and some texts, notably Noël de Fribois. Jean Gerson, the fifteenth-century theologian, quotes Saint Denis in a sermon entitled

---

[66] P. S. Lewis, 'The centre, the periphery and the problem of power distribution in later medieval France', *The Crown and Local Communities in England and France in the Fifteenth Century*, J. R. L. Highfield and R. Jeffs (eds.) (Gloucester: Alan Sutton, 1981), 34, 36, repr. in P. S. Lewis, *Essays in Later Medieval French History* (London: Hambledon, 1985), 152, 154. Cf. K. Daly, '"Centre", "power" and "periphery" in late medieval French historiography: some reflections', in *War, Government and Power in Late Medieval France*, C. T. Allmand (ed.) (Liverpool University Press, 2001), 124–44.

[67] Camille Serchuk, 'Images of Paris in the Middle Ages: Patronage and Politics' (Ph.D. dissertation, Yale University, 1997). For the ways in which the body politic was gendered in French thought, see Sarah Hanley, 'Mapping rulership in the French body politic: political identity, public law, and the *King's One Body*', *Historical Reflections* 23 (1997), 129–49.

'Vivat Rex' from 1405. 'Rien ne peust durer sans unité', he says: nothing can last without unity.[68] In his view, it would be unity that would permit France to endure. In the calamity of the fifteenth century, so great was the desire for unity that ancient traditions that had long shaped perceptions of France and French territory gave way to new formulations. Later generations returned to the model of a tripartite Gaul. But before they did, an ideology of unity found expression in many forms, textual, pictorial, cartographic, and were offered with passion and force, in a bid to heal the wounds of the Hundred Years War.

[68] Jean Gerson, '*Pour la réforme du Royaume*', in *Oeuvres complètes*, Palémon Glorieux (ed.) (Paris: Desclée, 1960–73), 7, 1149.

9   Passion and conflict
Medieval Islamic views of the West

Karen C. Pinto

At first glance the typical medieval Islamic map of 'the West' – Surat al-Maghrib – strikes us as nothing more than a quaint abstraction of circles, triangles and oblong shapes ornately adorned with vivid pigments.[1] Closer study presents a more complex image, however, of passion and conflict; of attraction and revulsion; of love and hate. Indeed the Maghrib map is by far the most dissonant image in the extant collection of medieval Arabic and Persian maps and, as such, one of the most engaging. Whereas all the other images have a veneer of harmony and balance, this one is – by deliberate design – passionately conflicted. It is the discord of desire inlaid within the Muslim pictographs of the Maghrib that is the focus of this chapter, the over-arching question being how did medieval Islamic cartographers settle on such a strange-looking image as a representation of the western Mediterranean – in particular, North Africa, Islamic Spain, and Sicily? Answering this question requires immersing ourselves in the map-image itself, and takes us through a series of subliminal messages ranging from intra-Islamic imperial ambitions to erotic and nostalgic Andalusian poetry.

The bulk of the medieval Islamic cartographic tradition is characterized by emblematic images of striking geometric form that symbolize – in Atlas-like fashion – particular parts of the Islamic world to the familiar viewer. They comprise a major carto-geographic manuscript tradition known by the universal title of *Kitab al-Masalik wa al-Mamalik* (*Book of Roads and Kingdoms*) that was copied with major and minor variations throughout the Islamic world for eight centuries.[2] It was a stylized

---

[1] In Arabic, 'al-Maghrib' literally means 'far west'. As a geographical term it refers to the western extremities of the known world – North Africa and southern Spain.

[2] For the sake of brevity, I refer to this carto-geographic tradition as the *KMMS* series, an acronym based upon the universal title of the most popular Arabic and Persian carto-geographic manuscript in the series, namely al-Istakhri's work known by the universal title of *Kitab al-Masalik wa al-Mamalik* (*Book of Roads and Kingdoms*). This mapping tradition dates back to the tenth century, although the earliest extant manuscript containing maps is from the eleventh century. For more detail on this manuscript tradition see Karen

amimetic vision restricted to the literati and, specifically, to the readers, collectors, commissioners, writers and copyists of the particular geographic texts within which these maps are encased.[3] The plethora of extant copies dating from the eleventh to the eighteenth centuries produced all over the Islamic world – including Iraq, Iran, Syria, Egypt, North Africa, Anatolia, and even India – testifies to the long-lasting and widespread popularity of a particular medieval Islamic cartographic vision.[4] Each manuscript typically contains twenty-one iconic maps starting with an image of the world, then the Arabian Peninsula, the Indian Ocean, the Maghrib (North Africa and Andalusia), Egypt, Syria, the Mediterranean, upper and lower Iraq, as well as twelve maps devoted to the Iranian provinces, beginning with Khuzistan and ending in Khurasan, including maps of Sind and Transoxiana.

An examination of the icon that Middle Eastern cartographers used to symbolize the Mediterranean reveals the incongruence between our textually based, historiographic perception of conflict between the Christian and Muslim halves of the Mediterranean, and the counterintuitive picture of perfect harmony that the Islamic maps seem to proclaim (Fig 9.1).[5] This disjuncture between text and representation arises because of the curious bulbous form that the Middle Eastern cartographic artists employed to symbolize the Mediterranean. It is a perfectly symmetric form that deliberately does not match up mimetically with the actual coastline of the Mediterranean. Instead it presents the northern flank of the sea as a mirror image of the southern, as if to suggest that the Muslims conceived of the other side of the Mediterranean as a reflection of themselves. This is a reading that flies in the face of the dominant discourse that asserts that the Mediterranean was a

Pinto, 'Cartography', in *Encyclopaedia of Islam and the Muslim World*, Richard C. Martin, Said Amir Arjomand, Marcia Hermansen, Abdulkader Tayob, Rochelle Davis and John Obert Voll (eds.) (New York: Macmillan Reference, 2003), 128–31.

[3] Although the recently discovered *Book of Curiosities* manuscript has received a lot of attention in the United Kingdom, it is, in fact, a later, single manuscript reflecting a medley of hybrid traditions. The scholars who originally discovered the *Book of Curiosities* advertised it as an early eleventh-century manuscript, but later revised their dating to thirteenth/fourteenth century. Exclusive focus on a unique manuscript diverts attention from the bulk of the tradition. In this chapter, and in my work generally, I focus on the most popular and widespread tradition that is – without doubt – the *KMMS* mapping tradition. While the *Book of Curiosities* manuscript contains some maps typical of the *KMMS* series it does not contain an exclusive map of the Maghrib and therefore does not warrant inclusion in this chapter.

[4] Karen Pinto, 'The maps are the message: Mehmet II's patronage of an Ottoman Cluster', *Imago Mundi* 63 (2011), 155–79.

[5] See Karen Pinto, '"Surat Bahr al-Rum" (Picture of the Sea of Byzantium): possible meanings underlying the forms', in *Eastern Mediterranean Cartographies* (Athens: National Hellenic Research Foundation, 2004), 223–41.

Figure 9.1 Map of the Mediterranean
Source: Leiden: Bibliotheek der Rijkuniversiteit, Or. 3101, Eastern
Mediterranean (589 AH/1193 CE), f. 33r.

heavily contested zone and that Muslims and Christians were constantly at loggerheads across it.[6]

Comparison with other maps of the Mediterranean, the World, Egypt and Syria

In their world maps the Muslim cartographers employed a similar motif – an elongated tear-drop shape – to represent the Mediterranean. In these, the Mediterranean nestles like a bird between the landmasses of Africa, Asia and Europe, while the long wings of the Bosphorus and the Nile spread out perpendicularly from either side. Albeit elongated and of smaller scale, the Mediterranean retains its characteristic iconic bulbous form (Fig 9.2). But when the cartographer's lens zooms into the larger-scale depictions of individual regions around the Mediterranean one finds that the forms of representation change dramatically (Fig 9.3). Gone is the trope of the bulbous form. Instead we are faced with either a marginalized Mediterranean or a disruptive one, so while in the maps of Egypt and Syria, the Mediterranean is just an incidental boundary serving as trimming for the central focus that lies within the land, in the Maghrib map the emphasis is completely different: it plays a central – albeit disruptive – role in the image (Figs 9.4 and 9.5). The veneer of symmetric harmony has vanished. We are faced instead with two sides of the sea that are no longer mirror images of each other. The difference between this image of the Maghrib and the image of the Mediterranean is so stark as to leave no doubt that the disjuncture was deliberately imposed.

To a modern audience concerned with mimesis and accuracy in cartography, such a deliberate break in form will come as a surprise. How could the larger-scale, close-up view of the Mediterranean not match up with the smaller-scale view of the entire sea? Resolving this visual anomaly is quite simple: if one has no intention of providing a mimetically accurate map for navigation, but intends instead to create an easily memorized, cognitive picture of key geographical spaces, then that intention is better served through stark contrast rather than mimicry.[7] As the anonymous author of *Ad Herennium* (one of the most

---

[6] See, for instance, the work of Henri Pirenne, *Mohammed and Charlemagne* (n.p.: Barnes & Noble, Inc., 5th impression, 1968), 156–62, and Samuel Huntington, *The Clash of Civilizations* (New York: Simon Schuster, 1996). For a novel approach to the discourse on the Mediterranean, see Miriam Cooke, 'Mediterranean thinking: from Netizen to Medizen', *Geographical Review* 89 (1999), 279–90.

[7] Evidence for this can be found in psychological analyses of the recollection of shapes using colour and contrast. See, for instance, Richard N. Wilton, 'The structure of memory: evidence concerning the recall of surface and background colour shapes', *The Quarterly Journal of Experimental Psychology* 41A (1989), 579–98.

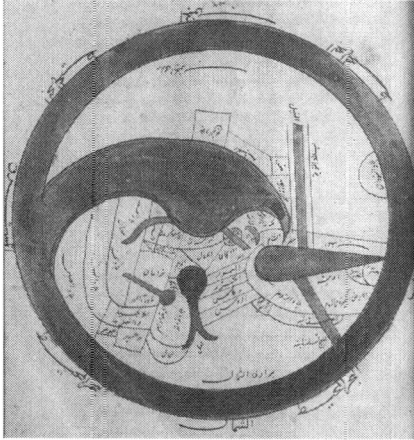

Figure 9.2 'World Map'.
Source: Istanbul: Sülemaniye Camii, Aya Sofya 2971a (1473), f. 3r.)

Figure 9.3 'Map of Egypt'.
Source: Istanbul: Topkapı Sarayı Müzesi Kütüphanesi, A3348,
Cairo (1258), f. 26r.)

popular books on rhetoric and memorization in the Middle Ages and
early Renaissance) puts it:

We ought, then, to set up images of a kind that can adhere longest in the memory.
And we shall do so if we establish likeness as striking as possible; if we set up

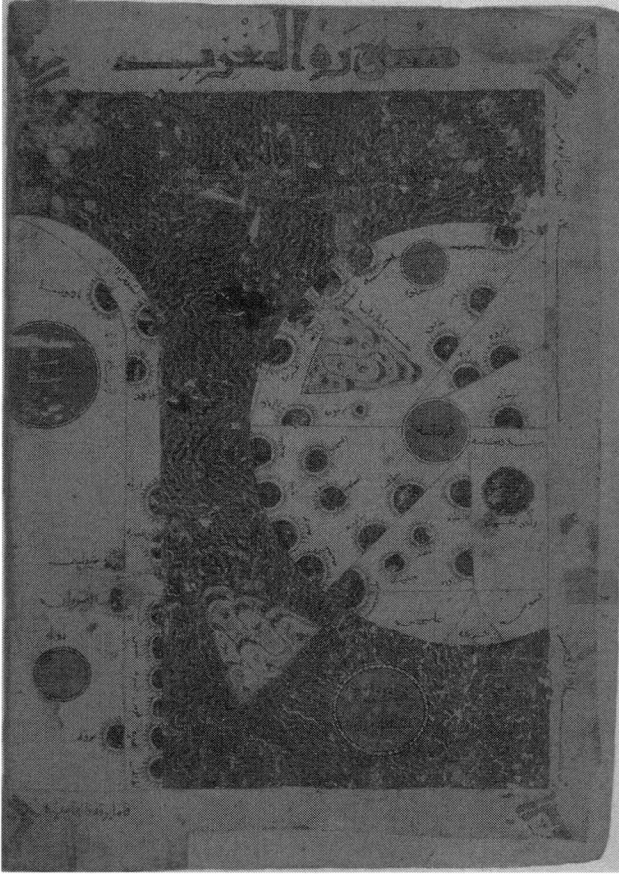

Figure 9.4 'Picture of the Maghrib'.
Source: Leiden: Bibliotheek der Rijkuniversiteit, Or. 3101, Eastern
Mediterranean (589 AH/ 1193 CE), p. 20.

images that are not many or vague, but doing something; if we assign to them
exceptional beauty or singular ugliness; if we dress some of them with crowns or
purple cloaks, for example, so that the likeness may be more distinct to us; or if
we somehow disfigure them, as by introducing one stained with blood or soiled
with mud or smeared with red paint, so that its form is more striking, or by
assigning certain comic effects to our images, for that, too, will ensure our
remembering them more readily.[8]

[8] *Ad Herennium*, Harry Caplan (tr.) (New York: Loeb Classical Library, 1968), 221.

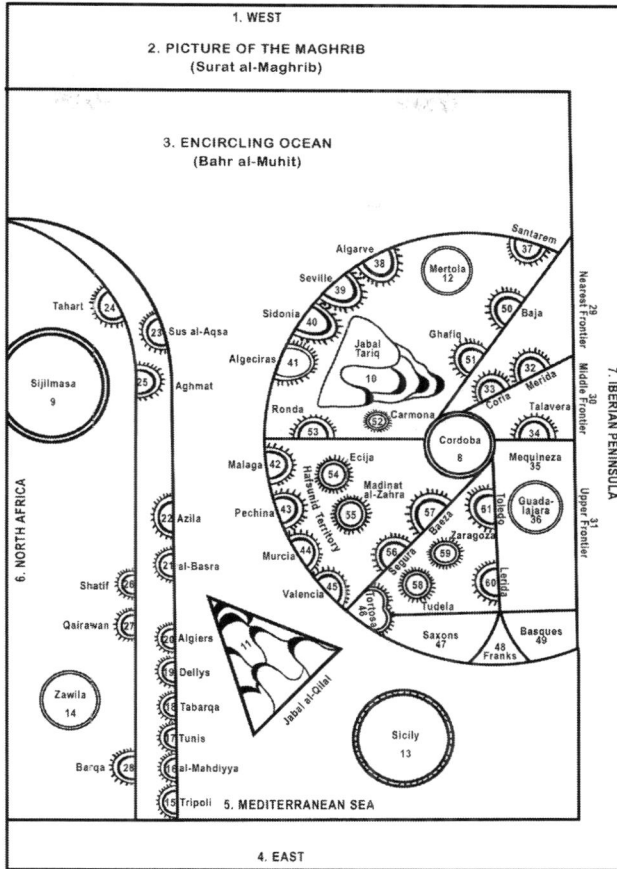

Figure 9.5 Line drawing of Figure 9.4 with transliterated names and numbers.

It is to the subtleties underlying the startling image of the Muslim West that this chapter turns through three levels of visual engagement, or 'gaze', focused on the Maghrib map. The aim is to go beyond the 'carto-' on the map's material surface through to the graphic that lies behind, or beneath, which conceals within it a series of multi-layered subliminal messages. These levels of gaze provide a sense of the map's powerful hold upon its viewers, as well as helping us expose the multiplicity of iconic meanings that lies within.[9]

---

[9] For visual art theories on different levels of gaze, see, for example: Roland Barthes, 'Rhetoric of the image', in Roland Barthes, *Image, Music, Text*, Stephen Heath (tr.)

Analysing the gaze: Maghrib maps in cartographic perspective

This Muslim map of the Maghrib is not a simple rendering of geopolitical space.[10] Rather it is a pictograph of very deliberate design made from the perspective of the eastern heartlands of the medieval Islamic world (Egypt, Syria, Anatolia, Iraq and Iran) graphically representing the reigning view among the eastern Muslim elite of the far West (Figs 9.4 and 9.5). Unlike most of the other *KMMS* maps, which face South, this map has West on top [1]. This deliberate switch in orientation serves to emphasize that the eye of the cartographer is from the bottom of the map [4] looking up: i.e. from the East looking West. One is struck immediately by the dramatic contrast between the forms that make up the image: the perpendicular North African landmass [6] jostling for attention with the semi-circular Iberian peninsula [7]. In between the two, the Mediterranean [5] intervenes, seizing centre stage. The Muslim map of the Maghrib represents a distinct departure from the symmetry of the full Mediterranean map (Fig 9.1). Whereas the Mediterranean map emphasizes a symmetrically harmonious balance of forms, the image of the Maghrib is asymmetric and unbalanced. The eye is drawn to a multitude of conflicting, off-centre focal points that form a patchwork of triangulated gazes laid haphazardly one over the other. Here I will focus on decoding each layer of 'gaze', starting with the layer of the most prominently marked sites and working through progressively more subtle planes demarcated by locations of lesser significance. The empirical basis of this analysis comes from the well-known *KMMS* example of MS. Or. 3101 housed at the Rijksuniversitat library in Leiden (Fig 9.4).[11] Firmly dated to 1193 CE and of eastern Mediterranean provenance,[12] this Maghrib map of Leiden's Or. 3101 is chosen here as the focus for discussion because of its dramatic stylized forms, which usefully facilitate analysis of how symmetry works in these portraits of 'the West'. Once the basic form

---

(New York: Hill and Wang, 1977), 32–51; and Norman Bryson, *Vision and Painting: The Logic of the Gaze* (New Haven: Yale University Press, 1983).

[10] Numbers in square brackets refer to a template of the Maghrib map with translations of the place names: see Figure 9.5.

[11] The maps from this manuscript in Leiden are the best known of the *KMMS* manuscript series. The world map was first printed in Leo Bagrow, *Die Geschichte Der Kartographie* (Berlin: Safari-Verlag, 1951). Subsequently more maps from this MS were reprinted in J. Brian Harley and David Woodward (eds.), *The History of Cartography: Cartography in the Traditional East and Southeast Asian Societies* 2:2 (University of Chicago Press, 1994).

[12] The precise provenance of this map is still open to question. Stylistically it appears to be an eastern Mediterranean product – possibly Ayyubid given the 1193 CE date of the manuscript although Sicilian or Andalusi provenance is also a possibility.

and its successive layers are decoded it is then possible to fathom fully the implications of variations in other maps.

In looking at the Leiden Maghrib map, our eye is initially confused and uncertain and wanders between the three big red circles (Cordoba [8], Sijilmasa [9], Sicily [13]) and then the two smaller ones (Zawila [14] and Mertola [12]), hovering in the blank blue-coloured neck-like space (entrance/exit of the Mediterranean from/to the Atlantic) between the edge of the semi-circular shape (Iberia [6]) and the long tubular one (North Africa [7]). We are drawn along the trajectory of this vision to the curious scallop-adorned triangular shape (Jabal al-Qilal [11]) located right below the measured centre of the image. From the triangle our eye continues on to the second-largest red circle (Sicily [13]) of the image, nestled in the lower right-hand section of the blue space. From this our eye rises naturally to the alternative focus of this image: the prominent red circle (Cordoba [8]) at the centre of the large white/cream semicircular form that juts in from the right-hand edge of the map.

The label of 'Qurtuba' [8] indicates that it is the marker for the city of Cordoba, one of the most famous cities of medieval al-Andalus. As the capital of the rebel Umayyad Emirate (later Caliphate), established by 'Abd al-Rahman I in 756, Cordoba came to be ranked as one of the ornaments of the Muslim world.[13] 'Abd al-Rahman was the only Umayyad prince to escape the massacre that followed the Abbasid overthrow of the Umayyad Caliphate in 747 in Iraq, Iran and Syria – the heartland of the early Muslim empire. Cordoba, which had been languishing economically until the first Muslim invasions of 711, reached its zenith under the Umayyads in the tenth century. The prominence accorded to Cordoba in this late twelfth century map is, however, a curious feature. This map was made during the period of the Almohads, one of the two North African Berber dynasties that governed Islamic Spain, two centuries after the city's heyday.[14] By this time, other places, such as Seville [39], had superseded Cordoba in importance.

---

[13] Maria Rosa Menocal, *The Ornament of the World: How Muslims, Jews, and Christians Created a Culture of Tolerance in Medieval Spain* (Boston: Little Brown and Company, 2002), 32–5.

[14] For background information on Islamic Spain during this period see David Wasserstein, *The Rise and Fall of the Party-Kings: Politics and Society in Islamic Spain 1002–1086* (Princeton University Press, 1984); Hugh Kennedy, *Muslim Spain and Portugal: A Political History of al-Andalus* (London: Addisson Wesley Longman Limited, 1996); Anwar G. Chejne, *Muslim Spain: Its History and Culture* (Minneapolis: University of Minnesota Press, 1974).

One explanation for Cordoba's exaggerated prominence on this twelfth-century map is that the *KMMS* image of the Iberian peninsula was frozen in time: possibly during the mid-tenth century or even earlier.[15] Another explanation is that a romanticized memory of the greatness of Cordoba lingered on in spite of the fact that the city was no longer the centre of power in Andalus. At its apex during the tenth century, Cordoba had a population of around 100,000, some 900 baths, thousands of shops, seventy libraries, including the Caliphal library which is said to have housed around 400,000 books, and a reputation as one of the most cosmopolitan cities in the Islamic world. It was the cultural capital of the Muslim West. Poets and artists flocked to the city, and it gained a reputation as the third most splendid city and cultural centre in the Muslim world after Baghdad and Cairo. It was an emporium for gold and goods that was envied in the heartlands of the Muslim East and came to be known in European circles as 'the brilliant ornament of the world [that] shone in the west'.[16]

The complex matrix of the Maghrib map does not, however, permit us to linger on Cordoba. Other prominent markers draw our eye to locations all over the map. There is, for instance, the very large reddish-brown semicircle [9] beckoning from the upper left-hand section of the map. Even though it is incomplete and tucked away along the southern margin of the North African landmass, the marker signifying the southern Moroccan entrepôt, Sijilmasa, commands attention. Its exaggerated size and singular deep reddish-brown hue suggests that the cartographer considered Sijilmasa [9] one of the most important sites in the West. This raises the question of why the better-known North African sites, such as Algiers [20], Tunis [17], Tripoli [15] or al-Mahdiyya [16] – all bordering on the Mediterranean – were not accorded as much or more importance than a place like Sijilmasa at the edge of the Sahara desert, along the margins of the Maghribi world. The answer lies in the lucrative West African gold trade and the crucial role that Sijilmasa played as the gateway to this gold. The route to West Africa via the southern Sahara was dangerous and plagued with sandstorms. Thus the primary access to the gold-rich kingdoms of

---

[15] Other places on the Maghrib map and the Mediterranean map reinforce my reading of the 'freezing of time' on *KMMS* maps. Take, for instance, the town of Nakur in North Africa, located opposite Malaga, in the present-day area of Morocco. Even though it is marked on the Mediterranean maps, according to al-Idrisi (1184), Nakur had ceased to exist by the late twelfth century.

[16] Menocal, *Ornament of the World*, 32–5.

Awdaghust and Ghana was along a curve that skirted the northern Sahara from Cairo to Fez and then dropped down to the West African Gold Coast via Sijilmasa.[17] Even during its twilight years in the mid-fourteenth century, Ibn Battuta used Sijilmasa as the central hub of his travels in the western Sahara.[18]

West African gold was prized as the highest-quality gold of the medieval period.[19] Ronald Messier, a numismatist and reigning expert on Sijilmasa, notes that 25–30 per cent of the gold that circulated in the Middle Ages passed through Sijilamasa, and during the city's zenith, under the Almoravids (1056–1147), there was an abundance of gold coin mints.[20] During eleventh–thirteenth centuries, Sijilmasa was the financial and business capital of the Islamic West: 'it was a focal point where Africa, Europe, and the Middle East met during the Middle Ages.' Through Sijilmasa large quantities of African gold and slaves made their way to Europe and the eastern Muslim world.[21] The map of the Maghrib makes no mistake about this. Sijilmasa may have been on the edge of the West, but it is still represented on the map as one of the most important places of the Maghribian realm. But how can we equate the signification of Sijilmasa on the map with the prominent presence of Cordoba?

As many of the Muslim chroniclers and geographers tell us, control of the city-state of Sijilmasa, ideally located at the head of the trade routes across the Sahara from West Africa, was the object of intense competition between the Umayyad rulers of Cordoba and the Shi'ite Fatimids of Egypt and Tunisia. Throughout the latter half of the tenth century, the Umayyad Caliphate of Cordoba controlled Sijilmasa, either directly or indirectly through client Berber tribes, wrestling constantly with the Fatimids for control of the Maghrib and the gold

---

[17] Ronald A. Messier, 'The Almoravids, West African gold and the gold currency of the Mediterranean Basin', *Journal of the Economic and Social History of the Orient* 17 (1974), 31–47, map at 32.

[18] In his travelogue, he repeatedly notes his need to go back to Sijilmasa to find a caravan to his next destination. See Ibn Battuta, *Tuhfat al-Nuzzar*, Said Hamdun and Noel King (eds. and trs.), *Ibn Batutta in Black Africa* (Princeton: Markus Wiener Publisher, 1975; reprinted 1994), 29–30, 36, 73–4.

[19] In spite of the fact that the median fineness of West African gold was less than their counterparts from Fatimid Egypt, Sudanese gold was still praised by the Muslim geographers as the best. See Messier, 'Almoravids', 36–7.

[20] John Lynch, 'Sijilmasa exhibit to tour USA', http://www.mtsu.edu/~proffice/Record/ Rec_v07/rec0716/body.html. Messier notes that 'the Almoravid dinar was used eventually as a unit of currency within Christian Europe itself': see Messier, 'Almoravids', 30–3.

[21] Lynch, 'Sijilmasa exhibit to tour USA'.

trade route.[22] The Shi'ite Fatimid Caliphate, centred in Cairo, ruled a vast area from Tunisia to Palestine for two centuries from 910 to 1171. Being an Ismaili Shi'ite state they vigorously opposed the orthodox Sunni Caliphate of the Abbasids in Iraq and Iran and the Umayyads in Spain. Indeed, it is said that the Cordoban Umayyads made the transition from Emirate to Caliphate as a counterstroke to the imperialism of the Fatimids.[23]

The first level of gaze could be read as emblematic of this tenth-century Fatimid-Andalusi struggle for control of Sijilmasa and the gold trade. This interpretation of the dominance of the Andalusi Umayyad and Fatimid Caliphate rivalry in the matrix of places on the *KMMS* Maghrib map is further reinforced by the third prominent place marker, Sicily [13]. Located in the lower right-hand section of the map, at the oblique centre of the Mediterranean Sea, it was also a heavily contested site. It was long admired and sought after by both Muslims and Christian conquerors, who recognized Sicily's pivotal position as one of the key islands of the Mediterranean whence traffic on the sea could be controlled. The prominence accorded to the island makes it clear that the illustrator of this map considered Sicily extremely significant. In 652 CE, barely twenty years after the death of the Prophet Muhammad, the first Muslim naval ships began raiding Sicily, then under Byzantine control. By the ninth and tenth centuries this turned into a full-blown struggle for control between North African Muslim ruling groups: in particular, the Aghlabids (800–909 CE), the Fatimids (909–1171 CE) and the Kalbids (948–1053 CE), who struggled first with the Byzantines, then among themselves, and later with the Normans for control of this key Mediterranean island. Following the Norman conquest of Sicily in 1061 a number of Sicilian Muslim families emigrated to Andalus, including the famous Sicilian poet, Ibn Hamdis, and Andalusi poetry was extremely popular in Sicily even after

---

[22] In fact, the first Fatimid ruler, al-Mahdi, declared himself Caliph in Sijilmasa in August 909. Heinz Halm, *The Empire of the Mahdi: The Rise of the Fatimids*, Michael Bonner (tr.) (Leiden: E. J. Brill, 1996), 128–42 and 280–4. And, when the Cordoban Umayyad ruler 'Abd al-Rahman III declared himself Caliph in 929 he ordered the minting of special coins to commemorate the event that were made out of West African gold. See Hugh Kennedy, *Muslim Spain and Portugal: A Political History of al-Andalus* (London: Longman, 1996), 91.

[23] By the early tenth century the Abbasid Caliphate was crumbling in Iraq and the Fatimids had announced their Caliphate in North Africa. 'The matter was made more pressing by the growing influence of the Fatimids in the Maghreb: if the Umayyads were to counter this expansion, they too would have to boast an equal title': Kennedy, *Muslim Spain*, 90. See also, Menocal, *Ornament of the World*, 31–2; Halm, *Empire*, 281.

the Norman takeover.[24] The Sicilian fondness for and close cultural affiliation with Muslim Spain is reflected in the affectionate epithet 'daughter of Andalusia', which the twelfth-century Muslim Andalusi traveller of the Mediterranean, Ibn Jubayr, uses to describe Sicily in his travelogue.[25]

Subliminal levels of gaze: visualizing worlds within the world

Seen together, Cordoba, Sijilmasa, and Sicily [8, 9, 13], form an obtuse triangle of gaze laid upon the basic template of the image, amid an oppositional balance of mountains and other site markers, which suggest alternative subliminal levels of gaze that conflict with the primary axis of Cordoba, Sijilmasa and Sicily (indicated by a triangle on Fig 9.6). Prominent among these other 'oppositional' site markers is the enigmatic mountain island marked on the maps as Jabal al-Qilal (Mountain of Qilal) [11] – a symbol for the Pillars of Hercules of yore adopted into the Muslim cartographic repertoire as a warning to readers of the dangers that lay beyond it.[26] They are presented in the Islamic geographical literature as an island containing pillars inscribed with warnings to sailors not to venture into the forbidding darkness of the Encircling Ocean where monstrous fish and the Devil and his helpers lurk.

Pushing its way into the narrow neck of water that separates Iberia from North Africa, the distinctly triangular form of the mythical island of Jabal al-Qilal disrupts the plane of the Cordoban–Sijilmasa–Sicilian–gaze outlined above. Through its odd placement and striking patterned form, it appears that the cartographer is suggesting that an alternative axis

---

[24] Alex Metcalfe, *Muslims and Christians in Norman Sicily* (London and New York: Routledge Curzon, 2003), 29–30; William Granara, 'Ibn Hamdis and the poetry of nostalgia', in *The Literature of Al-Andalus*, Maria Rosa Menocal, Raymond P. Scheindlin, and Michael Sells (eds.) (Cambridge University Press, 2000), 288; and Karla Mallette, *The Kingdom of Sicily, 1100–1250: A Literary History* (Philadelphia: University of Pennsylvania Press, 2005).

[25] *The Travels of Ibn Jubayr*, R. J. C. Broadhurst (tr.) (London: Jonathan Cape, 1952), 339. In a recent article about the geometric designs in the pavement of the Cappella Palatina in Palermo, Jonathan Bloom asserts that there were closer artistic and architectural connections between Norman Sicily and Almoravid Spain than have hitherto been investigated. Jonathan Bloom, 'Almoravid geometric designs in the pavement of the Cappella Palatina in Palermo', in *The Iconography of Islamic Art*, Bernard O'Kane (ed.) (Edinburgh University Press, 2005), 61–80.

[26] For more detail on the Encircling Ocean and its association with death and the Devil see Karen Pinto, *Ways of Seeing* (University of Chicago Press, forthcoming).

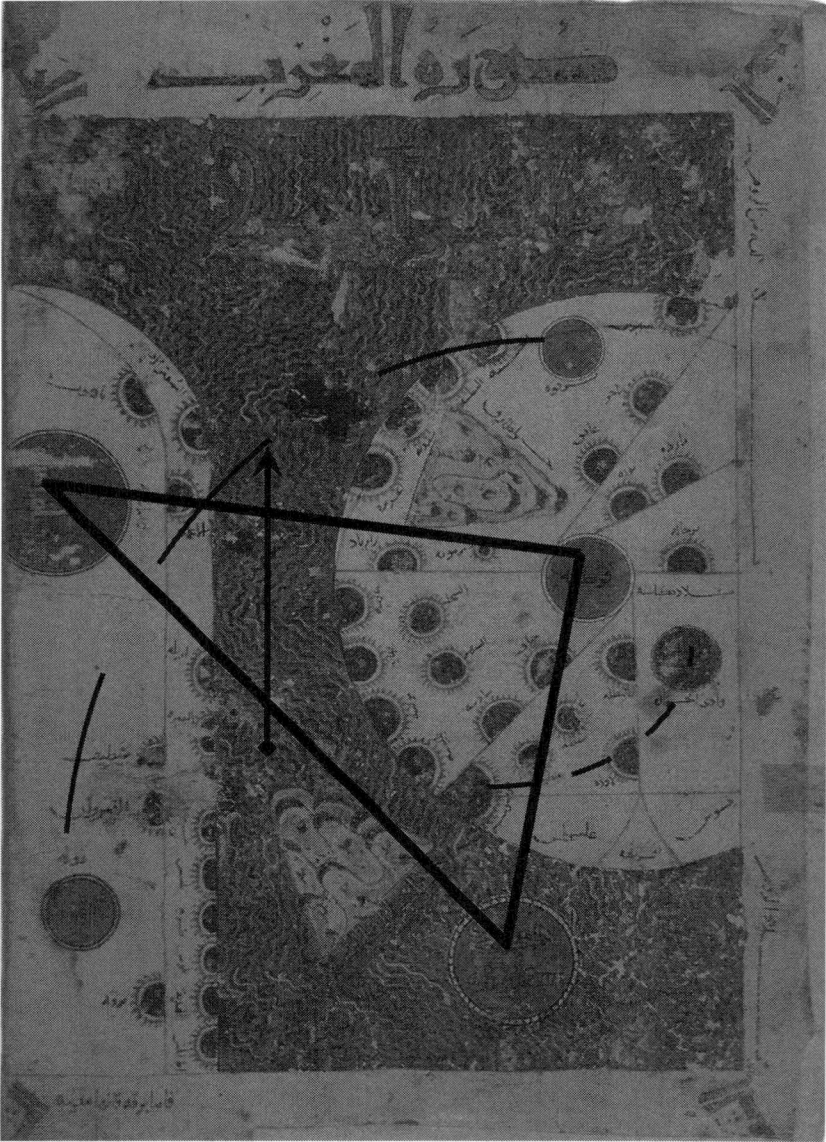

Figure 9.6 'Picture of the Maghrib', showing lines of gaze and emphasis.
Source: Leiden: Bibliotheek der Rijkuniversiteit, Or. 3101.

exists: that of Iberia and North Africa with Jabal al-Qilal as the linchpin. If the mountain island of Jabal al-Qilal is projected to the endpoint of its implied trajectory it would command the mouth of the Mediterranean and plug the sea gap that divides the two landmasses (indicated by dashed line with arrow on Fig 9.6). It is a visual suggestion that implies that whoever controls the island of Jabal al-Qilal will command the crucial sea route between the two landmasses and the entry and exit to and from the Atlantic Ocean (which is referred to in the Muslim maps and texts as the 'Encircling Ocean' – the 'Bahr al-Muhit'). What is strange about the placement of this mythical island of Jabal al-Qilal is that in the map of the entire Mediterranean it has achieved the endpoint of its trajectory and is centrally located between the two edges of the coast (see Fig 9.1). It is already centrally located at the mouth of the Mediterranean, equidistant between the North African and Iberian coasts. Why has this mythical island of Jabal al-Qilal achieved the location of its desire in the smaller-scale regional rendition of the entire Mediterranean and not in the larger-scale Maghrib map? Can this be read as a deliberate design by the artist to insert a conflicting paradigm into the map?

To reinforce this message of underlying conflict, Jabal al-Qilal's Andalusian twin, Jabal Tariq (Gibraltar) [12], is given a prominent berth on the southern flank of Spain. Together they create a subliminal line of gaze that cuts through the angles of the main Cordovan–Sijilmasa–Sicilian gaze, inserting an alternative visual axis into the image (indicated by a line with an arrow on Fig 9.6). What is this other conflicting paradigm? If we read the Cordovan–Sijilmasa–Sicilian gaze as representing the triangle of Andalusian Umayyad and Fatimid influence (possibly refracted through a Norman lens), then we can read the Jabal-al-Qilal–Gibraltar line as a contesting power forcibly inserting itself across the Umayyad–Mediterranean sphere. This is precisely what the overzealous North African Berber conquerors, in particular the Almoravids (454–541/1062–1147) and the Almohads (524–668/1130–1269), do. Fired up with messianic religious zeal and the temptation of a weak and fractured Andalus, conquerors from both groups crossed the sea and took control of most of Islamic Spain. This reading fits with the 1193 date of this map. The map can thus be read as reflecting the rise of North African Berber political aspirations against the backdrop of the Andalusian Umayyad heritage.

The main gaze of the image is deliberately unbalanced by additional site markers. Note, for instance, the prominent demarcation of two sites also indicated by red markers: Mertola [12] on the westernmost end of the Iberian peninsula and Zawila [14] on the easternmost end of the North African landmass. These need to be read as a further reinforcement of the conflict-ridden gold trading grid; superimposed – or rather 'sub-imposed' – below the layer of the main Cordoba–Sijilmasa–Sicily gaze. Mertola [12], a Berber stronghold, was protected from eastern Muslim interference by its location at the western extremity of the Iberian peninsula. It would have provided an alternative protected route for gold from North Africa to Cordoba.[27] Zawila [14], on the other hand, provided an alternative route from the Mediterranean through the Sahara to the gold reserves of the West African sites of Ghana and Kanem. The eleventh-century Andalusian Muslim geographer al-Bakri described Zawila as 'a town without walls ... situated in the midst of the desert ... the first point of the land of the Sudan ... [where] caravans meet from [and radiate out] in all directions'. These two smaller red sites locate alternative angles of gaze – 'key' routes of movement that exist at an intermediate level below the main triangle of focus but above the level of a plethora of smaller markers that dot the surface of the map.

What do the profuse array of smaller half- and full-moon markers outlined in red with ray-like hatchings radiating out from Cordoba on the Iberian mainland and hugging the coastal flanks of North Africa indicate? Within this third layer of sites the illustrator does not exploit the size of the marker to indicate the significance of places. That some places are provided a berth on the map at all, when many other sites were omitted, appears to be the extent of their visual privileging. For instance, even though al-Mahdiyya [16], founded by the Fatimids in 909 CE as their first capital in Tunisia, is a site of considerable importance in the history books, there appears to have been no attempt to distinguish its site marker on the Maghrib map. Its presence is acknowledged with a small semi-circular black marker squeezed in among other similar black markers on the North African flank. In deliberate contrast to the Iberian peninsula the map seems to suggest that there were no key towns in North Africa around which other places were situated. Instead a string of seemingly randomly selected coastal sites on the Atlantic coast[28] run along the Mediterranean coast or parallel to it, while significant historical sites such as Tangiers, Ceuta, Nakur and Malila are left out.

The inner line, running parallel to the coast, appears to be a demarcation for the chief commercial and military artery of the early Muslim

---

[27] There is very little information available on the site of Martula. Previous scholars who examined this map either left the site unidentified or misidentifed it as Merida. Merida is already marked on the map, further inland, in keeping with its actual location. Susana Gómez, a specialist on Martula, and her colleague Claudio Torres are working to resurrect the history of this forgotten site. See Claudio Torres, *Mertola Almoravide et Almoahade* (Mertola: Museum of Mertola, 1988).

[28] Starting with Tripoli [15], al-Mahdiyya [16], Tunis [17], Tabarqa [18], Dellys [19], Algiers [20], al-Basra [21], Azila [22], and ending with Sus al-Aqsa [23].

armies and traders. This route along the edge of the Sahara ran from Cairo to the far-west Moroccan city of Tahart [24]. It was preferred by early Muslim conquerors because it bypassed the coast where they were at the mercy of the Byzantine fleet. Known as the Qairawan corridor, the route is named for the city of the Qairawan [27], an important place of learning and culture, which began as a Muslim garrison and grew to become one of the central Muslim cities in the region.[29] Hence the naming of the route that ran all the way from Cairo, passing through Barqa [28], Shatif [26], to the far-western Moroccan city of Tahart [24] – with all except Cairo marked on the North African flank of the map. Tahart benefited from its location on the Qairawan corridor. Lying at the nexus of the West African gold route, and the North African point of trade with Muslim Spain, Tahart became one of the richest cities in the region. The sites on the Iberian peninsula, on the other hand, radiate out in all directions from the centre. In doing so, they stress the roundness of the form used to represent the Iberian landmass and weight the image towards a single central focal point: Cordoba [8]. All the places of Muslim space in Iberia are arrayed in triangular radial sectors around a central Cordoban node. The northernmost three sectors are a visual reference to the main frontier areas, which divided Muslim and Christian Spain. This was a crucial buffer zone between the Umayyad kingdom and the Rumiyya's – ie. Christians – lurking beyond the ever-shifting Duero line.[30]

Below the three triangles representing the Christian–Muslim frontier border zones lies the prosperous heart of Muslim Andalus, from Santarem [37] to Tortosa [46] on the eastern flank of the Iberian peninsula,[31] which is located right next to the three little triangles demarcating the land of the Saxons [47], land of the Franks [48], and land of the Basques [49]. We can presume that the selection of sites marked in this segment,[32] with Cordoba [8] at the centre, were the places that

---

[29] In spite of the importance of the city of Qairawan, noted by the naming of the route, the cartographer made no attempt to distinguish the site marker.

[30] Sites in this frontier region include: al-Thagar al-Adna ('the Nearest Frontier') [29], which contains site markers for Merida [32] and Coria [33]; al-Thagar al-Awsat ('the Middle Frontier') [30], which contains the site marker for Talavera [34]; and al-Thagar al-A'la ('the Upper Frontier') [31] containing markers for Mequineza [35] and Guadalajara [36].

[31] In between Santarem and Tortosa are markers for Algarve [38], Seville [39], Sidonia [40], Algeciras [41], Malaga [42], Pechina [43], Murcia [44], and Valencia [45].

[32] These sites are Baja [50], Ghafiq [51], Carmona [52], Ronda [53], Ecija [54], Madinat al-Zahra [55], Segura [56], Baeza [57], Tudela [58], Zaragoza [59], Lerida [60] and Toledo [61].

the cartographers considered most important in the Andalusi interior. They confirm that the map was modelled on a tenth-century geopolitical reality because Tudela was conquered by Christian armies in the 1110s, Zaragoza in 1118, Lerida in 1149, Toledo in 1185, and Madinat al-Zahra ceased to exist after 1030. The deliberate misplacement of Toledo [61] into the north-eastern quarter containing Tudela [61], Zaragoza [59] and Lerida [60], when it should have been placed in the region of the Middle Frontier just above Cordoba suggests that the Muslims saw the zone between Lerida and Toledo as a single but very large region. Note that two place markers in the interior of Andalus, in the north-eastern quadrant, are graced with larger black markers: Wadi al-Hijara (known today as Guadalajara) [36] and Turtusha (Tortosa – conquered in 1149) [46]. The two together create the line of yet another level of gaze (indicated by a line on Figure 9.6) tucked beneath the dominant Cordoba–Sijilmasa–Sicily gaze. The emphasis is related to the fact that both sites figured prominently as frontier towns with key fortresses protecting the northern boundaries of the Andalusi realm. Indeed, Tortosa is often referred to as the most important 'outlying' town, especially after the loss of Barcelona in 801.

Third gaze and beyond: the fundamental motif and the carnivalesque?[33]

The visual is essentially pornographic, which is to say that it has its end in rapt, mindless fascination; thinking about its attributes become an adjunct to that . . . all the fights about power and desire have to take place here, between the mastery of the gaze and the illimitable richness of the visual object; . . . history alone, however, can mimic the sharpening or dissolution of the gaze.[34]

In the end, after all the specifics of places and emphases, it is the underlying template of the map, upon which the landforms are conceived, that unites the disparate renditions of the Maghrib to provide a coherent conception of *the West* in the Islamic cartographic imagination: the perpendicular North African coast versus the circular Iberian peninsula with the Mediterranean gushing up in between them like a fountain. How should we read this fundamental motif? Is it nothing more than stylized geometry – a semicircle, a rounded rectangle, and a triangular bottleneck shape with a bunch of smaller circles thrown in for good

---

[33] Inspired by Roland Barthes's essay on 'The Third Meaning', in Barthes, *Image, Music, Text*, 52–68.

[34] Frederic Jameson, *Signatures of the Visible* (New York: Routledge, 1992), 1–2.

measure? If we strip the map of its clutter of place names, markers and islands, what do we see?

The iconic marker of the Maghrib map emerges: a decidedly phallic North Africa, jostling for attention with the semicircular, breastlike, Iberian peninsula. Between the two landmasses the Mediterranean intervenes, seizing centre stage between the two landforms. Once one has developed a familiarity with this mnemonic hallmark one can pick out the image anywhere, irrespective of embellishments, colours and other variations. This form is the essential motif at the heart of the classical Islamic *KMMS* cartographic expression of the Maghrib. It is in this form that the maps of the Maghrib reveal their innermost expression: that of competition and desire. The North African coast is overtly masculine in form, while Iberia is unquestionably feminine. This map can, therefore, be read in terms of the push-and-pull dynamics of lust and conquest and, ultimately, the necessity of rejection in order to retrigger the cycle of desire all over again. That the Muslims conceived of themselves as a masculine force is asserted by the phallic form that they ascribed to North Africa in their maps (as well as the form ascribed to the Arabian peninsula – the symbolic centre for Islam). What takes us by surprise is that they conceived of Iberia as female. Support for this reading comes from both Arabic poetry and history.[35]

The tumultuous and bloody history of Andalus and the Muslim presence in Spain is well attested. The early conquests of 711 were intense and far-reaching – extending all the way to the Pyrenees. The Visigoths were pushed past the Cantabrian mountains of northern Iberia. Only the famous battle of Poitiers in 732 against Charles Martel was able to stem the tide. Quick though the muslims were to conquer the Iberian peninsula, slow were they to colonize. Despite the fact that the conquests were followed by rapid conversions to Islam, external forces such as the Christian kings of Aragon, Leon and Navarre, and outside groups, such as the Berbers, proved impossible to quell. During the early years, Andalus was controlled by the central emirate through a series of governors. A formal state, however, was not established until the century after the last surviving Umayyad prince,

---

[35] Although there is no poetry embedded in the texts that accompany the *KMMS* maps, we do have a striking example of the interweaving of poetry and geographical text from the famous Arab navigator Ahmad ibn Majid al-Sa'di al-Najdi (866 AH/1462 CE), best known for guiding Vasco da Gama around the Cape of Good Hope. Ibn Majid liberally intersperses prose with nautical poetry in his *Kitab al-fawa'id fi usul 'ilm al-bahr wa al-qawa'id*, using poetry to elucidate geographical concepts. Of the 171 pages of his manuscript, approximately 70 are devoted to poems about themes such as navigating the Gulf of Aden, descriptions of the Persian Gulf and directions to Mecca. For more information, see Marina Tolmacheva, 'An unknown manuscript of the Kitab al-Fawa'id', *Journal of the American Oriental Society* 114 (1994), 259–62.

Abd al-Rahman, arrived in Andalus as a mark of dissent for the bloody revolutionary Abbasid overthrow of his family. The emirate was centred on the cities of Seville and Cordoba, and although they lasted for almost three centuries, and established Cordoba as a cultural capital of the Muslim world, they were never able to completely control the outlying provinces. Eventually, thanks to the growth in power of groups whom the Umayyads of Andalus had brought in as mercenaries to help control their state – specifically Berbers from North Africa – and the growing strength of Christian kingdoms in northern Spain, the Umayyad Caliphate of Andalus collapsed in 1031.

Thus began a period of intense political fragmentation in Andalus engraved in Spanish history as the 'Taifa Period' – i.e. the age of the Muluk al-Tawa'if (Reyes de Taifa). No less than thirty-nine short-lived principalities have been identified, some better known and longer lasting than others, during a cataclysmic period of two centuries (from the eleventh to the early thirteenth) in which control of the kingdoms frequently changed hands between rival emirs and Christian challengers.[36] Eventually, only the Muslim Kingdom of Granada remained, which survived until 1492. In between, two major Berber dynasties, the Almoravids (al-Murabitun) 1062–1147 and the Almohads (al-Muwahhidun) 1130–1269 tried but failed to hold sway over Andalus and rein in the warring Muslim principalities. As the Christian *Reconquista* gained momentum from the north and the Berber conquest from the south, the Muslim *Taifa* rulers were killed or found themselves prisoners and refugees in North Africa. There began an intense period of nostalgia and romantic love and hankering for lost Andalus that is reflected in Hispano-Arabic poetry. Most of the extant Islamic map manuscripts that date from this period reflect the prevailing political tensions and thwarted desires.

The repertoire of nostalgic Hispano-Arabic poetry and prose is steeped in longing for Andalus, sometimes expressly stated, at other times metaphorical. 'On Forgetting a Beloved', a discussion of youthful passion for a slave girl by Ibn Hazm (d. 1064) in his famous discourse on love, 'Ring of the Dove' (Tawq al-Hamama), has, for instance, been read as an allegory about the loss of Cordoba.[37] The analogy can be expanded to represent a longing for the Andalusian homeland in general:

---

[36] David Wasserstein, *The Rise and Fall of the Party-Kings: Politics and Society in Islamic Spain 1002–1086* (Princeton University Press, 1985).

[37] Peter Scales, *The Fall of the Caliphate of Cordoba* (Leiden: E. J. Brill, 1994), 28–30; Alexander E. Elinson, *Looking Back at al-Andalus: The Poetics of Loss and Nostalgia in Medieval Arabic and Hebrew Literature* (Leiden: Brill, 2009), 3. Menocal, *Ornament*, 114.

No hopes of easy conquest were to be entertained so far as she was concerned; none could look to succeed in his ambitions if these were aimed in her direction; eager expectation found no resting-place in her . . .

She revived that passion long buried in my heart, and stirred my now still ardour, reminding me of an ancient troth, an old love, an epoch gone by, a vanished time, departed months, faded memories, periods perished, days forever past, obliterated traces. She renewed my griefs, and reawakened my sorrows . . . my anguish was intensified, the fire smouldering in my heart blazed into flame, my unhappiness was exacerbated, my despair multiplied. Passion drew forth from my breast all that lay hidden within it . . .

If I had enjoyed the least degree of intimacy with her, if she had been only a little kind to me, I would have been beside myself with happiness; I verily believe that I would have died for joy. But it was her unremitting aloofness which schooled me in patience, and taught me to find consolation. This then was one of those cases in which both parties may excusably forget, and not be blamed for doing so: there has been no firm engagement that should require their loyalty, no covenant has been entered into obliging them to keep faith, no ancient compact exists, no solemn plighting of troths, the breaking and forgetting of which should expose them to justified reproach.[38]

Ibn Zaydun (d. 1070), a contemporary of Ibn Hazm, held by some to be the most outstanding poet of Andalus, is best known for his nostalgic love poetry. His fifty-two verse rhyming qasida, *Nuniyya*, is considered one of the masterpieces of eleventh-century Andalusi poetry:

> Morning came – the separation – substitute for the love we shared,
>   for the fragrance of our coming together, falling away.
> The moment of departure came upon us – fatal morning. The crier
>   of our passing ushered us through death's door.
> Who will tell them who, by leaving, cloak us in a sorrow not worn
>   away with time, though time wears us away.
> That time that used to make us laugh when they were near returns to
>   make us grieve.
> We poured for one another the wine of love. Our enemies seethed
>   and called for us to choke – and fate said let it be.
> The knot our two souls tied came undone, and what our hands
>   joined was broken.
>   . . .
> O fragrant breath of the east wind, bring greetings to one, whose kind
>   word would revive us even from a distance.
> Will she not, through the long pass of time, grant us consideration,
>   however often, however we plead?[39]

---

[38] Olivia Remie Constable (ed.), *Medieval Iberia* (Philadelphia: University of Pennsylvania Press, 1997), 77–8.

[39] For the complete version, see Michael Sell, 'The Nuniyya (poem in N) of Ibn Zaydun', in *Literature of Al-Andalus*, 77–8.

The fourteenth-century philologist, literary critic and biographer al-Safadi tells us that, 'the *Nunniyya* became so emblematic of longing and exile that anyone who memorized the poem, it was rumored, would surely die far from home'.[40]

Ibn Shuhayd (d. 1035), an older contemporary of the Cordoban poets cited above, adopts less romantic language in his nostalgic poetry, but the sentiment of Cordoba/Andalus as a seductive temptress remains:

> A dying hag, but her image in my heart is one of a beautiful damsel.
> She's played adultress to her men,
> Yet such a lovely adultress![41]

The famous Hispano-Arab poet Abu Ishaq ibn Khafaja (d. 1139), who is known for introducing themes of pastoral yearning into Arabic poetry, creating a whole new form, escaped to North Africa during the Spanish occupation of his home town of Valencia. There he wrote of his estrangement from his beloved homeland:

> A garden in al-Andalus has unveiled beauty and a lush scene.
> The morning glistens from its teeth and the night is overshadowed by
>   its scarlet lips.
> When the wind blows from the East
> I cry: O how I long for al-Andalus![42]

Two and a half centuries later, one of the viziers of Granada, Lisan al-Din ibn al-Khatib (d. 1375), exiled in Salé, Morocco, reminisces about al-Andalus employing similar bucolic natural motifs:

> May the clouds water you with their showers,
> O age of happy love in al-Andalus.
> Your fulfillment was but a sweet dream
> Of sleep or a thief's stolen pleasure.[43]

Even more direct is the nostalgic poetry of Abu 'l-Baqa' al-Rundi (d. 1091), the poet prince of Ronda who was better known by the poetic nickname of al-Radi. More at home with his books than in battle, and often berated by his father for 'preferring the pen to the lance', al-Radi suffered through one of the worst and most tragic ends of a Taifa kingdom. His father, al-Mu'tamid, known as the 'poet-king', was betrayed by the very Almoravid Berber rulers he called upon to save his

---

[40] Devin J. Stewart, 'Ibn Zaydun', in *Literature of Al-Andalus*, 311.
[41] Salma Jayyusi, 'Andalusi poetry: the Golden Period', in *The Legacy of Muslim Spain*, Salma Khadra Jayyusi (ed.) (Leiden: E. J. Brill, 1994), 338.
[42] Ross Brann, 'Judah Halevi', in *Literature of Al-Andalus*, 274.
[43] Yaseen Noorani, 'The Lost Garden of Al-Andalus: Islamic Spain and the poetic inversion of colonialism', *International Journal of Middle East Studies* 31 (1999), 243.

kingdom! After Seville fell and al-Mu'tamid was imprisoned, al-Radi was forced to put down his arms in Ronda, whereupon he was unceremoniously slaughtered by the Berber raiders along the ramparts of his own castle. But not before he penned some of the most famous nostalgic poetry of Andalus, of which the two examples below provide further corroboration of my interpretation of the fundamental image underlying Muslim cartographic representation of the Maghrib. In his 'Nuniyya' poem (based on the Nuniyya prototype established earlier by Ibn Zaydun), al-Radi opines:[44]

> So ask Valencia: What of Murcia and wherefore Játiva and Jaén
> ... Pillars of the country they were; why stay if the pillars are gone?
> Unblemished Islam weeps, as a lover would on separation,
> For lands of Islam bereft, gone to waste with the erection of
>   Unbelief,
> Where mosques have become churches containing only bells and
>   crosses.

In another poem, al-Radi is even clearer:[45]

> It is this home that deceives men and severs the ropes that bind us,
> We are distressed by it, without wine (to relieve us) and we choke
>   from it without cool water.
> Even though we love it even more, however our striving for it is a
>   delusion
> *(Like the striving) for a beloved female whose love does not last and whose
>   lover is ever wanting.*[46]

It is precisely this sentiment of nostalgia and Andalus as the lost lover that can be read into the fundamental iconic form underlying the typical medieval Islamic *KMMS* map of 'the West'. If we strip the medieval Islamic map of the Maghrib down to its pornographic essentials, Iberia appears akin to an attractive female who thrusts her breast seductively across the intercoursing waters of the Mediterranean, tempting the Muslims to come and get her. Like a fickle lover, Iberia has no loyalties, switching allegiance back and forth between Muslims and Christians; seeding rivalry and intrigue amidst her conquerors; providing an enclave for rebels. The desire for her is insatiable because she can never be fully

---

[44] Al-Maqqari, *Nafh al-tib min ghusn al-Andalus al-ratib*, I. 'Abbas (ed.) (Beirut, 1968), vol. 4, 487–8. See also Aziz al-Azmeh, 'Northerners in Andalusi eyes', in *Legacy of Muslim Spain*, 264.
[45] Usamah ibn-Munqidh, *Kitab al-manazil wa'l-diyar*, Mustapha Hijazi (ed.) (Cairo, 1968), 299–300. I would like to thank Avraham Hakim of Tel Aviv University for pointing out this poem to me and assisting with the translation.
[46] Emphasis my own.

conquered. For precisely this reason, she cannot be admitted as the central focus of the image. Rather, she must be obliquely situated as the indirect object of Muslim desire.

Analysing the *KMMS* Maghrib map by peeling away the layered gazes proves that there is more to medieval Islamic maps than meets the eye at first glance. Moving from the obvious, such as the centrality of Cordoba, to an analysis of the positioning of other places reveals a secondary matrix of gazes that incorporate elements of synergy and conflict. These can be decoded and understood when we match them up with what we know of the history of the period. Finally, stripping the map of all its surface markers and visual clutter leaves us with the outlines of the basic template that undergirds all *KMMS* maps of the Maghrib. These point us in a surprising direction that the creators of the *KMMS* maps viewed al-Andalus/Spain/Europe as female. This fits within the context of post-*Reconquista* nostalgic discourse. When examining the European cartographic record for self-representation we find another surprising parallel: Europe routinely depicted itself in a female form.[47] The *c.* 1330 map of Europe and Africa by Opicinus de Canistris is a particularly good example of this phenomenon,[48] as is the late sixteenth-century map of 'Europe as Queen' in Sebastian Münster's *Cosmographia universalis*.[49] Can we begin then to posit a relationship between medieval Islamic and medieval/Renaissance Christian cartographic imaginations through a shared envisioning of Europe as woman?

[47] For a broad range of visual examples depicting Europe as female from the medieval to the modern/post-modern eras, see, Michael Wintle, *The Image of Europe* (Cambridge University Press, 2009), 111–13, 117, 119, 127, 134–5, 136, 149, 176, 248, 250–1, 253, 261, 267, 293, 304, 323–4, 336, 343, 358, 367–8, 373, 375, 381, 429, 457, Plates 2, 3, 4, 5, 6, 9, 30, 31, 33. (Note: I have referenced both cartographic examples and those found in paintings and sculptures because these can be seen as another form of mapping.)

[48] Opicinus de Canistris's suggestive anthropomorphic maps are discussed, in Wintle, *Image of Europe*, 175–7; and Denis Hué, 'Tracé, écart: Le sens de la carte chez Opicinus de Canistris', in *Terres Médiévales*, ed. Bernard Ribémont (Paris: Klincksieck, 1993), 129–57. Note that Opicinus de Canistris produced both male and female versions of Europe, and vice versa for Africa. In both versions, the Mediterranean Sea is depicted as a devilish persona suggesting the enabling vehicle of intercourse between the two figures of the opposite gender. This too would fit with an interpretation of the Mediterranean in the *KMMS* rendition as a sea of intercourse between the Muslim and Christian worlds. The difference is that since Islam does not see sexual intercourse as something evil the sea will not be depicted in a persona of the Devil.

[49] Sebastian Münster's depiction of 'Europe as Queen' is a well-known, much-reprinted and much-discussed image. See, for instance, Trevor Barnes and James S. Duncan (eds.), *Writing Worlds: Discourse, Text and Metaphor in the Representation of Landscape* (London: Routledge, 1992), xiv, plate 1. Wintle, *Image of Europe*, 247–52.

10　Hereford maps, Hereford lives

Biography and cartography in an English cathedral city

Daniel Birkholz

When speculating about cartographic consciousness and map-reception, historians of medieval cartography have tended to deal in collectivities. Typically, we determine social meanings by reconstructing maps' initial display settings and institutional contexts. For maps bound in manuscript we investigate copying stemmatics, patronal contexts and other aspects of textual production – especially authorial intentionality. In a word, our heightened scholarly moments remain originary. We fetishize moments of map creation, or wend further backwards, tracing content genealogies and formal influences; data-gathering tactics and first-hand observational opportunities; sources and analogues. These are not negligible undertakings; all interpretive work depends on them. In place of asking, however, what a given medieval map-artefact means in and of itself, or as designed and displayed by its creators, or in relation to generic contemporary audiences, my goal in this chapter will be to explore how some *specific* medieval persons understood a cartographic text they each encountered – over consecutive centuries, but in the same English cathedral-city milieu: that of Hereford, on the borderlands of England and Wales. My essay thus approaches its work of mapping medieval geographies not from the standpoint of medieval culture writ large, but from an individuated and located perspective, elucidating three men's interconnected cartographic lives within a shared urban/ecclesiastical setting.[1]

As an institutional locale, Hereford Cathedral is notable for how it played host to a high incidence of medieval cartographic activity, of various orders. To explore Hereford's unusual standing in this respect, this chapter utilizes a single documentary artefact – an otherwise unknown map, usually described as a 'map of Europe', that is embedded in a *c.* 1200 copy of Gerald of Wales's *Topography of Ireland* (*c.* 1188) and *Conquest of Ireland* (*c.* 1192): Dublin, National Library of Ireland MS 700 (hereafter NLI 700). As we shall see, this particular manuscript ties

---

[1] For another essay attuned to individual and idiosyncratic, rather than generic or collective cartographic engagements, see Power, ch. 4, this vol.

together three Hereford lives, each connected with the cathedral and the city, and with cartography as a material practice. In one sense, then, my essay hinges, like many in this volume, upon a scholarly encounter with a standard (graphic) medieval map; but like some others, it also examines the (non-graphic) 'mapping' inherent in selected 'textual cartographies' from the period.[2] The essay takes as its launching point, however, yet another sort of medieval map, best described as scholarly 'mapping', a modern charting of medieval lives: in short a *medievalist practice* as opposed to data we encounter in extant medieval objects, whether visual or textual. At the crux of this inquiry are the documentary 'life-maps' I have assembled for three Hereford individuals associated with NLI 700. The first belongs to the famous man who wrote NLI's texts: a certain Welsh-born topographer, ethnographer, hagiographer, royal clerk and ecclesiastical polemicist known as 'Giraldus' or Gerald (*c.* 1143–1221), educated at St David's (Pembrokeshire), St Peter's Abbey (Gloucester) and the University of Paris.[3] Next comes a historical actor virtually unknown to modern scholarship, one Walter Mybbes or Mibbe, who enters the picture by virtue of an inscription at the back of NLI 700 that records its donation to Hereford Cathedral's College of Vicars Choral in 1438.[4] Between Gerald and Walter lies a third Hereford life intimately connected with the city and its mapping traditions: a Hereford-based cathedral clerk named Roger Breynton (*c.* 1290–1351). Breynton held various positions in Hereford's ecclesiastical hierarchy; his importance here lies chiefly in his roles, during a quarter-century tenure as Cathedral Canon, as custodian of the chapel shrine where the Hereford *Mappamundi* was displayed, and as overseer of scribal production, for example, as Hereford's designated examiner of episcopal notaries.[5]

---

[2] See the chapters by Kupfer and Pinto, and by Roland and Lavezzo, respectively. Several contributors, such as Serchuk and Small, combine visual and textual modes profitably.

[3] On Gerald's life and works, see Robert Bartlett, *Gerald of Wales: 1146–1223* (Oxford University Press, 1982).

[4] Inscription on f. 99r; map on f. 47r. For digital reproduction, see The Dublin IAS Irish Script on Screen website (http://www.manuscripts.cmrs.ucla.edu/Manuscripts_view.php? editid1=711). For texts, see James F. Dimock (ed.), *Giraldi Cambrensis Opera*, vol. 5 (London: Longman, Green, Reader and Dyer, 1867); cf. Gerald of Wales, *The History and Topography of Ireland*, John J. O'Meara (tr.) (Harmondsworth: Penguin, 1982 [1951]). For manuscripts and provenance, see Dimock (ed.), *Giraldi Cambrensis*, ix–xlvi, and A. B. Scott and F. X. Martin (eds. and trs.), *Expugnatio Hibernica: The Conquest of Ireland by Giraldus Cambrensis* (Dublin: Royal Irish Academy, 1978), xxxiv–xxxix.

[5] See Daniel Birkholz, 'Biography after historicism: the Harley Lyrics, the Hereford Map, and the life of Roger de Breynton', in *The Post-Historical Middle Ages*, Elizabeth Scala and Sylvia Federico (eds.) (New York: Palgrave Macmillan, 2009), 161–89. Emerging from Anthropology and Native Studies, 'map-biography' projects are human-subject research initiatives 'in which a person's life history is told in map form', or in which a community's

Gerald, Walter, Roger: famous topographer, mystery benefactor, ecclesiastical official; occupants, respectively, of the early thirteenth, early fifteenth and early fourteenth century, but all affiliated with Hereford Cathedral's cohort of landowning, book-perusing and map-viewing canons. As regards each man's institutional milieu and the manuscript map that connects them, the operative point is that denizens of the past like Gerald, Walter and Roger both *are* and *are not* the predictable product – historically, just a function – of their reconstructed times, artefact webs and social locations. Historicist analyses of medieval cartography (for decades now, the scholarly norm) have provided insight into map-reading communities, yet to insist upon a radically *individualized* vantage point offers, paradoxically, an expanded interpretive purchase. The documents I use to animate my three biographical subjects relate where possible to the (Hereford-affiliated) manuscript map they appear, sequentially, to have known. My inquiry, however, also highlights a series of notarized topographies, or 'written maps', in the form of legal instruments and property transactions – dozens of which connect to Walter Mibbe and (especially) Roger Breynton. As non-graphic texts of a territorial nature, documents such as these constitute a species of what Nicholas Howe has described as 'non-visual cartography'.[6] Whether written maps or figural ones, all cartographic texts at once contain *and* help produce their own meanings. Individual maps prompt their cultural interpretation by internal formal means as well as through negotiation with external environments. Yet without wildcard individuals to look at them – socially mobile eyes to 'poach upon' them unpredictably, according to diverse experiences – the meanings of extant maps remain frozen in accord with production and initial display contexts.[7]

---

characteristics and shared experiences are rendered graphically, according to how its members traverse, use, distribute, represent, remember, and otherwise engage with their physical surroundings; see Terry Tobias, *Chief Kerry's Moose: A Guidebook to Land Use and Occupancy Mapping, Research Design and Data Collection* (Vancouver: Ecotrust Canada, 2000), 12–18. For similar approaches in Geography, see David Seamon, *A Geography of the Lifeworld: Movement, Rest and Encounter* (London: Croom Helm, 1979); Allan Pred, *Lost Words and Lost Worlds: Modernity and the Language of Everyday Life in Late Nineteenth-Century Stockholm* (Cambridge University Press, 1990). In literary and historical life-writing, actual maps remain conspicuous by their absence, although metaphors of mapping are common: cf. Peter France and William St. Clair (eds.), *Mapping Lives: The Uses of Biography* (Oxford University Press, 2002). For an exception, see Sarah Ann Wider, 'The contour of unknown lives: mapping women's experience in the Adirondacks', *Biography* 25 (2002), 1–24.

[6] Nicholas Howe, *Writing the Map of Anglo-Saxon England: Essays in Cultural Geography* (New Haven: Yale University Press, 2007), 3–5. For the importance of non-graphic mapping practices to this volume, see Lilley, 'Introduction', this vol.

[7] See Daniel Birkholz, 'The vernacular map: re-charting English literary history', *New Medieval Literatures* 6 (2004), 11–77.

My goal in mapping three Hereford lives is to produce a historical 'snapshot' in which the medieval biographical individual, at some distinct moment and place, engages with a unique map-artefact – a document with a history of its own, formal, institutional and material. To stage such a scene (or rather, three such scenes) requires great particularity on the contextual historical front, but also, and more unusually, on the individual documentary front. The critical, textual and biographical details shall pile up, but collectively they matter because each of my three protagonists helps orient us to his fellow inhabitants of medieval Hereford. Roger Breynton's fourteenth century will be our destination, but in order for his case to acquire full meaning, it is to thirteenth-century Gerald and fifteenth-century Walter – their documentary traces, cathedral connections and cartographic lives – that we must first proceed.

Giraldus Herefordensis; or, mapping the absent canon

Some call him Gerald de Barri – for so he identifies himself, as in a letter 'To Master Albinus, Canon of Hereford' (*c.* 1209). More often Gerald calls himself 'The Archdeacon' – he was Archdeacon of Brecon *c.* 1175–1203, later deferring to an ungrateful nephew – but no scholars champion this epithet.[8] Those who choose Gerald de Barri prefer it to 'Giraldus Cambrensis', as most manuscripts name him, because they find Gerald of Wales does not adequately represent 'the variety of allegiances [he] bore during his complicated and contradictory career'.[9] De Barri does a better job, as it foregrounds the writer's affiliations with a Norman marcher aristocracy committed to the domination and conquest of vernacular Welsh and Irish territories and populations. Robert Bartlett calls Gerald's *Conquest of Ireland* 'in many ways, a family epic', noting also his 'active involvement [in] containment of the native Welsh' and his service in *The Journey through Wales* (1191) and *The Description of Wales* (1194) as 'frequently a spokesman for the marchers'.[10] Recent Geraldine

---

[8] Giraldus Cambrensis, *Speculum Duorum or A Mirror of Two Men*, Yves Lefèvre and R. B. C. Huygens (eds.), Brian Dawson (tr.) (Cardiff: University of Wales Press, 1974), 156. For reference to himself as 'The Archdeacon' see *The Autobiography of Giraldus Cambrensis*, H. E. Butler (tr.) (London: Jonathan Cape, 1937), 50ff. Some scholars ironically follow suit; see J. S. Brewer (ed.), *Giraldi Cambrensis Opera*, 1 (London: Longman, Green, Reader and Dyer, 1861) [hereafter Brewer, *Opera*], lxxx, lxxxvi. Cf. Richard M. Loomis (ed. and tr.), *Gerald of Wales (Giraldus Cambrensis), The Life of St. Hugh of Avalon, Bishop of Lincoln 1186–1200* (New York: Garland, 1985), xiii.

[9] Kathy Lavezzo, *Angels on the Edge of the World: Literature, Geography, and English Community, 1000–1534* (Ithaca: Cornell University Press, 2006), 158.

[10] Bartlett, *Gerald of Wales*, 20, 15, 65. Further complicating Gerald's 'ambiguous sense of national identity' (9–10) are his university years at Paris (1165–72; 1176–9), where he

scholarship, virtually all on his Irish and Welsh topographies, has focused increasingly on his works' proto-colonial effects: 'hybrid' bodies and 'mongrel' identities, 'moralized geographies' and dislocating 'marvels of the west'.[11] Yet if a certain postcolonial distaste marks recent analyses, 'the problem of national identity' also inflects older scholarly treatments; as when Gerald is half-claimed as a progenitor by a self-styled *Cambrian Plutarch* in 1834 or when England's Rolls Series (1861/7) apologizes that Gerald's *Topography* and *Conquest* 'ought to have been edited by an Irishman'.[12] The present essay will not resolve this time-honoured mess. Instead it proposes another epithet for multicultural Gerald, with attendant geographical identity: 'Giraldus Herefordensis', Gerald of Hereford.

Good biographical evidence supports this renaming – chief among them Gerald's position (from *c*. 1193) as Canon of Hereford. Mostly he drew income and prestige while a resident vicar fulfilled his duties in choir. But Gerald did dwell at Hereford when he compiled a life of its cathedral's patron saint – his *Vita Sancti Ethelberti* (*c*. 1194–6) – from material in the Chapter archives.[13] Canon Simon de Fresne had urged Gerald's relocation to Hereford Cathedral, 'where all the liberal arts are studied' and 'where [he] would receive his due honour as a scholar, from men of like interests'.[14] Considering Gerald's studies at St David's and Gloucester, Hereford's location along the route from Wales into England suggests he had a wayfarer's familiarity with the city from an early age. The autobiographical *De Rebus a se Gestis* (1208) documents Gerald's

absorbed 'the internationalism of twelfth-century culture' (5) before an early career as ecclesiastical reformer (1174–6) and multipurpose royal clerk (1184–94).

[11] Beyond Lavezzo, see Jeffrey Jerome Cohen, 'Hybrids, monsters, borderlands: the bodies of Gerald of Wales', in *The Postcolonial Middle Ages*, Jeffrey Jerome Cohen (ed.) (New York: St Martin's, 2000), 85–104; Asa Simon Mittman, 'The other close at hand: Gerald of Wales and the "Marvels of the West"', in *The Monstrous Middle Ages*, Bettina Bildhauer and Robert Mills (eds.) (University of Toronto Press, 2003), 97–112; and Rhonda Knight, 'Werewolves, monsters, and miracles: representing colonial fantasies in Gerald of Wales' *Topographia Hibernica*', *Studies in Iconography* 22 (2001), 55–86. For the *Topographia* as 'essentially a moralized geography', see Monika Otter, *Inventiones: Fiction and Referentiality in Twelfth-Century English Historical Writing* (Chapel Hill: University of North Carolina Press, 1996), 133.

[12] Bartlett, *Gerald of Wales*, 9; John H. Parry, *The Cambrian Plutarch, comprising Memoirs of Some of the Most Eminent Welshmen* ... (London: Simpkin and Marshall, 1834), 146: 'Had it been his lot to be born in a more enlightened age ... he would have bequeathed to posterity a splendid fame'. Mittman, 'The other close at hand', 105, notes Gerald's '[adoption]' as a national hero' by 'some modern Welsh authors'; Dimock (ed.), *Opera*, 5, lxxxvi.

[13] Bartlett, *Gerald of Wales*, 58, 217.

[14] Brewer, *Opera*, I, 378–80; 382–4; cf. Kathleen Edwards, *English Secular Cathedrals in the Middle Ages* (Manchester University Press, 1949), 192–3.

persecution at Hereford by political rivals, upon returning from Brecon in 1201, such that 'scarce any of the Canons or of his friends dared to speak with him for fear of the public power, and scarce any of the citizens would harbour him'.[15] Earlier – following a chapter on 'How the Archdeacon became a follower of the court' (1184) – *De Rebus* narrates a diplomatic adventure that occurred 'before great men' at breakfast and in the garden of the Bishop's House in Hereford, when via 'courteous jest' and rhetorical rejoinder he derailed a Welsh invasion 'single-handedly'.[16] However reliable, such anecdotes illustrate Hereford's transactional location and borderlands function in Gerald's geopolitical imagination. Similarly forceful from a cultural gazetteer standpoint is that Gerald's ethnographic, crusade-preaching *Journey through Wales* both departs from Hereford and returns us back there to conclude.[17]

Recent scholarship has demonstrated Gerald's implication in proto-nationalist English royal as well as universalist Roman ecclesiastical programmes of domination and conquest. As an English county seat, economic centre and cathedral city standing poised upon the Welsh border, Hereford occupies a key intermediary place in a centre–periphery system whose secular and ecclesiastical facets were mutually reinforcing.[18] The city's Anglo-Saxon toponym 'Hereford' (literally, 'the army's way') neatly translates this long-standing borderland function. Edward I would employ Hereford as a military base during his conquest of Wales, afterward using its charter of urban privileges as a model when establishing new towns in English-occupied Welsh territory.[19] In historical practice and not just Gerald's autobiography, then, and for both Norman England and ecclesiastical Rome, Hereford serves as cultural outpost, material way station and administrative staging-point. If 'places, like maps, are never neutral, but are always filled with meanings – culturally shared meanings as well as meanings based on personal experiences';[20] and if the normative realm of England serves in Gerald's

---

[15] Butler, *Autobiography of Giraldus Cambrensis*, 226–7.   [16] Ibid. 81, 82–5.

[17] Gerald of Wales, *The Journey through Wales/The Description of Wales*, Lewis Thorpe (tr.) (Harmondsworth: Penguin, 1978), 204: 'We thus described a full circle and returned once more to the place from which we had begun'.

[18] It was at Hereford's Chapter House, for example (with a visiting Welsh Bishop presiding), that Gerald's Archdeaconry dispute with his nephew was adjudicated in the latter's favour; see Lefèvre and Huygens, *Speculum Duorum*, 156–7; Loomis, *St. Hugh*, xvi–xvii.

[19] Daniel Birkholz, *The King's Two Maps: Cartography and Culture in Thirteenth-Century England* (New York: Routledge, 2004), 115, 121. On Edward's new towns in Wales, see Keith D. Lilley, Christopher D. Lloyd and Steven Trick, 'Designs and designers of medieval "new towns" in Wales', *Antiquity* 81 (2007), 279–93.

[20] Rhonda Lemke Sanford, *Maps and Memory in Early Modern England: A Sense of Place* (New York: Palgrave, 2002), 25.

topographical works as a buffer-zone, negotiating between a theologically authorized Rome and a unsettlingly marvellous, barbaric, spiritually monstrous Celtic fringe,[21] then frontier-located Hereford repeatedly stages and necessarily inflects such processes, impacting them on a logistical level while being imaginatively characterized by them in turn.

The key source on Gerald's relationship to Hereford is his *Epistola ad Capitulum Herefordensis* (*c.* 1218–20). Dispatched shortly before his death (presumably at Lincoln, though datable only through Hereford documentation), Gerald's *Letter to the Hereford Chapter* has been cited chiefly for its listing of his works, otherwise generating two lines of comment.[22] One concerns its reference to a mysterious *Mappa Kambriae* (Map of Wales) which Gerald has found, or made, or perhaps *had* made. This map (or proposed map – no copy is extant) Gerald promises to send to Hereford to accompany his Welsh *Description* and *Journey*, works already in its Chapter's possession.[23] The second issue concerns Gerald's *Letter*'s apparent reference to NLI 700; that is, the Irish *Topography* and *Conquest* manuscript at whose centre lies an otherwise unknown 'Map of Europe' – or 'Map of Western Europe', or *mappamundi* derivative, or map of England's relationship with Rome, depending on one's perspective (Fig 10.1).[24] It seems no accident that the map separating the *Topography* from the *Conquest of Ireland* in NLI 700 survives, uniquely, in an artefact with demonstrable links to Hereford Cathedral, just as, among all Gerald's voluminous writings, it is specifically here, in a stocktaking letter to Hereford, that the great auto-promulgator pauses to describe his *Mappa Kambriae*. A notable characteristic of Gerald's Hereford correspondence, not to put too fine a point on it, is how important *cartography* is, as epistolary topic and material transaction.

To associate Hereford Cathedral with cartography is now commonplace, given the significance of Richard of Haldingham's map (*c.* 1290–1305) for heritage tourists and academic medievalists alike. Judging by its

---

[21] See Lavezzo, *Angels on the Edge of the World*, 46–70; Cohen, 'Hybrids, monsters, borderlands'; and Mittman, 'The other close at hand', 97, 106–7.

[22] Brewer, *Opera*, I, 409–21; Loomis, *Life of St. Hugh*, xxiii, xlv.

[23] John J. Hagen (ed. and tr.), *The Jewel of the Church: A Translation of Gemma Ecclesiastica by Giraldus Cambrensis* (Leiden: E. J. Brill, 1979), xiv; Bartlett, *Gerald of Wales*, 134. J. C. Davies, 'The Kambriae Mappa of Giraldus Cambrensis', *Publications of the Historical Society of the Church in Wales*, 2 (1950), 46–60, attempts a reconstruction, using Gerald's description here and geographical data in the Welsh *Journey* and *Topography*. Interestingly, none of the fabulous material so important to recent Geraldine criticism appears on Davies's map, despite the precedent offered by *mappaemundi* in this respect.

[24] Scott, *Expugnatio*, liii–liv. Characterizations from Arthur Dürst, 'Manuscript map of western Europe from *c.*1200', *Cartographica Helvetica* 20 (1999), 35–8; G. R. Crone, *Early Maps of the British Isles AD 1000-AD 1579* (London: Royal Geographical Society, 1961), 7, 14; and Lavezzo, *Angels on the Edge of the World*, 66–8 (see also note 4).

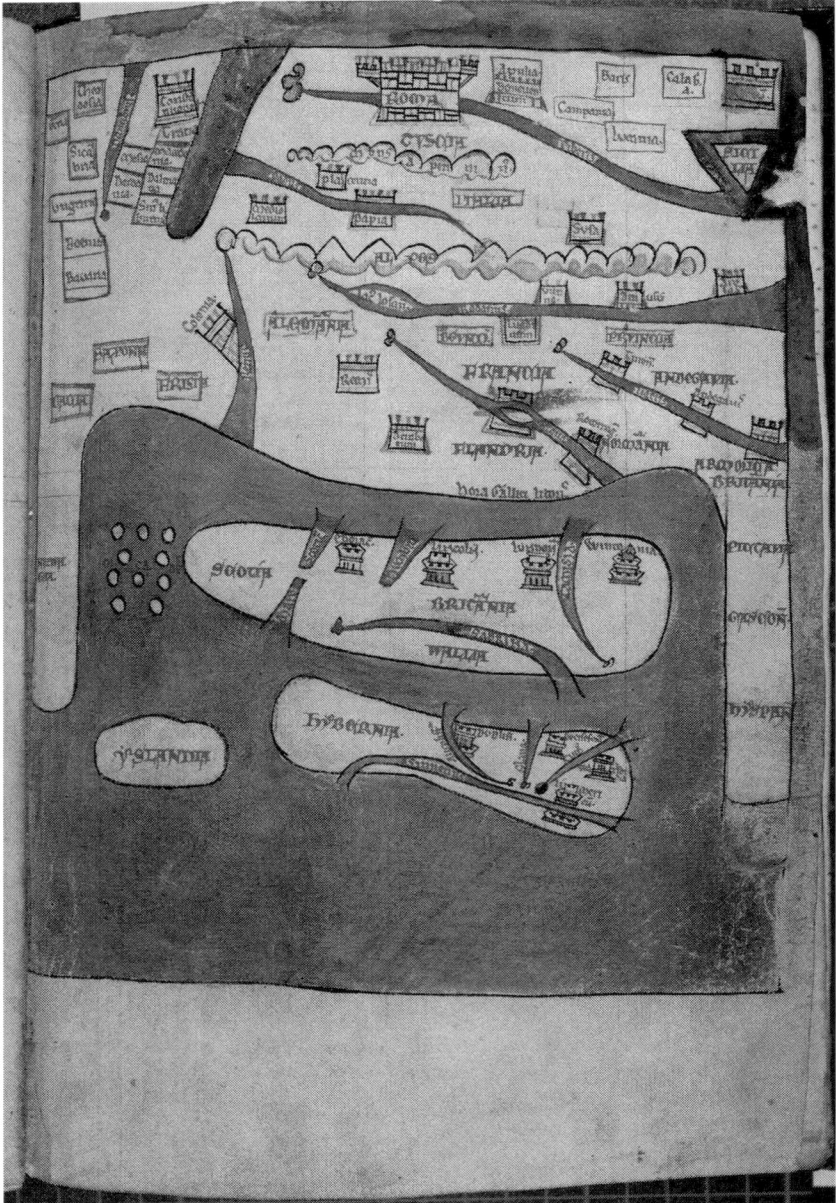

Figure 10.1 'Map of Europe', featuring an oversized Ireland and
Britain, with 'Roma' placed centre-top. This map sits between Gerald's
Topographia Hiberniae and the *Expugnatio Hibernica* in the Cambrensis
codex, NLI MS 700 f. 48r.
Source: Reproduced courtesy of The National Library of Ireland.

featured location in (former bishop) St Thomas Cantilupe's 'shrine-complex', however, the Hereford *Mappamundi* carried still more urgent meaning back in the day, as a reflection of the cathedral's devotional, artistic and academic prestige.[25] Long before Bishop Richard Swinfield (1282–1317) procured its eponymous map, Hereford Cathedral enjoyed a distinguished reputation in scholarly geography – as Canon Fresne enumerates, 'not only astronomy and astrology, but also geomancy' (*geomantia* = divination by means of lines and figures, or by geographic features). An enthusiast like Gerald, Simon promises, will study among those who 'share his tastes'.[26] Bartlett has highlighted the 'problematic phenomenon' of Gerald's 'naturalism' – his solution being Paris, that very incubator of theological and natural science in which the ecclesiastical *mappamundi* appears to have developed.[27] The study of natural history could be pursued elsewhere, however; England's cathedral schools had not as yet been eclipsed by the universities. Kathleen Edwards emphasizes the special '[expertise] in natural science that was maintained at Hereford'. Where Lincoln so specialized in theology that Gerald was urged to abandon his worldly topographical pursuits, Hereford's fame – already for William of Malmesbury – lay 'particularly [in] astronomy and the science of calculation'.[28] Edwards deploys the Hereford Map as *coup de grâce*, but the argument needn't be circular. The 1305 recruitment to his cathedral chapter of Richard Bello, along with his Lincoln-produced world map, by Bishop Swinfield – a noted patron of learning – may constitute academic reinvestment, the shrewd buttressing of an institutional strength.[29]

It may be too much to see in the emergent English *mappamundi* genre a blend of Lincoln Cathedral's distinction in theology with Hereford's strengths in natural history and geography. And yet, as Gerald's letter attests, Lincoln-produced maps were in circulation at Hereford Cathedral, subject to the scrutiny of its Canons, well before the arrival of

---

[25] Dan Terkla, 'The original placement of the Hereford *Mappa Mundi*', *Imago Mundi* 56 (2004), 131–51. For a revised view, see Thomas de Wesselow, 'Locating the Hereford *Mappamundi*', *Imago Mundi* (forthcoming).

[26] Edwards, *English Secular Cathedrals*, 193. Brewer, *Opera*, 1, 382–3. Swinfield was Cantilupe's protégé and successor.

[27] Bartlett, *Gerald of Wales*, 6; Scott D. Westrem, *The Hereford Map: A Transcription and Translation of the Legends with Commentary* (Turnhout: Brepols, 2001), xxxiv–xxxvii. For Gerald at Paris, see note 10, above.

[28] Edwards, *English Secular Cathedrals*, 192–3; Lefèvre and Huygens, *Speculum Duorum*, 168–9 ('To Master William, Chancellor of Lincoln'): 'You have said that we ought to write theological works and that this would be more becoming to our maturity'. As Bartlett, *Gerald of Wales*, 146, notes, 'dallying too long with the natural world was suspect'.

[29] For Richard Bello, Richard Swinfield, and the Lincoln origins of the Hereford Map, see Westrem, *Hereford Map*, xxii–xxv.

Lincoln Canon and Treasurer Richard of Haldingham's devotional/ encyclopaedic masterwork. Relevant here is the large-scale *mappamundi* form's baseline function as a negotiator between matters secular and spiritual; between communities lay and ecclesiastical; between the manifest and the universal.[30] Gerald's 'supernatural naturalism' may discomfit modern scholars, and so too his colonialist aggression, theological dogmatism, brazen self-aggrandizement and bombastic rhetoric. The Hereford *Mappamundi*, by contrast, has managed to offer, to modern approbation, a palatable means by which historical Others (medieval believers) could negotiate between the claims of this world and the next, which is to say, navigate between those very shoals (religious intercession and ideological didacticism) which have so eroded modern respect for Gerald as a writer of moralized geographies and divinatory topographies. This difference in critical reception – bracing critique of Gerald's topographica versus tolerance for the Hereford Map's incarnate theology – has consequences for our encounters with the map in NLI 700. It is not my purpose to 'close-read' the map in NLI 700. Instead it is to examine this topographical compendium's *gifting*-history, specifically its habit of being presented (*c.* 1218–20, by Gerald) and re-presented (*c.* 1438, by Walter Mibbe) to Hereford Cathedral. The mystery of this map's authorship remains as insoluble, I would suggest, as it is immaterial – that is, impossible to determine by material means.

Thomas O'Loughlin, by contrast, has advocated reading the map in NLI 700 as 'a window into the world of Giraldus Cambrensis' and in part '[a reflection of routes] taken by Gerald on his own travels'. For O'Loughlin, NLI 700's central-lying map, like its corrected Irish texts and commissioned illustrations, was categorically produced by his scribes and at his direction, indeed, 'lovingly executed for [Gerald's] own pleasure'.[31] As argued, however, by his Hereford and

---

[30] See David Woodward, 'Medieval *mappaemundi*', in *The History of Cartography Volume 1: Cartography in Prehistoric, Ancient, and Medieval Europe and the Mediterranean*, J. B. Harley and David Woodward (eds.) (University of Chicago Press, 1987), 286–370.

[31] O'Loughlin assiduously catalogues the map's textual and visual content, although it is not *from* any cartographical close reading that we arrive at a conclusion of its Geraldine provenance, but the other way around: his assertion 'that the map reflects [Gerald's] view of the world' remains tautological, a function of the strenuous effort to link it to him: Thomas O'Loughlin, 'An early thirteenth-century map in Dublin: a window into the world of Giraldus Cambrensis', *Imago Mundi* 51 (1999), 24–39, at 24–6, 29–30, 32, 34. O'Loughlin leans heavily on Scott, *Expugnatio*, lv. See also Mittman, 'The other close at hand', 97, 110, and Michelle P. Brown, 'Marvels of the West: Giraldus Cambrensis and the role of the author in the development of marginal illustration', *English Manuscript Studies 1100–1700*, vol. 10, A. S. G. Edwards (ed.) (London: British Library, 2002), 34–59, at 40, 47–8.

other correspondence, 'Gerald's own pleasure' lay not in manuscript hoarding – reserving his topographies' manifold wit and manifest wisdom for himself – but in enabling their conspicuous consumption (by the eager likes, he reports, of King, Pope, and Archbishop).[32] Whatever NLI 700's production dynamics, the points to be seized are, first, that Giraldus Herefordensis regards the Hereford Chapter as a fitting audience with whom to talk cartography; and second, that he pledges a map's delivery, from his present base to a previous academic home.[33] This early thirteenth-century journey of a Lincoln map to Hereford prefigures that more famous, late-century migration of a map across the English midlands. Scholarly consensus continues to link Richard Swinfield's acquisition of the Hereford Cathedral Map to the younger Richard Bello's 1305 appointment as a Hereford Canon.[34] In 1326 Roger Breynton would become the next occupant of Hereford Cathedral's Prebend of Norton – 'vacated by the death of Richard de Bello'.[35]

Nearly eight centuries on, we cannot know who – eying an exemplar now lost – traced out the contours, sketched out the icons and copied out the toponyms comprising 'Giraldus Cambrensis' Map of Europe', so-called. But we can assert that our Norman-born, Welsh-bred, Paris-trained, Lincoln-based Canon of Hereford – 'The Archdeacon' who answers to De Barri – arranged for this document's shuttle between cathedrals. How, for medieval Britain's pre-eminent ethnographer, do issues of cultural identity and territorial possession relate to cartography – mapping, that is, as a social-institutional, *and a* material graphic, *and a* documentary textual practice? To plot a route towards such issues requires a shift away from cartographic authorship, towards cartographic transaction – or put another way, territorial possession. Here, academic geography's favourite chestnut proves subject to reversal: the map *is* the territory.[36] A topographical diagram like that in NLI 700 – a spare if unique rendering of Ireland, Britain and parts of Europe – is at

[32] Hagen, *Gemma Ecclesiastica*, xvi; Scott, *Expugnatio*, liii.
[33] Such a scenario aligns with current thinking on NLI's illustrations – not that Gerald executed them (à la Matthew Paris, who drew maps and scenes for his *Chronica Majora*) but that as author of the book's recopied texts he (may have) arranged for its deluxe illumination. Brown, 'Marvels of the West', 35, proposes the M. Paris analogue, though '[his] contribution would remain equally valid if Gerald acted in a supervisory capacity'. Cf. Mittman, 'The other close at hand', 110; Knight, 'Werewolves, monsters and miracles', 60.
[34] Peter Barber, 'Medieval maps of the world', in *The Hereford World Map: Medieval World Maps and their Context*, Paul D. A. Harvey (ed.) (London: British Library, 2006), 1–44, 27; see also Westrem, *Hereford Map*, xxii–xxv.
[35] A. T. Bannister (ed.), *Registrum Ade de Orleton, Episcopi Herefordensis 1317–27* (London: Canterbury and York Society, 1908), 389.
[36] David Harvey, *Explanation in Geography* (London: Edward Arnold, 1969), 373–6.

once possessed territory (i.e. a material object unto itself) and a document of territoriality (i.e. a cartographic technology providing evidence of geographical possession and/or a means to its epistemological control). Either way, such a map promises its holders a privileged geo-identity.[37]

O'Loughlin wants to award this map to Giraldus Cambrensis, and labours to discover his cultural signature (his hybrid national experience, his episcopal ambitions) within its contours and toponyms; but as conscientious executors, we need to respect the expressed wishes and material acts of the man himself – Giraldus Herefordensis, that is. Canon Gerald already bequeathed this artefact, choosing as destination a body of colleagues to whom he appealed at moments of personal and professional crisis: his erstwhile scholarly cohort, his debtors in hospitality, his 'old' but he hopes 'not outdated' [*antique et non antiquate*] friends in the Hereford Chapter.[38] What is most important about NLI 700's map is not whose Euro-Hiberno travels and textual-authorial travails might underlie it, but that *it* travels as a material document. Our map's journey from an absent Hereford Canon to his residentiary fellows, however, is not the only traversal of a local landscape we can reconstruct. Its next move shall take us across town as well as across time.

Corporate self-confidence; or, mapping the mayor

I may be alone in declaring him 'Herefordensis', but others wish to see NLI 700 safely delivered unto Gerald's fellows at Hereford, of 'true and lasting friendship'.[39] A. B. Scott, for example (citing N. R. Ker), emphasizes that this manuscript 'belonged to the Vicars Choral of Hereford Cathedral, with ascription to Walter Mybbe' – the hand being '?14th century'. The redoubtable Ker, however, did not gesture at the *fourteenth* century; he specified '1438'.[40] Moreover, Ker ascribes not the book but its *gifting* to Mibbe. Implied here is the authorial restitution programme later advocated by O'Loughlin – NLI 700 as 'lovingly executed for Gerald's own pleasure'. As Scott ruminates: 'So, the fact that our manuscript belonged to a Canon of Hereford in the later Middle Ages has naturally set people thinking that perhaps it is the very manuscript sent to Hereford by Gerald'. Where others wish to demonstrate

---

[37] See Jeremy Black, *Maps and Politics* (University of Chicago Press, 1996).
[38] Lefèvre and Huygens, *Speculum Duorum*, 158–9.   [39] Ibid. 160–1.
[40] Scott, *Expugnatio*, xxxvii; N. R. Ker, *Medieval Libraries of Great Britain*, 2nd edn. (London: Royal Historical Society, 1964), 99, 268.

that NLI 700's re-edited Irish texts and unique map 'emanate from Gerald himself',[41] my questions concern *where* these topographical materials emanated *to*, and with what consequences. Judging by what he sends – copies of maps, letters about maps, topographical manuscripts – Giraldus Herefordensis seems to regard his former Chapter as a preferred audience, when it comes to his unusual and not always appreciated geographical interests. His (vast) output was generically varied, but at Hereford Gerald's legacy had a pronounced cartographic element.[42] What, then, did this medieval geographical compendium, with its Irish *Topography* and *Conquest* and intervening map of Europe – or more tellingly, the 'extracted' European 'portion of a *mappamundi*'[43] – mean for those who donated, received and otherwise encountered it, *after* its *c.* 1220 deposit at Hereford?

A biographer's windfall, Gerald has made certain that posterity knows him well. As for that other name asserting proprietary rights to NLI 700 – folio 99r's 'Walter Mybbe', he of the 1438 donation to Hereford's Vicars Choral – there is less information to hand. On the scent in 1978, Scott speculated that working this angle might produce 'a "break-through"' on NLI 700, while in 1999, O'Loughlin revives the search. Unfortunately, he is 'unable to find any other reference to Walter Mybbe'.[44] Whatever books and maps Gerald sent to Hereford must have been known to later cathedral clerks. My essay's concluding section will explore Gerald's cartographic legacy at Hereford by examining the legal-topographical profile of a later Canon, Roger Breynton – his property transactions, material-territorial bequests and other textual mapping practices: first, though, to find Walter Mibbe.

Penelope Morgan's typescript *Index to the Earlier Hereford Cathedral Muniments* (1957) gives fourteen entries for 'Walter Mubbe' and another for 'Mybbes'.[45] Dating 1404–early 1440s, these (and other) documents do not paint the picture of 'a Canon of Hereford' as Scott presumed. NLI 700's Walter Mibbe emerges, rather, as a leading citizen of Hereford. A frequent City Bailiff, Mibbe doesn't merely associate with Hereford Mayors – he serves in 1419–20 *as* 'The Right Worshipful

[41] Scott, *Expugnatio*, liv–lv.
[42] Hagen, *Gemma Ecclesiastica*, xlii, proposes 'four categories: hagiographic; doctrinal instruction and reform; historical; and autobiographical' – none of which apprehend Gerald's geographical importance.
[43] See n. 24, above.
[44] Scott, *Expugnatio*, liv–lv; O'Loughlin, 'Window into the world', 34.
[45] Penelope Morgan, *Index to the Earlier Hereford Cathedral Muniments* (HCL typescript, 1957); B. G. Charles and H. D. Emanuel, *A Calendar of the Earlier Hereford Cathedral Muniments* (Hereford Cathedral Library typescript, 3 vols., 1955).

the Mayor of Hereford'.[46] Walter appears as witness or principal in
numerous city land-deals, but also in major transactions outside urban
precincts. One sees him granted partial rights to the castle and manor of
Snodhull; in others, he and co-querents (urban worthies, county esquires
and the odd aristocrat) 'have the manor of Brimfield rendered to them'
or 'acquire rights to the manor of Talgarth' and accompanying rents –
typically in exchange for cash.[47] Mibbe's associates bear surname
descriptors like 'Chapman', 'Le Tailler', 'Tanner' and 'Mason' – featuring repeatedly is Mayor John Falke, Draper – but Walter's own line is
unclear.[48] Nevertheless, he is well enough off, financially, and well
enough placed, politically, to litigate against the estate of John Prophete,
Dean of Hereford Cathedral.[49] Walter Mibbe, in short, embodies
that burgeoning class of provincial merchant-gentry whose capital and
initiative were transforming late medieval English society. Economic
and political demographics aside, such men had cultural impact,
particularly – from the local cathedral's standpoint – in the realm of
material-devotional patronage.

   Holding aside NLI 700's identity as an artefact of Anglo-Irish colonial
ambition, the question for 1438 becomes: why does an eminent city
father, five years before his death, present a deluxe, antique
ethno-topographical manuscript to Hereford Cathedral's *College of
Vicars Choral*? Some have envisioned Mibbe's bequest as the *return* of a
cathedral treasure; but if so, whether legally alienated or surrepti-
tiously removed, why not give such a book back to the Cathedral *Chapter*
(as Scott assumed), for repatriation to its Canons' famous Chained
Library, with its reputation in scholarly geography? In short, what are
the consequences of Mibbe's choice of philanthropic venue? Our answer
to this may lie in the vexed relationship between these symbiotic cathedral
subgroupings – Canons and Vicars Choral – as compounded by the
commerce of each with city-burghers like Walter Mibbe.

   Cathedral Canons and Vicars Choral stood together at a social-
institutional nexus where the competing pressures 'of administrative

[46] Hereford Cathedral Muniments [hereafter HCM] #130 [Jan. 1419/20].
[47] Feet of Fines: CP 25/1/83/55, #47 [Nov. 1437]; CP 25/1/83/51, #10 [April 1404]; CP 25/1/83/51, #22 [May 1409].
[48] HCM #709 [Feb. 1415/16]; #177 [Dec. 1410]; #168 [Dec. 1410]; #145 [Dec. 1403]. Could 'Mibbe' [nib of pen?] indicate a profession related to scribal practice or material writing technologies?
[49] Feet of Fines: CP 25/1/83/51, #22; *CPR* 13 April 1404. Among his 'high position[s] in church and state', John Prophete was Prebendary of Lincoln; Dean at Hereford (1393–) and York (1407–); keeper of the privy seal; and king's executor (1408). For his will (April 1416), see *Testamenta eboracensia: Wills from the Registry at York*, III (Durham: Surtees Society, 1865), 53–5.

needs and diocesan pastoral care, of this world and the next', revealed medieval Christianity's fundamental tensions.[50] But where the former enjoyed comfort, income and status, to the latter fell a range of administrative and sanctuary duties, with 'long hours', 'poor stipends' and 'little hope of career advancement'.[51] Vicars and chaplains proliferated as cathedrals developed into 'increasingly complex prayer factories', until carefully regulated and strategically located Colleges of Vicars Choral became every see's 'indispensable adjunct'.[52] Ironically, the creation of buildings to contain them 'enormously enhanced the corporate self confidence of the Vicars Choral just as it distanced them from the Canons whose deputies they formally were'. Vicars' subsequent busyness in 'acquiring and consolidating substantial blocks of urban property' made them familiar real-estate partners for city residents.[53] If Vicars' increasing corporate identity relates to their acquisition of communal property, such legal-territorial, real-geography initiatives served to integrate local clerks, as much professionally as socially, within Walter Mibbe's world of civic entrepreneurs. As evinced by testamentary bequests, especially of books, Vicars were 'intensely loyal to the history and traditions' of their cathedrals, colleges and cities;[54] but as Mibbe's gift attests, such bonds extend beyond College walls.

As a leader in Hereford's merchant-gentry, Walter Mibbe pursues manorial property in his countryside dealings, and with them a gentleman's privileges; but within city limits, Mibbe trades parcels with lesser personages. His litigation against the Dean notwithstanding, he has smooth real-estate dealings with four decades of cathedral clerks. The fields leased and domiciles exchanged invariably lie near ecclesiastical properties (for instance, 'a messauge with a dovecote and storehouse built over it ... extending as far as the land of the Lord Bishop of Hereford'), but those involved are not dignitaries.[55] Instead, Mibbe witnesses the 'release of a tenement in the city' between two local rectors and a chaplain, or the grant of another, 'with garden adjoining', to an urban chaplain; or himself leases a house and curtilage 'to two chaplains and a clerk', or a garden

---

[50] Barrie Dobson, 'The English Vicars Choral: an introduction', in *Vicars Choral at English Cathedrals: Cantate Domino: History, Architecture and Archaeology*, Richard Hall and David Stocker (eds.) (Oxford: Oxbow Books, 2005), 1–10, at 1.

[51] David Lepine, *A Brotherhood of Canons Serving God: English Secular Cathedrals in the Later Middle Ages* (Woodbridge: Boydell Press, 1995), I, 4; Julia Barrow, 'The origins of Vicars Choral to 1300', in *Vicars Choral*, Hall and Stocker (eds.), 11–16, at 13–14.

[52] Richard Hall, 'Preface and Acknowledgements', in *Vicars Choral at English Cathedrals*, Hall and Stocker (eds.), i–xiv, at vi–vii; Dobson, 'English Vicars Choral', 3–4, 6.

[53] Dobson, 'English Vicars Choral', 5–6.

[54] Ibid. 9–10; Barrow, 'Origins of Vicars Choral', 15.   [55] HCM #185 [May 1441].

'to two clerks', including 'one of the Vicars in the choir of the Cathedral Church of Hereford'.[56] Such dealings comprise most of Mibbe's extant documentary activity. Yet whether Walter presides (as Mayor/Bailiff) or participates himself in these transactions, the 'lesser' local cathedral clerks appearing with him are, notably, those whose institutional autonomy and economic standing these years saw burgeoning.

Considering their closer match along a hierarchical social continuum, Walter Mibbe likely had better relations with his city's College of Vicars Choral than with its august Cathedral Chapter. Judging by property deeds, Mibbe had regular encounters with local vicars and clerks as a consequence of their shared urban lives – beyond any devotional inter-action. It seems relevant here that beyond sanctuary duties, Vicars Choral were 'expected to help with [local] fundraising', surely because the 'great proportion' of such clerks (unlike outsider Canons) 'were recruited from the cathedral city' itself.[57] But if vicars were 'local boys', 'trained in the cathedral school choir', not all Hereford Cathedral School enrollees were intended for ecclesiastical careers. A significant proportion returned to lay life, their cathedral sojourn reinforcing pat-ronage ties with prominent families and city guilds. Youths themselves received literacy training, marking them out as future city bureaucrats – precisely the sort to occupy bailiff roles à la Mibbe and his ilk. That numerous Hereford mayors appear earlier as notaries (for example, eight-times mayor 'John Mey, clerk') bespeaks a practical if not institutionalized relationship between chartered city officials and the cathedral's lower-tier clerks – as likely as not, their former fellow students.[58] In 1395, Richard II officially incorporated Hereford Cathedral's College of Vicars Choral; a decade earlier (1384) he had instituted by charter the office of Hereford City Mayor. The former group's growing 'corporate self-confidence' may connect with the burgeoning fortunes of the class represented by the latter. Something of the two bodies' social-professional intercalation may be seen in how a *city*-generated deed mentioned above (HCM #55) is dated not '6 October' but 'Thursday after the Feast of St Thomas of Hereford' – local miracle-working bishop Thomas Cantilupe, that is, at whose shrine the Hereford Cathedral *Mappamundi* initially stood. In a single documentary-devotional clause, a (cathedral educated?) city

---

[56] HCM #107 [Jan. 1411/12]; #137 [Jan. 1425/26]; #185 [May 1441]; #55 [Oct. 1429]. See also #709 [Feb. 1415/16]; #706 [June 1428].

[57] Barrow, 'Origins of Vicars Choral', 14–16; Dobson, 'English Vicars Choral', 8.

[58] Dobson, 'English Vicars Choral', 8; Lepine, *Brotherhood of Canons*, 130.

notary neatly encapsulates the enfolded nature of official cathedral and urban mercantile interests.

To read Mibbe's donation of NLI 700 as the return of an alienated treasure is to pit Hereford's Chapter of Canons against its College of Vicars Choral. Mayor Walter's 1438 bequest reprises Canon Gerald's initial gifting of the book to Hereford Cathedral – but with a difference. Walter's philanthropy, it may be, commemorates his youthful experience in the cathedral fold, the education that prepared him for his career in Hereford's mercantile bureaucracy. So understood, the act reads as an expression of preference for one Hereford Cathedral sub-corporation above another – whether payment of personal debt or statement of social coalition. Viewed through Chapter–Vicar relations, Bailiff Walter's bequest of Canon Gerald's antique geography of Ireland helps to map the social-institutional landscape around Hereford Cathedral. This fifteenth-century bibliographic transaction also has resonances, however, of a more far-flung, bio-topographical nature – its prequel tremors extending beyond documentary names (Gerald; Walter; Roger) to historical addresses and architectural remains. And so we move toward 29 Castle Street, original site of Hereford's College of Vicars Choral.[59]

Mapping the residentiary canon; or, an archivist of the landscape

His quarter-century in the Hereford Chapter – the last sixteen years in residence – makes Roger de Breynton an ideal reader for Giraldus Herefordensis, the absent Canon so concerned with posterity, with British and Irish Topography, and with his reception at Hereford. Breynton's positions were various, but his identity as a homegrown Hereford Canon (1326–51) dominates his profile. Such was his effectiveness in the employ of Bishops Richard Swinfield (1316–17) and Adam Orleton (1317–33) that his culminating canonry seems inevitable. Before entering episcopal service, however, and especially before his first benefice (1313) and study leave, Breynton languished in professional flux, his future distinction no given. Unlike his friend Thomas Talbot – of comparable age, but promoted years earlier – Roger was

---

[59] Ron Shoesmith, '"A Brave and Ancient Priviledg'd Place": The Hereford Vicars Choral College', in *Vicars Choral*, Hall and Stocker (eds.), 44–60, at 46–7; Ron Shoesmith, 'The close and its buildings', in *Hereford Cathedral: A History*, Gerald Aylmer and John Tiller (eds.) (London: Hambledon Press, 2000), 293–310, at 306.

obscurely born, making advancement unlikely. This uncertainty is reflected in the shifting naming conventions with which Hereford notaries attempt to place him prior to 1313.[60] But however fluid his early identity, to map Breynton in this nominative respect helps to establish his urban positioning over the course of his career. Such localized identity conditions in turn inflect his experience of a map like Canon Gerald's.

To study medieval urban cartography is necessarily '[to study] imagination and language', as Daniel Lord Smail observes, since 'late medieval cities were rarely represented in their own time by means of graphic maps or other visual images'. Yet 'urban space was in fact mapped – by the record keeping bureaucracy of the late Middle Ages'.[61] A crucial feature of medieval urban documentation's 'implicit linguistic cartography' – for Breynton's Hereford no less than for Smail's Marseille – is its direction towards neither state territorial nor theological-devotional ends (as in emergent regnal mapping and main-line church cartography).[62] Instead, cartographic lexicons and grammars were used, on behalf of a cross-section of city-residents, 'to map out possession and identity'. Those who shaped this process were not ecclesiastical or governmental elites, but public notaries, who 'conspicuously favored a street-based cartographic template that was to become the urban norm'.[63] This documentary technology, in which the 'map' of the city is regularized according to a single architectonic form, fostered their interests because it made necessary the act of cartographic 'translation' notaries provided. Confronted during property transactions

---

[60] Born shortly after Cantilupe's miracles began (1287), Breynton attended Hereford Cathedral School, then apprenticed with Nicholas Reigate, Cathedral Treasurer (1302–8) and Bishop's Registrar, from whom he learned documentation and accounting, and received instruction in the inventory, handling, upkeep and display of cathedral 'treasures' (like the Hereford Map). Cathedral records describe him as manorial reeve (1306), subdeacon (1307), 'clerk' (1309, 1310), and 'Vicar in the Church of St. Ethelbert' (1311); HCM #R352 [Sept. 1306]; #1733 [Dec. 1309]; #61 [June 1310]; #203 [June 1311]. My thanks for the 1307 reference to Susan Ridyard, who is currently editing Cantilupe's canonization dossier (Vatican Library MS Lat. 4015); personal communication, 8 May 2009.

[61] Daniel Lord Smail, *Imaginary Cartographies: Possession and Identity in Late Medieval Marseille* (Ithaca: Cornell University Press, 1999), xi, 7. For such a concept's importance to this volume overall, see Lilley, 'Introduction', this vol. For textual mapping in another medieval English cathedral city, although in a hagiographic rather than bureaucratic context, see Lavezzo, ch. 11, this vol.

[62] Smail, *Imaginary Cartographies*, 8–10, xiii–xiv; Birkholz, *King's Two Maps*, xxi–xxviii, 65, 95, 166.

[63] The basis for their influence lay in the shifting nature of medieval identity clauses, especially the geographic component of documentary formulae. Smail, *Imaginary Cartographies*, xi–xiv, 8–10.

by a 'considerable gap between vernacular and notarial cartographic languages', notaries translated their clients' habitual modes of cartographic discourse into the documentary grammar of their own preferred form.[64] As a result, 'notaries *became cartographers*':

> Working for clients ranging from common men and women to great landlords, they acted unconsciously as map-makers or archivists of the landscape, recording toponyms, locating buildings, and identifying streets, plazas, markets, squares. This cartographic role emerged because the legal requirements of the acts drawn up by notaries demanded that real estate be mapped and that the acts themselves be validated by reference to a time and place.[65]

A corollary of their development into so many mapmakers is that professional notaries, though amorphous as a collectivity, became *de facto* custodians of cartography – 'keepers of the city's authoritative linguistic map'. Passed down between generations of city notaries, Marseille's *ad hoc* notarial casebooks thus constitute, in Smail's memorable coinage, 'a dispersed cartographic archive'.[66]

Roger Breynton occupies multiple positions within Smail's framework of medieval urban notarial culture, as he does within Hereford's social-ecclesiastical landscape. His early years – pre-benefice and study leave (1313–16), pre-episcopal *familia* (1316–33) – see him progress from nondescript lesser clerk to small-time dealer in agricultural and residential property. Sometimes Roger's brothers Hugh Topesleye (a layman) or William Breynton (a fellow clerk) accompany him; trading partners come from their mixed-estate, suburban Breinton vicinity. Occasionally he serves in neighbours' deeds as agent of record himself.[67] The point is, we see 'Roger de Breynton, clerk', mixing freely with Hereford citizens of various professions, classes, ages, genders and neighbourhoods. In acquiring properties in 'Breinton and Upton', then increasingly in the city core, near cathedral environs where assorted lesser clerks and chaplains dwelt, Breynton straddles multiple communities – city vicinities that coalesce around his suburban childhood vill (adjacent to Canon's Moor, the Cathedral Treasurer's designated manor) and his urban professional present (in the cathedral close). Entry into Swinfield's *familia* (with its coterie pressures, administrative workload and constant perambulation) brings to a halt his side dealing in local property, with later years under Orleton being especially taken up with travel and political volatility. Future career stages find Breynton multi-tasked, transforming him – if we regard his bureaucratic work as so much

---

[64] Ibid. 12, 40–1.   [65] Ibid. 39. My emphasis.   [66] Ibid. 39, 29.
[67] HCM #553 [Feb. 1311/12]; cf. #12–14 [June 1314]; #17 [Oct. 1313]; #117 [Dec. 1314]; #1733 [Dec. 1309].

dictation and retention of written maps – into a medieval cartographic archive virtually unto himself.

Considering his plurality of roles, any attempt to fix Breynton geographically must founder. Yet two features of his documentary life-map bear emphasizing. First, once his geographical ambit expands (and urban transactions cease) with entry into episcopal service, Breynton's territorial dealings follow standard benefice-acquisition patterns for secular clerks. The second notable aspect of his bio-cartography we have already encountered. Early on Roger had been a notarial client and sometime documentary mapmaker (i.e. a commissioner of 'textual cartography' in the form of property deeds), but his prevalent role in career maturity is as overseer of scribal production. In short, Roger Breynton emerges as an exceptionally placed, cathedral-based 'archivist of the [Hereford] landscape'. His cartographic portfolio extends across urban and rural, secular and ecclesiastical, familial and institutional boundaries, touching everything in between – from city tenements; to suburban houses, buildings and gardens; to rural crofts, meadows and arable land; back finally to Roger's double-sized canonical house in Cabache Lane (now 20 Church Street).[68] If, as Smail asserts, there are telling differences 'between how a city is mapped by its bureaucracy and how it is mapped by its residents',[69] the collected records (or city 'maps') of resident bureaucrat Roger Breynton, as much cathedral dignitary as native son of its suburban fringe, insist on illustrating how thoroughly geographical identity categories intermingle.

If the Hereford properties he acquired early (1309–16) still remained during his hectic middle years, in late career Roger Breynton begins, like Walter Mibbe, to liquidate real estate holdings, converting them into assets less material, though no less valuable for being intangible. Consider this pair of written maps from 15 April 1336:

Licence for the alienation in mortmain by Roger de Breynton, prebendary of Norton in the church of St Ethelbert, Hereford, of two messuages, a virgate and a half of land, 4 acres of meadow and 10s. of rent, in Breynton and Upton, to a chaplain to celebrate divine service daily in the cathedral church of Hereford, for his soul and the souls of William de Breynton and Nicholas de Reygate, late treasurer of the said church, and their fathers and mothers.

---

[68] HCM #203 [June 1311]; #218 [Sept. 1311]; #17 [Oct. 1313]; #23 [before 1313]; #17 [Dec. 1314]; #61 [June 1310]; #1733 [Dec. 1309]; W. W. Capes (ed.), *The Register of Thomas de Charlton, Bishop of Hereford 1327–44* [hereafter *HRC*] (Hereford: Wilson and Phillips, 1912), 2 [Oct. 1328].

[69] Smail, *Imaginary Cartographies*, 12.

Licence for the alienation in mortmain by Roger de Breynton, clerk, to the dean and chapter of the church of St Ethelbert, Hereford, of a messuage in Castle Street, Hereford, for a habitation for the vicars and clerks of the church.[70]

These documents show Breynton picking his way between personal and institutional interests, and urban and ecclesiastical spaces, as carefully as in younger days – negotiating thereby a re-grounded relationship to his professional past and local origins. Smail would interrogate such notarial mappings via their identity-clause choices: how Breynton's honorific in the first document ranks him institutionally and grounds him geographically ('Prebendary of Norton in the Church of St Ethelbert, Hereford') where the second styles him merely 'Roger de Breynton, clerk' – this despite his being Vicar-General (acting bishop) at this very moment, and despite his conspicuous position as benefactor, in a bequest providing new 'habitation' for the underendowed 'vicars and clerks of the church'. Roger's generic designation here, strangely a-hierarchical for a documentary culture that thrives on status differentiation (not since 1313 has he appeared so unadorned), may constitute an expression of solidarity by a former Hereford Vicar, the manifestation of a levelling impulse that places all Hereford clerks on a single status-plane.[71]

Identity clauses aside, what glimpse into Breynton's textual cartographic profile do these bundled documents offer – a pair of commissioned maps 'taken at Hereford' but enrolled together by royal notaries at the Tower of London? Lying within hailing distance of the Bishop's Garden where a certain Archdeacon Gerald did diplomatic wonders, as well as of narrow tenements leased by Walter Mibbe to waves of local chaplains, the alienated Breynton 'messauge in Castle Street' that is now Number 29 would eventually house Hereford's College of Vicars Choral in splendid communal style.[72] The operative figure for our second record resides in the first, wherein Roger alienates 'two messuages, a virgate and a half of land, 4 acres of meadow and 10s. of rent, in Breynton and Upton' to endow a perpetual mass: 'a chaplain to celebrate divine service daily in the cathedral church of Hereford'. Roger's action here benefits

---

[70] *Calendar of Patent Rolls 1334–38*, 247 (15 April 1336).

[71] After years in episcopal service, Canon Breynton has taken up residence at his home cathedral, where, especially during Bp. Thomas Charlton's absence as Justiciar of Ireland (1336–9), he performs leading diocesan roles. He typically appears as '*Dominus* Roger de Breynton, Archdeacon of Gloucester', or 'Roger de Breynton, Canon of Hereford and Vicar-General'. R. M. Haines, *Register of Simon de Montacute, Bp. of Worcester 1334–37* (Worcestershire Historical Society, n.s. 15, Kendal, 1996), 44, 943.

[72] As Shoesmith, 'Hereford College', 46–7, notes, 'their [1395] foundation charter indicates that the vicars already had a communal residence', in Castle Street – 'parts of which still survive within the present Number 29' – for which 'evidence indicates a [mid-to-late 14th-c.] constructional date'.

souls: his own, his parents', his brother William's (d. 1332), and that of 'Nicholas de Reygate [d. 1308], late Treasurer of the said church'.[73] But what is long-dead Nicholas doing here, consorting in 'daily divine service' with Breynton's immediate family? Breynton donates the Castle Street messuage because Hereford's lesser clerks needed housing,[74] but what testamentary mapping does invocation of Reigate's memory perform, on behalf of our 'archivist of the [Hereford] landscape'?

If 'metaphoric models of place', maps and places themselves all 'shape how people conceptualize and experience their world' – especially '[as] these texts become inscribed in memory' – then the practice of 'thinking cartographically' is what allows pre-modern subjects 'to take virtual possession of their topographic surroundings'.[75] Still, it is practical, legal, testamentary control of urban territory that counts, as notaries 'map out possession and identity' on behalf of diverse clients but with increasingly standardized textual practices. Pre-modern maps and mapping discourses facilitate acts of remembering not only systematically (as in spatial memory systems). Cultural codings aside, maps conjure intense personal memories for individuals.[76] Scholars continue to privilege early maps' 'relevan[ce] to matters of state' – how 'the medium of the map comes in very handy for the exercise of power'.[77] Yet since maps, like places, trigger not just shared meanings but private ones, 'meanings based on personal experiences', Roger's encounter with 'Gerald of Wales' Western Europe' in NLI 700 must engender responses of an individuated sort, beyond the 'armchair travel', 'colonial fantasy' and 'power-politics' generally operative for this map.[78]

Like Walter Mibbe's, Roger Breynton's gift to the Vicars of Hereford combines community patronage with personal nostalgia. He directs a favour to his former self: a young clerk of uncertain prospects, facing a dead-end career. *Treasurer and former registrar* Nicholas Reigate's inclusion in Breynton's family mass-bequest acknowledges his mentor's profound intervention, which enabled Roger (so unusually) to jump career tracks, from generic lesser clerk to eventual Cathedral Canon. In 1336, Roger does his bit for coming generations of Hereford Vicars.

---

[73] W. W. Capes, *Charters and Records of Hereford Cathedral* (Hereford: Wilson and Phillips, 1908), 215–16.

[74] Chantry culture's growth meant more chaplains generally, but in 1327 the cathedral secured a major grant, '[providing] for statelier celebration' (ten additional chaplains) of cathedral masses, especially at Cantilupe's shrine; Phillip Barret, 'The College of Vicars Choral', in *Hereford Cathedral*, Aylmer and Tiller (eds.), 441–60: 444–5. For the bequest, see Capes, *HRC*, 34–40.

[75] Sanford, *Maps and Memory*, xiii, 3.   [76] Ibid. 18, 21, 22–3, 139.   [77] Ibid. 18.

[78] Ibid. 25, 139, 18; Knight, 'Werewolves, monsters and miracles', 55; O'Loughlin, 'Window into the world', 33.

In a neat territorial twist, Breynton appears to have acquired the city-core property he donates as 'a habitation for the cathedral's clerks and vicars' – Number 29 Castle Street – through his executorship of Reigate's estate.[79] And so too the land-parcels 'in Breynton and Upton', adjacent to Canon's Moor (in the Treasurer's portfolio), that Roger uses to endow his mass.[80] Breynton could be financially liberal (forgiving debts to Bp. Orleton's estate, for example),[81] but his Castle Street bequest was an unusually concrete act of generosity. Of course, Canon Roger had reason to be grateful in turn.

Mapping NLI 700; or, convergent Hereford lives

Medieval streets do not simply carry traffic and house buildings, Rhonda Lemke Sanford observes, but also operate as 'loci in a memorial system that combines maps and memory'.[82] The textual mapping we have examined bears unique period characteristics – that is, its functions are essentially medieval – in that it works not to render a place objectively knowable (in an emergent architectonic sense). Instead, medieval mapping technology offers to transform one kind of substance – one order of territorial possession: material, economic, this-wordly – into another. Roger's parents, brother and mentor all partake in the purgatorial benefits that accrue to those for whom masses are performed. Cartographically, the point is that by converting economically productive agrarian land into spiritually beneficial clerical labour, Canon Breynton's documents reproduce, with local topographical material and on a micro-social scale, the universalist, geographical-theological transformation that is the signature operation of ecclesiastical *mappaemundi*.

Local boy, modest birth: by rights, 'Roger de Breynton, clerk' (1310) should have remained 'Roger de Breynton, clerk' (1336). Treasurer and Registrar Nicholas Reigate, however, presider over England's 'most complex accounting system' during a period of documentary revolution,[83] plucked a low-born but highly capable vicar, making him reeve for his manor Canon's Moor; training him in accounts and artefact inventory; and socializing him to a secular cathedral's ways and means. Institutionally speaking, Reigate gave his protégé a map of the place – an administrative key to Hereford Cathedral – as well as a place on its map,

---

[79] HCM #2873 [April 1311]; #218 [Sept 1311]; #203 [June 1311]; #609 [Jan. 1307/8].
[80] HCM #469 [April 1305]; #1489 [July 1308]; etc.
[81] R. M. Haines, *The Church and Politics in Fourteenth-Century England: The Career of Adam Orleton c.1275–1345* (Cambridge University Press, 1978), 52.
[82] Sanford, *Maps and Memory*, 142.   [83] Edwards, *English Secular Cathedrals*, 263–4.

a foothold in its hierarchy. The unusualness of Breynton's trajectory, in sheerness of ascent, bears repeating. Vicars Choral did not rise to become Canons, let alone papal envoys and royal escorts. That Breynton did underscores not just his own talents but the immensity of Reigate's intervention. Our 1336 deeds map Roger's tangible sense – still strong a quarter of a century onward – of professional gratitude, even spiritual obligation, to his mentor as well as institution. Appropriately, Breynton's personal obligation receives documentary grounding – it gets mapped memorially and becomes imprinted architecturally – within the social-topographical matrix of urban Hereford's property grid. Here, where ecclesiastical and civic, devotional and entrepreneurial, lay and clerical lives intertwine, and where a cathedral city's streets themselves are active memorial 'loci', Reigate's memory endures. From Number 29 Castle Street it is only a short way to where, at its liturgical-architectural core, the cathedral goes about its perennial business of negotiating between matters mundane and transcendent. In its triangulation of cartographic, economic and spiritual, the arrangement balances well with the Registrar's world-view that Roger gleaned from Nicholas and put to good use down the years.

If Canon Roger's documentary mappings carry larger weight, as 'personal' texts that are revelatory of medieval urban cartographic culture in its thick social-encrustment, what may be made, finally, of Canon Gerald's Hereford Map – that 'European' cartographic hinge lying between NLI 700's Irish topographies, in material and testamentary pursuit of which we began? What of the map itself? In the vein of previous scholarship, the temptation is to chase down medieval Ireland's relations – culturally, politically, economically – to Hereford families, institutions and elites. To address this enormously complex history for the Mibbe and Breynton eras would augment what Geraldine scholars have done for NLI 700's topographical texts and manuscript context; such that – presumably – the laconic cartographic data of NLI 700 would, finally, compellingly, acquiesce.[84] Ultimately, though, my reading places less emphasis on matters Irish than on this map's always conceded, but always underplayed status *as a mappamundi*: its definitive condition less as a textual derivative (or descendant) than as a fragment, extract or portion of a document whose formal qualities and essential meanings lay *squarely within* the genre to which the great bulk of medieval maps

---

[84] For a provocative model, see John J. Thompson, 'Mapping points of West Midlands manuscripts and texts: Irishness(es) and Middle English literary culture', in *Essays in Manuscript Geography: Vernacular Manuscripts of the English West Midlands from the Conquest to the Sixteenth Century*, Wendy Scase (ed.) (Turnhout: Brepols, 2007), 113–28.

belong. Adopting such a viewpoint cancels out this map's usual fetishiza-
tion in terms of geographical scale, content or context – its presumptive
identity as a rendering 'of Europe', in a book 'on Ireland'. Reinforced
instead is the primacy of medieval cartography's signal transformative
process: from mundane/ephemeral into spiritual/eternal; from this-
worldly into next-worldly; from terrestrial striving to universal stability.
Such a movement defines Roger Breynton's documentary mappings, as
his paired CPR records of 1336 demonstrate, as surely as it characterizes
his cathedral's treasured map.

Effectively, then, both ecclesiastical *mappaemundi* and ecclesiastical
acta offer to transform available earthly capital (wealth, dominion,
scholarly achievement) into usable spiritual capital. It is more than
coincidental, surely, that those who produce and archive notarial maps –
Cathedral Treasurers, Residentiary Canons, Registrars: the likes of
Roger Breynton and Nicholas Reigate – occupy the same institutional
positions as those (Gerald of Wales, Richard of Haldingham) connected
with known *mappaemundi* at these sites. If this chapter has a bio-
cartographical arrival point, it is that all of Breynton's 'mappings' –
commissioned and encountered; notarial and graphic; from loose
Hereford Cathedral Muniments to Chapter Library manuscript (NLI
700) – partake in similar generic processes. Arguably, they might be seen
as contributing to a single, overarching, transformative undertaking, in
equal measure social, geographical and theological. If this is so, it may be
because the *mappamundi* at Thomas Cantilupe's Hereford shrine looms
so large in Roger Breynton's professional life, devotional career and
personal experience[85] – a situation fully in keeping with how the *mappamundi* genre dominates medieval cartographic discourse overall. Written
or drawn, all of Roger's maps are, in essence, Hereford Cathedral
maps.[86] The earlier and later cases of Gerald and Walter, his fellow
stakeholders in cartography at this English cathedral city, help us to
recognize this essential affinity.

---

[85] For more on the Hereford Map and its host cathedral, see Daniel Birkholz, 'Mapping
medieval utopia: exercises in restraint', *The Journal of Medieval and Early Modern Studies*
36 (2006), 585–618.

[86] Interestingly, although he concedes the 'extraordinary significance' and 'near universal
interest in producing' *mappaemundi*, Smail himself, *Imagined Cartographies*, 2, 6–7,
separates medieval graphic and textual mapping practices categorically – finding 'little
to indicate that the vast body of people . . . were at all aware of the many ways in which
maps configure the world'. Breynton occupies an opposite extreme.

11 Shifting geographies of antisemitism

Mapping Jew and Christian in Thomas of Monmouth's *Life and Miracles of St William of Norwich*

Kathy Lavezzo

The object of geographical inquiry in this chapter is not a conventional map or chorography but the earliest written blood accusation narrative, the *c.* 1155 tale by Benedictine monk Thomas of Monmouth of how Jews ritually tortured and killed a twelve-year-old boy, William of Norwich.[1] As its title indicates, Thomas's *Life and Miracles of St William of Norwich* explicitly asserts its status as a hagiography: an account of the sanctity of the boy 'martyr' William, who was killed in Norwich in 1145 and later venerated as a saint. About a decade after the youth's death, Thomas, a monk who became the sacristan of William's shrine, began working on the *Life*, which makes the case that William was a martyr of Jewish ritual violence. But, as critics such as Gavin Langmuir and Jeffrey Jerome Cohen have made clear, Thomas's text offers a myth of Jewish animosity toward Christians.[2]

While Thomas's text speaks in many and various ways to its historical moment; above all, it presents us with an antisemitic fantasy about Jewish wrongdoing. Crucial to that fantasy is its imagined urban geography, its stress on the spatiality of medieval Norwich, the city where the action of Thomas's narrative occurs.[3] Norwich is very much at issue in Thomas's

---

[1] Hannah Johnson provides an overview of how scholars have dated the composition of Thomas's book in 'Rhetoric's work: Thomas of Monmouth and the history of forgetting', *New Medieval Literatures* 9 (2008), 64, n. 2. All of the scholars mentioned by Johnson date the start of Thomas's composition of the *Life* as between 1150 and 1155. More recently, scholars have questioned this dating and pondered the prospect that the monk wrote not only the later books of Thomas's seven-book text but the entire volume at a later date, around the time of Becket's 1170 martyrdom.

[2] Gavin Langmuir, *Toward a Definition of Antisemitism* (Berkeley: University of California Press, 1990), 210–36; Jeffrey Jerome Cohen, *Hybridity, Identity and Monstrosity in Medieval Britain: On Difficult Middles* (New York: Palgrave, 2006); Johnson, 'Rhetoric's work', 63–91. For many years, scholars examined Thomas's text with an eye towards the question of its validity – of answering the question, in other words, of whether or not Norwich Jews committed the crime: Langmuir, *Definition of Antisemitism*, 210–11.

[3] The term 'antisemitism' is used in this chapter to refer to negative fictions about Jews and Judaism that emerged in the Middle Ages. While it postdates the medieval period, the term's reference to fantasies about Jewish identity that had at times real-life motivations

text, and especially its first two books, which describe William's murder and its aftermath. In his account of William's death, discovery and ultimate veneration as a martyr, Thomas takes his readers on a tour of the city and its environs, through narrow passageways leading to the domestic site of William's murder, outside of town to the nearby forest where his corpse is deposited, to the urban thoroughfares through which his body is taken in a great procession to Norwich Cathedral, where a requiem mass occurs followed by a burial in the Monks' Cemetery. Later the body is transported to the chapter house and then, to its final resting place, the cathedral itself.

The *Life* of William, in other words, demonstrates how, as Nicholas Howe, Denis Cosgrove and other critics have shown, literary texts provide us with mental mappings that tell us much about the imagined geography of locations such as the medieval city.[4] Thomas's text offers a revealing imagined geography of the city of Norwich and its intersection with other sites, including England, Europe and the whole of western Christendom. In what follows, I consider the geographies imagined in Thomas's *life* of William, not only for what they tell us about medieval concepts of the city and other sites, but especially for the role those imaginative mappings play in the production of identities. As critics such as Patricia Yaeger point out, imagined geographies allow for a particularly nuanced consideration of identity formation.[5] As much emerges in Dan Birkholz's contribution to this volume, in which analysis of the 'life-map' of Roger de Breynton leads to the destabilization of 'lay/clerical, financial/spiritual, civic/ecclesiastical, local/universal' and still other oppositions commonly applied to medieval culture. Similarly, close analysis of Thomas of Monmouth's imagined geography of Norwich, England and the Christian West offers a means of problematizing the opposition of Jew and Christian.[6]

At the forefront of Thomas's antisemitic geography are two architectural sites: the Jewish house where William is killed, and the Christian church where his corpse ultimately rests. Through those structures and

and, unfortunately, real-life anti-Jewish consequences applies to medieval English representations of Jews. See Anthony Bale, 'Fictions of Judaism before 1290', in *The Jews in Medieval Britain: Historical, Literary, and Archaeological Perspectives*, Patricia Skinner (ed.) (Woodbridge, UK, and Rochester, NY: Boydell and Brewer, 2003,) 129–44, at 129.

[4] Nicholas Howe, *Writing the Map of Anglo-Saxon England: Essays in Cultural Geography* (New Haven: Yale University Press, 2008); Denis Cosgrove (ed.), *Mappings* (London: Reaktion, 1999).

[5] Patricia Yaeger, 'Introduction: narrating space', in *The Geography of Identity*, Patricia Yeager (ed.) (Ann Arbor: University of Michigan Press, 1996), 1–38, at 15. On the intersection of Jew and Christian in medieval culture, see Bale.

[6] See Birkholz, ch. 10, this vol.

related spaces, Thomas distinguishes Jews from Christians by associating the former with private and profane spaces and the latter with communal and sacred locations: in stark contrast to the 'dangerous' Jewish home where William secretly is killed is Norwich cathedral, the public site of William's veneration.[7] This chapter aims not merely to map out such literary oppositions of Jewish and Christian spaces but also to probe how they are inadvertently or intentionally queried and undermined. That is, in Thomas's 'city text', a Christian church does not simply contrast with but surprisingly is similar to a Jewish home. It is precisely through such ironic resemblances that Thomas shifts from demonizing Jews to acknowledging problems in his own Christian society.

In its interplay of public and private urban spaces Thomas's text thus negotiates not only the recent arrival of the Jewish diaspora to his city, but also new challenges posed by Norwich itself as one of many medieval cities undergoing massive changes during the twelfth century. Among the larger concerns informing this tale of covert Jewish operations is the question of what any city dweller does under the cover of a dark alley or behind a closed door. Conversely, open spaces emerge in Thomas's text less as an unproblematic alternative to the urban underworld than as its equally threatening obverse. Ultimately, analysis of the cultural geography of Thomas's text confirms how, even if the Jews left Norwich (which they eventually were forced to do, in 1290), problems linked to the spatiality of that city – to its commercial mappings and flows, to the public and private circulations of its inhabitants – would remain.

Augustine, diaspora and shifting medieval geographies of antisemitism

Thomas's urban, national and global 'mappings' of Jewish and Christian identities reflects the negative view that Christians notoriously began to assume towards Jews around the turn of the twelfth century. Before that time, the single most influential take on Jews and their spatiality was that found in Augustine of Hippo's writings. Central to Augustine's conception of Jews is their exile from Israel and consequent dispersal into foreign lands. Born of repeated historical catastrophes, the diaspora that

---

[7] That opposition of Jewish private space and Christian public space figures not only in Thomas's text but also serves as a kind of deep structure for a wide array of antisemitic literatures in England, from Marian miracle tales like the Jewish boy and Chaucer's *Prioress's Tale*, to host desecration narratives such as that performed in the *Croxton Play of the Sacrament* and post-Reformation dramas like Marlowe's *Jew of Malta*. This chapter comprises part of a larger book project that analyses the interplay of Jewish and Christian space in those and related narratives.

served as a source of deep suffering for Jews also proved troubling for Christians in ways that similarly pertained to the relationship between place and identity. By, for example, settling in Christian lands and thus rendering those territories at once Christian and Jewish, Jews prevented Christians from enjoying a coincidence of territory and identity, from possessing a place that they could claim as theirs alone. Compounding the problematic presence of the Jews in Christian lands was their fraught temporal status as a people out of whom Christianity grew – Jesus, after all, was Jewish – who, however, rejected that system of belief. Arguably the most charged religious others of the medieval Christian West, Jews confronted the Christians they encountered with a disturbing paradox: the sheer ongoing presence of God's original chosen people who nevertheless denied Christ's divinity.

Augustine registers the wide reach of the diaspora in Christendom, observing of the Jews in *The City of God* 'that there certainly is nowhere in the world where they are not present', and that they 'are scattered throughout all the nations, wherever the Christian Church extends itself'.[8] But, instead of viewing Jews as others who disturbingly mix and mingle with Christians, Augustine renders them reassuring symbols of Christian identity. According to his influential doctrine of witness, by exiling and scattering the Jews, God punishes them (by depriving them of a stable dwelling and of liberty) and provides Christians with living embodiments of the Old Testament and its prophecies regarding Christ. Blind to the manner in which their religious books affirm the ancient roots of the Christian message, dissenting Jews become in Augustine's thought less sentient beings than an insensible form of providential technology. The global scattering of the Jews constitutes in effect God's universal advertising campaign. 'For if that testimony of the Scriptures existed only in the Jews' own land,' writes Augustine, 'and not everywhere, then, clearly, the Church, which is everywhere, would not have it to bear witness in all nations to the prophecies which were given long ago concerning Christ.'[9] Jews are omnipresent, public billboards proclaiming the Christian message, 'milestones' who 'point the way to travelers walking along the route' to salvation who themselves remain 'inert and unmoving' until their final conversion during the Last Judgment.[10]

---

[8] *De Civitate Dei*, 18.46, Corpus Christianorum Series Latina (CCSL) 48:644; *The City of God against the Pagans*, R. W. Dyson (tr. and ed.) (Cambridge University Press, 1998), 891–2. My discussion of Augustine is indebted to the incisive account of his writings on Jews in Jeremy Cohen, *Living Letters of the Law: Ideas of the Jew in Medieval Christianity* (Berkeley and Los Angeles: University of California Press, 1999), 19–67.

[9] *De Civitate Dei*, 18.46, CCSL 48:644; *City of God*, 892.

[10] Augustine, *Sermo* 199.1.2, PL 38:1027, *The Works of Saint Augustine: A Translation for the 21st Century*, pt. III, vol. 6: *Sermons*, Edmund Hill (tr.), John E. Rotelle (ed.)

Augustine's doctrine of witness enjoyed currency among Christians until the 'long twelfth century', when relationships between Jews and Christians began to worsen.[11] Emblematic of that shift were the tragic pogroms that occurred in 1096 during the First Crusade.[12] The massacre of whole communities of Rhineland Jews by European crusader knights and local commoners horrifically evinces how the Augustinian plea for tolerance no longer held sway for certain Christians.[13] The question of why that unprecedented animosity towards Jews erupted during this period in the history of the medieval West has received much attention in recent decades by scholars including Robert Chazan and R. I. Moore. Such scholarship has pointed to a variety of reasons for that anti-Jewish behaviour, among them the racialist elements of Urban II's crusade rhetoric and socio-economic tensions between Christians and their Jewish neighbours. According to Chazan, the primary reason for the pogroms lies in the emergence of an extremist version of crusader doctrine that saw addressing Jewish unbelief as a core component of the crusader ideal. Geography played an essential role in that radical revision of crusader ideology. Jews, their attackers claimed, were, like the Muslim objects of crusader warfare, enemies of Christianity. But unlike Saracens, who lived far away in Jerusalem, Jews were spatially proximate to European Christians. Answering the call to, as the monk Ralph put it in a sermon for the Second Crusade, 'avenge the Crucified

---

(New Rochelle, NY: New City Press, 1993), 80. In that sermon, which is for the epiphany, Augustine is specifically referring to the directions to Jerusalem given by the Jews to the Magi. On the manner in which this passage exemplifies one of the six arguments offered by Augustine to explain the function performed by the Jews, see Cohen, *Living Letters*, 35–6.

[11] Among the many important works written about the rise of intolerance towards not only Jews but also other 'outsiders' in the medieval West, see: R. I. Moore, *The Formation of a Persecuting Society: Authority and Deviance in Western Europe, 950–1250* (New York: Blackwell, 1987); John Boswell, *Christianity, Social Tolerance, and Homosexuality: Gay People in Western Europe from the Beginning of the Christian Era to the Fourteenth Century* (University of Chicago Press, 1980); David Nirenberg, *Communities of Violence: Persecution of Minorities in the Middle Ages* (Princeton University Press, 1996).

[12] As Robert Chazan points out, we should avoid making a 'facile contrast' between the lives of Jews before and after the First Crusade: Robert Chazan, *European Jewry and the First Crusade* (Berkeley: University of California Press, 1987), 203. The 1096 attacks did not initiate a sudden downturn for European Jews, who in many respects flourished in economic, demographic and cultural terms during the twelfth century. According to Chazan, it is only at the end of the thirteenth century that extensive and disastrous violence began to occur, in a context that in many respects differed from that of 1096.

[13] See Chazan, *European Jewry*; and Robert Chazan, *God, Humanity, and History: The Hebrew First Crusade Narratives* (Berkeley: University of California Press, 2000); J. C. Riley-Smith, 'The First Crusade and the persecution of the Jews', *Studies in Church History* 21 (1984), 51–72; J. C. Riley-Smith, *The First Crusade and the Idea of Crusading* (London: Athlone, 1986).

upon his enemies who dwell among you', the crusaders attacked their Jewish neighbours.[14] Crusading, the attackers believed, begins at home, with the Jewish enemies in their midst.

On the question of precisely why Jews constituted enemies of Christ, crusader extremists cited the deicide allegation, as well as Jews' ongoing rejection of Christian doctrine. But it is not until Thomas of Monmouth's pseudo-hagiography, composed some sixty years or more after the First Crusade, that we find the earliest extant myth of Jewish conspiracy.[15] Versions of the ritual murder fantasy would provide in subsequent decades a spurious rationale for anti-Jewish attacks in England, most notoriously, the state-sponsored execution of nineteen people for their supposed involvement in the 'martyrdom' of Hugh of Lincoln in 1255.[16] Within *The Life and Miracles of St William of Norwich*, Thomas's ritual murder assumes its most explicit form when the monk offers the fifth in a series of pieces of 'evidence' affirming that Jews killed William. In testimony from one 'Theobald, who was once a Jew, and afterwards a monk' (*Theobaldo, quondam iudeo et monacho postmodum*), Thomas offers a tale of Jewish conspiracy as fraudulent and offensive as the modern era's Protocols of the Elders of Zion.

in the ancient writings of his fathers it was written that the Jews, without the shedding of human blood, could neither obtain their freedom nor could they ever return to their fatherland. Hence it was laid down by them in ancient times that every year they must sacrifice a Christian in some part of the world to the most high God in scorn and contempt of Christ, that so they might avenge their sufferings on Him; inasmuch as it was because of Christ's death that they had been shut out from their own country and were in exile as slaves in a foreign land. Wherefore the chief men and Rabbis of the Jews who dwell in Spain assemble together at Narbonne, where the Royal seed resides, and where they are held in the highest estimation, and they cast lots for all the countries which the Jews inhabit; and whatever country the lot falls upon, its metropolis has to carry out the same method with the other towns and cities, and the place whose lot is drawn has to fulfill the duty imposed by authority. Now in that year in which we know that William, God's glorious martyr, was slain, it happened that the lot fell upon the Norwich Jews, and all the synagogues in England signified, by letter or by message, their consent that the wickedness should be carried out at Norwich.[17]

[14] Shlomo Eidelberg (tr.), *The Jews and the Crusaders* (Madison: University of Wisconsin Press, 1977), 121; Chazan, *European Jewry*, 170.

[15] See Johnson, 'Rhetoric's work', 65, n. 4.

[16] See Geraldine Heng, 'Jews, Saracens, "Black Men", Tartars: England in a world of racial difference', in *A Companion to Medieval English Literature and Culture, c. 1350–c. 1500*, Peter Brown (ed.) (Malden, MA: Blackwell, 2007), 252–5.

[17] Cambridge University Library Ms. Add. 3037, f. 26v; Thomas of Monmouth, *The Life and Miracles of St. William of Norwich*, A. Jessop and M. R. James (trs. and eds.) (London: C. J. Clay; Cambridge University Press Warehouse, 1898), 93–4. Hereafter

Through Theobald's fictional 'report', Thomas affirms Augustine's stress both on exile as a punishment and on the geographic omnipresence of Jews. Thomas indicates the spread of the Jews from their erstwhile home of Jerusalem to the western territory of Occitania and far western terrain of England. More precisely, Thomas reflects how the problem of diaspora engaged by Augustine in the late fourth century had come home to twelfth-century Norwich.

But other aspects of the report starkly differ from Augustine's conception of Jewish dispersal. Instead of signifying the Old Testament prophecies, Thomas's Jews are locked into a twisted reading of the New Testament or, more precisely, the idea of Jews as Christ-killers that medieval Christians teased out from the Gospels.[18] Far from this being a universal ad campaign that publicizes the supersession of Judaism by Christianity, the Jews scattered throughout the world annually '[sacrifice] a Christian ... in scorn and contempt of Christ', thanks to a well-organized communication network that systematically links multiple geographic scales, from the nodal capital of Narbonne, to country, metropole, city and town. If Augustine's diaspora publicly circulates Christian meanings that are concealed from their Jewish bearers, Thomas's diaspora describes a network of conniving persons whose machinations are hidden from their Christian neighbours. If Augustine's witness doctrine recommends a kind of tolerance, Thomas's myth of Jewish conspiracy and animosity urges ridding Christendom of the enemy in its midst.

When Thomas began writing, in the mid-1150s, Jewish communities, which first entered England shortly after the Norman Conquest, had been living in Norwich for about twenty years, rendering Augustine's claim that they 'are not lacking anywhere' palpable to Christians such as Thomas.[19] Norwich during the mid-twelfth century was a radically different city than the Anglo-Scandinavian borough that existed prior to 1066. As James Campbell puts it, 'Norwich was more changed by the Norman Conquest than it had ever changed before (in so short a time);

Life. All citations and translations adapted from this edition, in consultation with Miri Rubin's transcription of the manuscript (http://yvc.history.qmul.ac.uk/passio.html). On the Protocols of the Elders of Zion, see Stephen Eric Bronner, *A Rumor about the Jews: Reflections on Antisemitism and the Protocols of the Learned Elders of Zion* (New York: St Martin's, 2000).
[18] On the history of the deicide allegation as well as Augustine's complex relationship to the claim that the Jews killed Christ, see Jeremy Cohen, *Christ Killers: The Jews and the Passion from the Bible to the Big Screen* (New York: Oxford University Press, 2007).
[19] V. D. Lipman, *The Jews of Medieval Norwich* (London: Jewish Historical Society of England, 1967), 4–5.

or than it was ever to change again.'[20] Many of those changes affected Norwich's urban landscape. Not long after 1066, the new Norman elite initiated a brutal colonial reconstruction programme that entailed razing native Anglo-Scandinavian homes and churches to make way for: a new borough for the *Franci de Norvic* or French of Norwich, established between 1071 and 1075; a new stone cathedral priory, whose main building was completed by 1148; and a new stone castle, completed by *c.* 1120.[21] Campbell cites the new French borough, cathedral priory and castle as three of four interlocking elements of the 'revolution' that took place in post-Conquest Norwich; the fourth element identified by Campbell is the Jews.[22] Forming less than 1 per cent of the town's population, the Jewish community of Norwich nevertheless constituted a visible presence in the reconfigured town, largely because of their habitation of neighbourhoods near the market located in the new French quarter, 'in the most populous part of the city'.[23] The Jews' habitation of the commercial centre of Anglo-Norman Norwich – the new market superseded the pre-Conquest market at Tombland – reflects their vital role in the economic life of the city.[24] The geography of Jewish settlement probably served for the native Anglo-Scandinavian inhabitants of Norwich as, in Jeffrey Jerome Cohen's words, a distressing 'visual reminder of the shift in the city's economic and social gravity'.[25]

In a move that reflects the general pattern of urban Anglo-Jewish habitation during the Middle Ages, the Jews of Norwich lived near the marketplace, the source of their livelihood, and also within close reach of the castle, where they could turn to seek protection and rights from the crown and its representatives (Fig 11.1).[26] Compounding the problems posed by the presence of Jews in the privileged centre of Norwich was the proximity of their homes to those of Christians. As Vivian Lipman warns, 'one must not fall into the error of thinking of the Jewish quarter as a closed ghetto: there is some evidence for Jews ... living elsewhere

---

[20] James Campbell, 'Norwich before 1300', in *Medieval Norwich*, Carole Rawcliffe and Richard Wilson (eds.) (London: Hambledon and London, 2006), 39.

[21] Campbell, 'Norwich before 1300', 39–41. See also Keith D. Lilley, *City and Cosmos: The Medieval World in Urban Form* (London: Reaktion, 2009), fig. 58.

[22] Campbell, 'Norwich before 1300', 39.   [23] Lipman, *Jews of Medieval Norwich*, 3.

[24] While not supplanted altogether by the new market, the Anglo-Scandinavian market at Tombland ceased to function as the main commercial centre of Norwich during the twelfth century: ibid. 16–17.

[25] Cohen, *Hybridity*, 159.

[26] On the central location of Jewish communities in other medieval European towns, see Keith Lilley, *Urban Life in the Middle Ages 1000–1450* (New York: Palgrave Macmillan, 2002), 246–8.

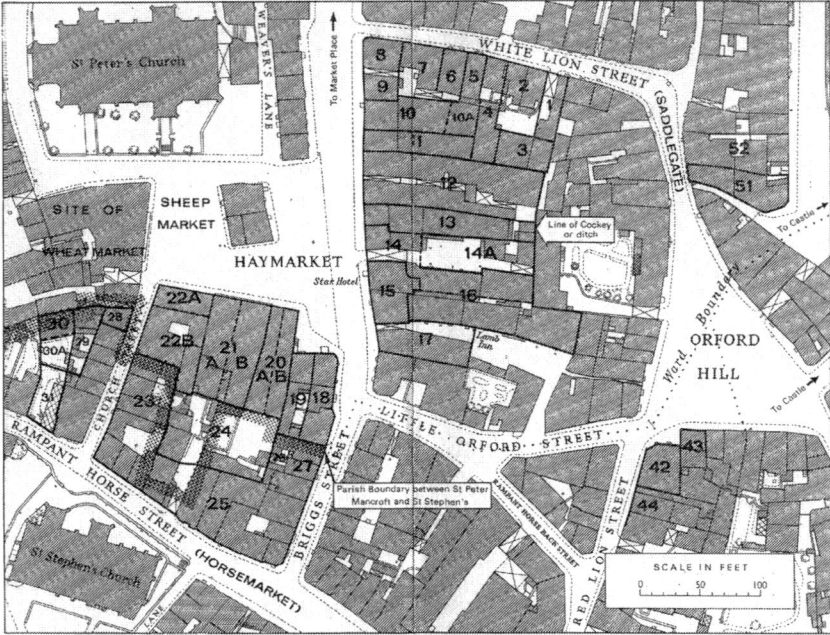

Figure 11.1 Section of Norwich Jewry. Numbers 1–2 and 9–17 refer to
properties owned by Jews; numbers 4–8 refer to properties owned by
Christians.
Source: V. D. Lipman, *The Jews of Medieval Norwich* (London: Jewish
Historical Society of England, 1967). Reproduced with kind permission
of the Jewish Historical Society of England.

in the city and there is plenty of evidence of Christians living side by side
with Jews in what was in fact only a partly Jewish quarter'.[27] Other Jews
lived beyond the market. For example, the prominent financier Jurnet
owned a home about 500 metres south-east of the market on King Street
near the river Wensum. Scattered throughout Norwich, contiguous
to their Christian neighbours, the Jews presented on an urban scale a

---

[27] For example, during the 1250s and 1260s, medieval properties located east of
Haymarket and south of Saddlegate (now White Lion Street) included, starting from
the corner of what is now Gentleman's Walk and White Lion Street, five properties
owned by Christians, followed by two properties owned by Jews, while to the south of the
Christian-owned corner property lay several Jewish-owned homes: see Lipman, *Jews of
Medieval Norwich*, 116–20, fig. 6.

distressing version of their global diaspora. Omnipresent but not exiled from the 'Jerusalem' or centre of the city, the Jews of medieval Norwich thus more than made Augustine's claim that they 'are not lacking anywhere' palpable to Christians such as Thomas of Monmouth.

Urban thresholds and geographies of suspicion

If the Jewish community of Norwich was small yet geographically open and prominent, the Jews of Thomas of Monmouth's Norwich are imagined in altogether different spatial terms. Central to Thomas's antisemitic discourse are notions of place, whereby the hidden and bounded nature of Jewish homes signifies Jewish hostility to Christians. This geography of suspicion has its narrative dimensions. As Langmuir points out, the *Life* is in many respects a detective story bent on tracking down and uncovering – in at times a voyeuristic manner – the nefarious activities that purportedly transpired in the homes of Jews.[28] As much emerges early in Book One, when Thomas recounts 'evidence' gathered by two female 'detectives'. The first is William's female cousin, who is sent by her mother Liviva to trail the boy and his abductor. The girl

ran out to explore the way they were going; and she followed them at a distance as they turned about through some private alleys, and at last she saw them entering cautiously into the house of a certain Jew, and immediately she heard the door shut. When she saw this she went back to her mother and told her what she had seen.[29]

Here as elsewhere, Thomas doesn't name the particular streets in which the Christian and Jewish persons depicted in the *Life* live. But we might surmise that Liviva's willingness to allow her daughter to wander the streets of Norwich, as well as the absence of any comment on how far the girl travelled, reflect the proximity of Jewish and Christian homes in the city. At the same time, Thomas asserts the separation of Christians and Jews by situating the latter in a closed domicile in a private alley. Later, in Book Two, Thomas underscores the bounded nature of the Jewish home when he reminds us that 'the little girl, [William's] kinswoman ... actually saw the door shut behind him', and offers additional evidence from a Christian maid employed at the Jewish home.[30] While passing a pot of boiling water into the place of William's torture, the maid 'through the chink of the door managed to see the boy fastened

---

[28] Langmuir, *Definition of Antisemitism*, 209.   [29] *Life*, 19; MS Add. 3037 6rb.
[30] *Life*, 89; MS Add. 3037 25v.

to a post. She could not see it with both eyes,' Thomas informs us, 'but she did manage to see it with one.'[31]

Thomas's account of urban intelligence-gathering places special weight on the door, whose closure the girl experiences through dual sensory registers of sight and sound, and whose slight opening enables the maid to glimpse the crime in process. Given the pattern of Jewish settlement, it's no surprise that the *Life* stresses doors. In the absence of substantial geographic boundaries separating Jews and Christians, domestic thresholds gained a special resonance. As Mary Douglas observes, thresholds such as building entrances or – shifting scales downward – bodily entrances like mouths are charged and vulnerable sites symbolizing 'points of entry or exit to social units' in a culture.[32] We can glean how medieval people viewed doors as apertures enabling inward and outward flows by looking at the entry on the mouth in Isidore of Seville's *Etymologies*, which states that 'the mouth (*os*) is so called because through the mouth as if through a door (*ostium*) we bring food in and throw spit out, or else from that place food goes in and words come out'.[33] Entities whose spatial affinities appear in the classical Latin *ostium*, which signifies door and also any kind of entrance, doors and mouths threaten destabilizing circulations and mixings.[34] Indeed, in Thomas's text not only doors but also mouths abound. To name only two of many salient instances, the Jews attack William while the unsuspecting youth is eating, and his torture commences with the placement of a painful gag in his mouth.[35] Thomas's text turns to the boundaries of both the Christian body and the Jewish domicile in order to produce and fix Jews as demonized others. While mouths and doors could serve as conduits of hospitable connection (i.e. the entry through a door of a home to share, say, a meal or an amicable exchange of words, or to work out the terms of a loan), here they indicate hostility.

---

[31] *Life*, 89–90; MS Add. 3037 25v. The status of the door as a charged threshold emerges also in Thomas's recounting of the Jews' eventual departure from the house, when they 'threw open their doors, and free entrance was granted to any that came' (*undique hostiorum patefactus est introitus, ac liber deinceps cuilibet datus est accessus*) enabling the maid to investigate the scene of the crime while performing her usual household work: *Life* 91; MS Add. 3037 26r. I am indebted to Jeffrey Jerome Cohen, who, while corresponding to me about the spatiality of medieval antisemitic literature, first alerted me to Thomas's text and its stress on the Jewish domicile.

[32] Mary Douglas, *Purity and Danger: An Analysis of Concepts of Pollution and Taboo* (London: Ark, 1984), 4.

[33] *Etymologies* XI.i.49; Stephen A. Barney, W. J. Lewis, J. A. Beach and Oliver Berghof (trs. and eds.), *The Etymologies of Isidore of Seville* (Cambridge University Press, 2006), 234.

[34] P. G. W. Glare (ed.), *Oxford Latin Dictionary* (Oxford: Clarendon Press, 1982), s.v. *ostium*.

[35] *Life*, 20; MS Add. 3037 51r.

Mirroring the probable exclusion of Jews from the Benedictine hospitality provided by the Norwich cathedral precinct, Thomas imagines those religious outsiders as similarly inhospitable. William, a poor English boy who moves from the country to the city to develop his skills as a leather-worker, is lured by the promise of a better job into the Jewish domicile, where he is then fed dinner. But what ultimately occurs in the home – and what ultimately enters William's mouth – neither nourishes the boy's body nor fosters his commercial ascent. In the case of William's gag, it is worth noting that Thomas tells us that the device used by the Jews to stop up William's mouth is *uulgo Teseillun dicitur* or 'called by the commoners a teasel'.[36] In Norwich's vital cloth-making industry, teasels – a kind of prickly thistle – were cultivated for use by fullers to raise the nap on cloth.[37] By having the Jews use a teasel not for its intended manufacturing use but as an instrument of torture, Thomas imaginatively inverts their historical relationship to the Norwich economy. The Jews of medieval Norwich held various occupations, and worked as educators, physicians, household servants, vintners, cheesemongers, fishmongers, traders, pawnbrokers, and, above all, financiers.[38] As Jeffrey Jerome Cohen affirms with respect to that final role, as moneylenders, Jews were 'the lifeblood of Norwich's commercial prosperity'.[39] But Thomas's account of William's murder suggests something altogether different. The urban population that infused capital into that city, thereby creating and stimulating economic flows in Norwich, is depicted as gagging the mouth of one of the city's young workers.

The door is similarly constitutive of Jewish animosity. When he describes how William's cousin both hears and sees the door close and how the maid peers into the room when the door is momentarily ajar, Thomas urges his readers to imagine the spatial phenomenology of Jewish aggression: the sealing off of the Jewish domicile from Christian eyes and ears in the manner of a barricade dividing one enemy camp from another.[40] That conception of the door as an impediment emerges

---

[36] Ibid.

[37] Evidence of the importance of the cloth trade in Norwich emerges in how that city and other English cities paid 'fines in 1202 to avoid complying with the regulations concerning the size of cloth': Penelope Walton, 'Textiles', in *English Medieval Industries: Craftsmen, Techniques, Products*, John Blair and Nigel Ramsay (eds.) (London and Rio Grande: Hambledon Press, 1991), 347. She writes that 'Teasels that "longyn to the office of fullers" were cultivated on a large scale from at least the early-thirteenth century in England': Walton, 'Textiles', 332.

[38] Lipman, *Jews in Medieval Norwich*, 79–94.   [39] Cohen, *Hybridity*, 158.

[40] In the case of the simultaneous closing of the Christian maid's eye with the Jewish door, Thomas indicates how the desire to separate themselves is mutual on the part of Jews and Christians.

elsewhere in medieval culture. In his entry on doorways, Isidore stresses how a door can impede, and plays on the Latin for doorway (*ostium*) and impeding (*obstare*). Isidore, moreover remarks upon the martial valences of doors, writing 'Others say doorway is so called because it detains an enemy (*ostis*, i.e. *hostis*) for there we set ourselves against our adversaries.' In that passage, Isidore plays on the similarities between the Latin terms for door, *ostium*, and *hostis*, meaning foreigner or enemy, to indicate the adversarial relationship demarcated by the object.[41] That slippage between architecture and aggressivity, between domestic seclusion and nefarious action, is realized, perhaps, in Thomas's medieval Latin, which spells door as *hostium* and repeatedly refers to the Jews as the *christiani nominis hostes* or enemies of the Christian name.[42] More generally, Thomas's stress both on doors and on dark alleys shores up the status of the medieval city as a site with a dark underbelly, a place that attracts a criminal element engaged in spurious actions in secluded locations. Such an anxiety over secrecy and private life may well have been endemic to medieval cities, where diverse populations gathered and cohabitated in close quarters.

This is not to say that Thomas's text ignores the public side of Norwich, far from it. Events such as judicial proceedings, ecclesiastical synods, and Christian rituals bring to life the civic aspects of the city. But among all such representations, Thomas's depictions of crowds especially connote Norwich's public side. On the face of it, crowds provide a corporate notion of Christian identity that opposes images of Jewish secrecy and criminality. Consider Thomas's first reference to crowds, where he tells us that after killing William, his abductors are forced to delay transporting the boy's corpse to Thorpe Wood because 'That day was the Absolution day [Holy Thursday], on which the penitents of the whole diocese were accustomed to assemble in crowds in the Mother Church at Norwich, and the streets of the whole city were crowded with an unusual multitude of people walking about'.[43] Having dwelt extensively upon the sordid details of William's torture in the Jewish domicile, Thomas opens his text outward, to the open thoroughfares or *platée* of Norwich. Thomas very likely alludes in that passage to Tombland (Scandinavian for 'open land'), just west of the Cathedral Close and the site of the old Anglo-Scandinavian marketplace over which the monks

---

[41] *Etymologies* XV.vii.4, 311.
[42] Cf. *Life*, 28, 44, 63, 71; MS Add. 3037 8v, 13r, 18v, 20v. It is impossible, however, to confirm Thomas's intentions, since medieval writers often would add an h to Latin words which began with an o.
[43] *Life*, 26; MS Add. 3037 8r.

now asserted their jurisdiction.[44] In a move that may reflect the Norman Christian remapping of Tombland, Thomas depicts both that open area and the rest of Norwich's streets as the setting for the performance of Christian ritual. Claiming that an exceptionally large group of Christians (*plus multitudine*) take up the city's streets, Thomas affirms the dominant religious identity of the city over and against that of its Jewish minority. Through its sheer size, the Christian public of Norwich opposes Jewish mobility and intentionality, preventing the Jews of the city from leaving their homes and taking the next step in their criminal plan.

Thomas's use of crowds to distinguish Christians from Jews culminates in his account of William's funeral in Norwich Cathedral and the subsequent interment of the boy's corpse in the Monks' Cemetery. Thomas tells us that when the monks sent by Bishop Eborard returned from the forest, where they had retrieved William's corpse,

so vast a concourse of the common people met them that you would have thought very few had stayed behind in the city. So that precious and exquisite treasure was carried with immense delight of clergy and people and brought by the venerable convent of the monks with a procession and introduced into the Cathedral Church and placed with its bier before the altar of the Holy Cross ... The mass of requiem was solemnly sung by the monks and the whole Church was filled from end to end with the crowds of citizens.[45]

The crowd attending the funeral infuses that Christian ritual with a public nature that contrasts with the secrecy surrounding his murder. The boy's abduction involved clandestine movements in dark alleys, but his translation entails a procession in broad daylight. While a few Jewish men torture the boy in the seclusion of a home, a multitude of monks sing William's requiem mass before an audience so great that it 'fills the whole Church from end to end'; later, during William's burial, we are told that 'the cemetery was filled by thousands of men' so that 'the area was hardly large enough for those who kept coming in.'[46] The opposition of public and private space, of Cathedral and home, of ritual veneration and ritual murder, all set the Christian inhabitants of Norwich apart from their Jewish counterparts. The supersession of Judaism by Christianity emerges in the sheer public spectacle of a multitude of Christians revering a dead boy. The very scale of Christian practice, heightened all the more by its public performance in and around

[44] Brian Ayers, 'The urban landscape', in *Medieval Norwich*, Rawcliffe and Wilson (eds.), 1–28 at 5, 14.

[45] *Life*, 50–1; MS Add. 3037 15r-v.

[46] *Impletur cimiterium milibus hominum alio de latere per portam introeuntium et intrantibus uix loci iam sufficiebat capacitas*: *Life*, 54; MS Add. 3037 16v.

Norwich cathedral, overshadows the isolated, diminutive enactment of ritual murder in the close confines of a Jewish home.

Between the universal church and the crowd of local worshippers: Thomas's cenobitic geography

Thomas's celebration of Christian collectives might reflect the larger historical moment in Church history of which he is a part. The twelfth-century witnessed an unprecedented wave of consolidation and codification of power on the part of the western church.[47] From the time of the Conquest to the end of the twelfth century, the papacy gradually extended its influence in England, heightening the sense that Christians belonged to a large community united by their shared relationship to Rome.[48] Particularly given the important role played by monasticism in bolstering the universal church, Thomas's emphasis on the sheer size of the Christian community in Norwich might reflect the monk's investment in the totalizing might of the universal church within the world. But a passage a bit later in the *Life* reveals Thomas's striking resistance to the authority of the universal church. At the opening of the second book of the *Life*, Thomas responds to a critic who claims 'it is very presumptuous to maintain so confidently that which the church universal does not accept and to account that holy which is not holy':

if they are accused of presumption who keep up the memory of saints whom the whole world does not know, or whom the church universal does not celebrate, you will find very few or none who will not incur the same blame. And to say the truth, saving only the glorious Virgin mother of God and John the Baptist and the Apostles, of few of the saints can it be said that the knowledge of them is spread abroad over all the earth whereon the religion of the Christian name prevails. In truth is it the fact that all those whom Rome herself honours Gaul and Britain accept as equally worthy of renown? Is it the fact that the famous name of the most blessed King and Martyr Eadmund or of the glorious Confessor Cuthbert, renowned in every part of England, is equally well known among the people of Greece or Palestine? Or, to sum it up, in the case of those whom Asia or Africa counts as famous, does all Europe pay them all a customary reverence?[49]

---

[47] The Church, of course, wasn't the only area of medieval society to become more authoritarian and centralized. Secular government, in England and elsewhere, also consolidated its power during the period. For two accounts of religious and secular consolidations as they pertained to the persecution of Jews and other disenfranchised persons, see Boswell, *Christianity*, and Moore, *Persecuting Society*.

[48] See Charles Duggan, 'From the Conquest to the Death of John', *The English Church and the Papacy in the Middle Ages*, C. H. Lawrence (ed.) (New York: Sutton Publishing, 1999), 63–116.

[49] *Life*, 59–60; MS Add. 3037 17v.

Here, Thomas addresses the problem posed by the fact that Rome has not yet canonized William with a striking account of Christian heterogeneity and geographic separation. Far from offering a picture of a united Christendom, Thomas describes a church in which only a few saints enjoy universal renown and the vast majority of holy men and women receive more local forms of veneration. This section represents one of two instances in the *Life* when Thomas's narrative assumes a distinctly global perspective, the other moment emerging in Theobald's testimony on the Jewish cabal. In the same way that the converted Jew's testimony conjures a map of the medieval world, from its holy centre of Jerusalem to its far western border in England, Thomas's account of the cult of saints creates a verbal *mappa mundi*, complete with its three continents of Asia, Africa and Europe.[50] But in stark opposition to the organized communication network of Jews, the Christian cult of saints lacks an effective means of spreading information.

Querying notions of a unified *universalis ecclesia* or universal church over the vast territories of Christendom, Thomas lays stress on the absence of concord or even effective communication when it comes to the matter of venerating saints. While Norwich in the year 1244 becomes a charged place for a global Jewish cabal that is at once cognizant of and amenable to the selection of William as its victim, both the boy and his city enjoy no such connection to the universal Christian church. All Jews agree to kill William, but not all Christians venerate or furthermore are aware of him. Thomas conjures an image of the Christian world in which geographic distance prevents the spreading of a saint's reknown from one continent to another.

What are we to make of Thomas's association of a global communication network with Jews but not Christians, particularly given the centralization of the Church during the monk's lifetime? The Jewish cabal, we might speculate, serves in the *Life* as a kind of dark double for the universal church, through which the monk can register his own ambivalence over forms of consolidated authority. Thomas's account of hagiographic practice suggests his investment in geographic localism as opposed to geographic universalism, his lack of interest in the idea of Rome extending its power throughout Christendom. His mythic Jewish cabal suggests an even more critical perspective on universal authority on the part of the monk, who implies that global forms of power are in some way antithetical to Christian practice. Thomas suggests as much when he writes of the sceptic seeking Rome's approval that 'a stony hardness

---

[50] See Lilley, 'Introduction', this vol. See also Figs. 1, 2, 13, this vol.

besets the passages of his brain' and that 'this is the way that a blind man strikes out.'[51] For Augustine and other Christian writers, contemporary Jews blindly and stubbornly resist the Christian message. Stubbornness and blindness, in other words, famously comprise Jewish insufficiency. Thomas's sceptics thus become Jewish in their insistence upon papal authority, and instead of opposing the Jewish cabal to its universal Christian counterpart, Thomas contrasts an implicitly Jewish worldliness with local and national forms of Christian community: with national saints Eadmund and Cuthbert, who are 'renowned in every part of England' but not elsewhere in Christendom, and with the new saint William, through which God has shown his mercy to 'the parts about Norwich, or rather to the whole of England'.[52]

While William's murder is produced via an international global network of widely *scattered* Jews, the Christian veneration of William involves the dense geographic *convergence* of persons in a single urban location in England. We can understand Thomas's regionalism partly in terms of the unique ability of dense urban groupings to provide a visible spectacle of concord, to publicize a united show of force. As Jeffrey Jerome Cohen affirms in his important reading of Thomas's text, the monk's ideological programme becomes especially clear in those moments when he imagines all of the disparate elements of the Norwich Christian population disappearing as they unite over William's martyrdom. The boy's funeral offers the most salient of such instances. Of the throng of clergy, monks and laity present, Thomas proclaims: 'though they who were present differed in grade and in sex, they were all of one mind in wishing to see the sight.'[53] It is of course a geographic and logistical impossibility for the whole of Christendom to gather together to perform its rituals as one body. But, for Thomas, it is conceivable that on the smaller scale of a single city such as Norwich, all the Christians living in that urban location could come together, within its public structures and thoroughfares, as a multitude united in their love of William or – as the *Life* elsewhere indicates – their hatred of Jews.[54]

Yet Cohen also goes on to note how Thomas troubles the image of urban unity in his text by admitting how 'not every citizen of the town was easily convinced that William was entitled to the *cultus* spreading under

---

[51] *Life*, 59; MS Add. 3037 17v.   [52] *Life*, 10; MS Add. 3037 3r–v.
[53] *Life*, 54; MS Add. 3037 16v. Cohen writes of the episode, 'Thomas's narration transforms a hagiographical commonplace (the *adventus* of a saint's relics for burial) into a culminating moment of civic unity': Cohen, *Hybridity*, 163.
[54] Cf. *Life*, 42; MS Add. 3037 13r.

his name.'[55] I would argue that we can go further in querying the idea of community in Thomas's text not only by looking at his references to sceptics but also by registering the manner in which the very image of Christian union depicted by Thomas is itself fraught. That is, the very multitude of Norwich Christians who are 'all of one mind' in their veneration of William, who oppose the imagined geography of Jewish danger with a public spectacle of Christian faith, present problems of their own. As much emerges after the requiem mass for William, when, Thomas tells us,

the body was laid up between the choir screen and the monks' choir lest the crowds of people that were pressing in desiring to kiss the bier, and if possible to rush forward to see the body, should be a hindrance rather than a help to the brethren who were performing the proper ministry of washing the corpse.[56]

Threatening to impede the monk's ministrations, the Christian multitude no longer figures as a reassuring sign of local Christian union but instead signals discord in Christian Norwich. The *turba* or moving press of Christians in the cathedral evinces an agitation that contrasts with the deliberate calm of the monks, by whom the mass was *solenniter cantabatur* or solemnly sung. Regardless of the good intentions indicated by their desire to kiss the bier, the crowd opposes the aims of the monks so much so that they deploy the *pulpitum* or choir screen to protect themselves and the corpse.

We have seen barriers, of course, before in Thomas's narrative. Both the crowd's status as potential *impedimentum* or obstacle to the monk's ministrations, as well as Thomas's stress on the *pulpitum* as a means of hindering the multitude from pressing upon William's corpse, resonate with that boundary so stressed in Thomas's account of the Jewish domicile, the door. Not unlike the Jewish door, which hides William and his attackers from the Christian public, the *pulpitum* serves to barricade William from his lay admirers. While elsewhere in the *Life*, the open spaces of public worship offer a Christian location that contrasts with the danger and secrecy of the Jewish domicile, here Cathedral and home converge. And it is precisely in that slippage between Christian cathedral and Jewish home, in the similarities that obtain between the Jewish door and the Christian choir screen, that we can understand how the *Life*'s antisemitism serves not only to imagine Christian union but also to acknowledge problems within Thomas's own Christian community.

---

[55] Ibid. On the manner in which issues of scepticism and doubt inform the *Life*, see Johnson, 'Rhetoric's work'.
[56] *Life* 51; MS Add. 3037 15v.

In part, the use of the choir screen as a means of separating William from his many admirers speaks to contradictions regarding the public and private aspects of monastic life in Norwich or indeed any medieval cathedral-priory.[57] The monks of Norwich cathedral-priory were uniquely obliged to open their doors to the secular world. In Roberta Gilchrist's words, 'The urban market was at its doorstep, and everyday pilgrims, merchants, and secular servants mingled at its thresholds.'[58] But in tension with that injunction to hospitality was the stress on seclusion, isolation and contemplation, that was at the very centre of monastic existence. Surrounding at least partially the precinct were 'massive walls' that served symbolic, jurisdictional and defensive functions.[59] Even more than the walls and gates of the precinct, the cloister demonstrates the geographic isolation at the core of monastic identity. A space from which the public was barred, the cloister provided the monks with a geographic sequestration indicated by its very terminology.[60] Clearly, the cloistered life that defined monasticism stood in deep tension with the public engagements of the Norwich cathedral-priory. As Isidore asks, after showing how the Latin *monachus*, monk, derives from the Greek word for oneness, 'if the word for monk means "a solitary", what is someone who is alone doing in a crowd?'[61] To paraphrase Isidore, if solitude epitomizes monasticism, what is a monk like Thomas doing celebrating crowds in his text?[62]

Situated at the centre of the precinct, the cloister affirms the isolating aspects of cenobitic existence; the *pulpitum*, on the other hand, functioned as a charged interface between the monks and the laity. In Norwich

---

[57] Thomas's image of the crowd also reflects the ambivalence over crowds evinced by the Church itself. As Elias Canetti has pointed out, as early as the second century, the Church generally has been averse 'from anything violently crowd-like'. As early as the Montanist heresy the swift spread and unruliness of crowds, and above all their release from ecclesiastical hierarchies, rendered them the prime enemy of the Church, so that an 'unshakeable conviction' regarding the threat posed by crowds informs 'the whole substance of the faith, as well as the practical forms of its organization': Elias Canetti, *Crowds and Power* (New York: Viking, 1963), 155.

[58] Roberta Gilchrist, *Norwich Cathedral Close: The Evolution of the English Cathedral Landscape* (Woodbridge: Boydell, 2005), 236.

[59] Ibid. 44–6.

[60] 'Cloister' is derived from the Latin *claustrum*, meaning 'enclosure', 'barrier', 'bolt' or 'gate': Gilchrist, *Norwich Cathedral Close*, 241.

[61] *Etymologies* VII.xiii.1, 172.

[62] English crowds appear in other twelfth-century monastic texts, such as the Augustine canon William of Newburgh's *History of English Affairs*, which describes both mobs of anti-Jewish rioters and 'martyr' worshippers. But those multitudes are unambiguously dangerous and misguided. William of Newburgh, *Historia rerum anglicarum Wilhelmi Parvi: The History of English Affairs*, P. G. Walsh and M. J. Kennedy (eds. and trs.) (Warminster: Aris and Phillips, 1988), 4.1, 5.21.

cathedral, a heavily restored late medieval choir screen stands on the site of the now lost original Romanesque structure. Following a plan typical of Benedictine conventual churches of the Middle Ages, the *pulpitum*, located at the third pier west of the crossing in Norwich cathedral, divided the lay and monastic sections of the nave. The *pulpitum* is a medieval monument that has received increasing attention over the last decade by scholars who have shown how choir screens 'fulfilled a wide variety of incorporative functions' during the medieval Christian liturgy.[63] But in the case of Thomas, we find evidence supporting the older view of the *pulpitum* as an architectural feature that separated the laity from the clerical performers of church rituals.[64] By stopping the flow of Christians towards William's corpse, the screen evinces how, even for a cathedral-priory enjoined to welcome into its precinct the secular world, even within a cathedral whose ample size admits within its borders what seems to be the whole urban Christian community, hospitality has its limits. For all Thomas's celebration of the local Christian multitude, he acknowledges how the mob opposes monastic identity. Troubling Thomas's image of a Norwich multitude united in its love of William (and its hatred of Jews) are divisions between Norwich's religious and lay populations, divisions given concrete reality through the *pulpitum*.

Thomas's text, then, exhibits a distinctly ambivalent stance towards English crowds and queries its very project of elevating an English market town to global prominence. While Theobald's report on a Jewish cabal conjures up a *mappa mundi* where Norwich enjoys prominence, the crowds rushing towards William's body imply the status of the city – and more generally England – as a backwater populated by rough primitives who require clerical management. Indeed, the public behaviour of the Christian majority comes close to mirroring the dangerous actions of the Jewish minority. But even as the potential for violence lurking in the crowd authorizes monastic intervention, the *Life* doesn't go so far as to affirm larger-scaled, papal-based forms of management. Although Thomas doubts the viability of the crowd he resists any wholesale legitimization of a newly consolidated universal church. Ultimately a cloistral ideology trumps any notion of Christian geographic expansiveness in Thomas's text. If the public character of William's

---

[63] Jacqueline Jung, 'Beyond the barrier: the unifying role of the choir screen in Gothic churches', *Art Bulletin* 82 (2000), 622–57. My thanks to Marcia Kupfer for bringing this essay to my attention.

[64] See the argument offered by Dorothy Gillerman, as summarized by Jung, 'Beyond the barrier'.

Christian worship suggests an open and communal Christian mapping of the town that opposes its secretive Jewish locations, other elements of the *Life* reverse that opposition. Anxiety over not only Jews but also the Christian laity and the universal *ecclesia* results in a monastic cultural geography in which Thomas and his brothers, like their supposed Jewish 'others', hide William from others. Thus we might ponder how the Christian doubling of Jewish violence shifts geographic scales in Thomas's text, from the similarities between the Christian cathedral and Jewish domicile, to the mirroring of the sequestered religious orders and the cloistered Jews, to the resemblances between the global Jewish cabal and universal Christian absolutism.

12 Gardens of Eden and ladders to heaven
Holy mountain geographies in Byzantium

Veronica della Dora

In the *Topographia Christiana*, a sixth-century illustrated cosmographical treatise, we are presented with an unusual world image (Fig 12.1). Here the earth is not depicted in a spherical shape, as ancient Greeks envisaged it, nor as the round 'world island' of western medieval *mappaemundi*. Instead, Cosmas Indicopleustes, assumed to be the author of the treatise (a traveller-geographer from Alexandria who later in his life became a monk in the monastery of Saint Catherine on Sinai), envisaged the cosmos in the shape of a tabernacle. The Sinaite monk considered a spherical earth and any other inheritance from the pagan world inappropriate for a Christian audience; a truly Christian cosmography, he maintained, should only come from the Scriptures.[1]

On Sinai, Cosmas believed, Yahweh did not reveal to Moses only the archetypal pattern of the tabernacle of the Temple of Jerusalem, but also the structure of the entire universe.[2] On the desert mountain God disclosed to the prophet 'an interlocking correspondence of spiritual and terrestrial geographies', which Cosmas literally mapped out in his treatise.[3] The higher part of the structure corresponded to the vaulted heavenly chamber inhabited by God. The rectangular lower prism, 'its length [being] twice its breadth like the table of Shewbread', enshrined the terrestrial realm.[4] This contained a rectangular οἰκουμένη γῆ (inhabited world) surrounded by the ocean and entirely dominated by a huge mountain 'as high as the breadth of the land towards the northern and western regions' and beyond which the sun, carried by two angels λαμπαδοφόροι, would disappear every

---

[1] A. Scafi, *Mapping Paradise: A History of Heaven on Earth* (London: British Library, 2006), 160–2.

[2] See Ex. 25–6.

[3] B. Lane, *The Solace of Fierce Landscapes: Exploring Desert and Mountain Spirituality* (New York: Oxford University Press, 1998), 142.

[4] E. Winstedt, *The Christian Topography of Cosmas Indicopleustes* (Cambridge University Press, 1909), 6.

Figure 12.1 Cosmas Indicopleustes' tabernacle-shaped cosmos.
Source: Reproduced with kind permission of the Biblioteca Apostolica
Vaticana.

evening.[5] The eastern part of the lower prism was occupied by the
terrestrial paradise, the Garden of Eden lost to mankind after Adam's
fall and located to the east, beyond the ocean, on 'a distant land', whose
outer reaches Cosmas attached to the lower extremities of heaven
(Fig 12.2).[6]

---

[5] *Topographia Christiana*, 4: 185. For a modern edition see Wanda Wolska-Conus (ed. and
tr.), *Cosmas Indicopleustès, Topographie chrétienne*, 2 vols (Paris: Les Editions du Cerf, 1968).
See also W. Wolska, *La Topographie Chrétienne de Cosmas Indicopleustès: theologie et science au
VI siècle* (Paris: Presses Universitaires de France, 1962). For its contextualization within
Byzantine cartography see O. A. W. Dilke, 'Cartography in the Byzantine empire', in *The
History of Cartography Volume 1: Cartography in Prehistoric, Ancient, and Medieval Europe and
the Mediterranean*, J. B. Harley and D. Woodward (eds.) (University of Chicago Press,
1987), 261–2; M. Kominko, 'New persepectives on Paradise: the levels of reality in
Byzantine and Latin medieval maps', in *Cartography in Antiquity and the Middle Ages*,
R. J. A. Talbert and R. W. Unger (eds.) (Leiden: Brill, 2008), 139–53.

[6] See Scafi, *Mapping Paradise*, 160–2.

Figure 12.2 Orthographic view of Cosmas's cosmos, with terrestrial paradise in the East.
Source: Reproduced by kind permission of the Biblioteca Apostolica Vaticana.

One of the very few examples of Byzantine mappings and the only illustrated Byzantine cosmography that has survived to us, Cosmas's treatise did not have a large impact on medieval thought, which continued to superimpose biblical ideas on ancient Greek geographies.[7] Even so, the *Topographia* offers precious insights into Byzantine sacred geographies and into the relationship between Creation and the Kingdom of Heaven. In Cosmas's and the desert fathers' geographical imagination, terrestrial paradise was a self-enclosed garden separated by an unsurpassable extent of ocean, or of desert, and thus out of the reach of mankind. The closest that three Mesopotamian monks managed to get to its gates, after a long and difficult journey to the East, was twenty miles, we read in the life of Saint Macarius of Rome the Mesopotamian (fifth century).[8] But as the desert fathers knew and Cosmas's main

[7] Dilke, 'Cartography in the Byzantine empire', 261–3.
[8] 'The life of Saint Macarius of Rome, a servant of God who was found to be near Paradise by Theophilus, Sergius and Hyginus', in *Vitae Patrum*, ch. 16, text available at: http://www.vitae-patrum.org.uk/page34.html. See also H. Maguire, 'Paradise withdrawn',

diagram seemed to remind the reader, the focus of every Christian should not be a peripheral and perhaps bygone (or unreachable) terrestrial Eden, but rather the Kingdom of Heaven – in other words, not a physical place, but a condition of the soul which could be attained through spiritual ascent.[9]

By the time the *Topographia* was being compiled, the self-enclosed Edenic garden and the mountain stretching to the vault of heaven were powerful symbols. For over three centuries, they had been signposting the writings of the Church's fathers as both physical and metaphorical spaces – as pathways to Salvation. Fourth-century holy hermits, for example, washed the barren land of the Upper Thebaid with their tears of repentance and made it blossom with 'fruit a hundred-fold' (echoing Yahweh's ability to transform the desert into a fertile land),[10] whereas sixth-century monks of Saint Catherine carved the hard rock of Mount Sinai into the Stairway of Repentance, a 3,750-step route beginning from south-east of the monastery and leading up to the summit – a physical reminder of both Moses' climb and the hardness of spiritual ascent.[11] Or again, like Peter, James and John, later desert hesychasts claimed to have 'reached the top of Mount Tabor' and seen the Uncreated Light – this time, however, through contemplative prayer, without ever leaving their cells.

Whether as figurative or physical spaces (or both), the garden and the mountain continued to mark biographies of saints and sacred geographies during the entire life of the Byzantine Empire (and after its Fall). This chapter shows how these two apparently so different spaces came to converge in what became one of the most characteristic features of Byzantine Orthodox Christianity: the holy mountain. Between the fifth and the eleventh centuries, a number of 'new' (non-biblical) holy peaks hosting hermits and subsequently monastic communities started to emerge in regions of the Empire well beyond Sinai and the Holy Land. Their creation and maintenance has much to do with the ability of their founders and monks to tame their inhospitable terrain and to transform them, literally or figuratively, into gardens of

in *Byzantine Garden Culture*, Antony Robert Littlewood, Henry Maguire, Joachim Wolschke-Bulmahn (eds.) (Washington, DC: Dumbarton Oaks, 2002), 23.

[9] D. Kyrtatas, 'Seeking paradise in the Egyptian desert', paper presented at the conference 'The Cosmography of Paradise', Warburg Institute, London, 2009.

[10] See for example, 'Apolytíkion tou Aghíou Antōníou', *Menaion*, 17 January. See also C. Rapp, 'Desert, city and countryside in the early Christian imagination', in *The Encroaching Desert: Egyptian Hagiography and the Medieval West*, J. Dijkstra and M. van Dijck (eds.), Church History and Religious Culture 86 (Leiden: Brill, 2006), 93–112.

[11] J. Hobbs, *Mount Sinai* (Austin: University of Texas Press, 1995), 110.

Eden.[12] Byzantine holy mountains can be envisaged as physical and imaginative nodes of extensive spiritual networks. This chapter focuses on the geographies of such networks. It considers their 'mountain-nodes' both relationally as 'stations', or *loci memoriae* within hagiographical accounts, and in their local specificities, through their founders and inhabitants' physical encounters and poeticization as potential gardens of Eden and privileged sites for encountering God – as ladders to heaven.

Biblical mountains

The Judaeo-Christian tradition is signposted by mountains: barren desert peaks and quiet verdant elevations; inaccessible mountain tops and sweet hills made higher by wondrous happenings and by human imagination; mountains hosting dark caves and prophets; mountains identified and accurately mapped; and finally mountains destined to remain invisible, or simply allegorical. Both Moses and Elijah encountered God on mountain peaks. In Matthew's Gospel a sequence of mountains signals different stations in the life of Christ, functioning as a 'chain' of iconic mnemonic landmarks. The drama starts to unfold on the top of the 'highest mountain' where Christ is tempted by the Evil with 'all the kingdoms of the earth and the glory of them';[13] it then continues on the Mount of the Beatitudes, the Mount of Feeding, the Mount of Transfiguration (Tabor), the Mount of Olives, to culminate on the Mount of Crucifixion (Golgotha), where through his sacrifice, Christ, the New Adam, redeems the sins of the old one, whose bones are buried under the mountain (as shown on Byzantine icon paintings).[14]

Biblical peaks are landmarks that are at once physical and imaginative. In the Judaeo-Christian tradition, the seductive, ambivalent half-seen mountain is a metaphor for speaking to God. In the earliest Hebrew cosmology Yahweh was deemed to dwell in the Upper Chambers above the firmament and coming down to meet with the faithful, thus giving mountains a special significance as the closest earthly places to God's abode.[15] Biblical mountains are thus privileged theophanic sites, or *axes mundi* through which 'the transcendent might enter the immanent'.[16] They are ladders that link earth to heaven. In the Old Testament, Yahweh was encountered by Elijah in a cave on Mount Horeb beyond

---

[12] R. Greenfield, 'Galesion: opposition, disagreement and subterfuge in the creation of a holy mountain', paper presented to the 21st International Congress of Byzantine Studies, London, 2006, Panel VI.6 Monastic mountains and deserts.

[13] Matt. 4: 8.   [14] Lane, *Fierce Landscapes*, 45.   [15] Psalm 104: 13.

[16] M. Eliade, *Patterns in Comparative Religion* (New York: Meridian Book, 1963), 231.

words and understanding, and by Moses on the top of Mount Sinai in a dense cloud of smoke.[17] In the New Testament, God revealed Himself to Peter, James and John on Mount Tabor in a cloud of light.[18] The two latter mounts, utterly different in their settings and in their physical appearance, stand out as the scriptural holy peaks par excellence, and thus as the archetypes and privileged terms of comparison for non-biblical Byzantine holy mountains.

Mount Sinai is a 2,288 m-high peak located in the southern part of the desert peninsula after which it is named, one of the most arid regions of the world. Mount Tabor is a 575 m-high hill located in Lower Galilee, at the eastern end of the Jezreel Valley, 17 kilometres west of the Sea of Galilee. Surpassed almost 400 m by the nearby peak of Mount Catherine, the highest on the Sinai peninsula, the mountain of Moses does not possess any particularly striking physical characteristic that enables one to differentiate it from the several other surrounding peaks. By contrast, in spite of its remarkably lower altitude, the Mount of Transfiguration pops up from a flat landscape, justifying its biblical epithet of 'high mountain'.[19] Sun and barren red granite rock, so impervious to moisture, are dominant features of Mount Sinai and inflict the drought that is the norm in the region.[20] In the sixth century, Justinian's court historian, Procopius, described the mountain of Moses as a 'precipitous and very wild mountain'.[21] By contrast, Elisaneus, a seventh-century Armenian pilgrim, portrayed Tabor in almost Edenic terms, as 'a charming place' around which are 'wells of water and many densely planted trees, which blossom from the rain of the clouds and produce all kinds of sweet fruits and delightful scents; there are also vines which give wine worthy for kings to drink.'[22]

The juxtaposition of Sinai and Tabor has been exploited through the centuries to illustrate the dual nature of eastern Christian doctrine: as kataphatic (or 'positive') and apophatic (or 'negative'). Sinai symbolizes the 'aniconic power' of the latter tradition, whereby God is defined as uncreated, unbound, unattainable, thus inviting us to recognize our *agnosia*, or 'knowledge about unknowledge'.[23] On Mount Sinai Yahweh revealed Himself to Moses in a scarcity of images, in absence of clarity – 'ἐν νεφέλῃ πυκνῇ' – in a thick cloud.[24] The prophet asked to see Yahweh's face, but was shown only His back. The apophatic impulse of Sinai, a desert mountain made of hard barren rock, is to empty the

---

[17] Ex. 19: 9; I Kings 19: 8–13.   [18] Matt. 17: 5.   [19] Matt. 17: 1.
[20] Hobbs, *Mount Sinai*, 19.   [21] Ibid. 2.   [22] Cited in Lane, *Fierce Landscapes*, 131.
[23] See Solrunn Nes, *The Uncreated Light: An Iconographical Study of the Transfiguration in the Eastern Church* (Grand Rapids, MI and Cambridge, Mass.: Eerdmans, 2007), 57.
[24] Ex. 19: 18.

faithful of 'all inadequate images, to destroy idolatries, to cut through all false conceptions of the holy'.[25]

Tabor, the mountain of Transfiguration, by contrast, symbolizes 'the iconic, imaginative power of the kataphatic tradition' – the approach through which we comprehend the natural world and translate it into knowledge of the divine.[26] On the top of Mount Tabor, a gentle verdant hill covered with trees, God manifested Himself to Peter, James and John in a sharpness of lucidity of image.[27] There Jesus of Nazareth was transfigured before them. 'His face shone like the sun, and his clothes became as white as the light.'[28] The three disciples knew themselves to have encountered the living God in human flesh. What was not given to see and know to Moses, was made manifest in the New Testament.[29]

Sinai and Tabor are interpenetrating poles. In Byzantine iconography, they are also interpenetrating images. The Mountain of Transfiguration is itself transfigured into the mountains of the Old Testament's revelations: Mount Horeb (topped by Elijah, who stands for 'the Prophets') on the left and Mount Sinai (topped by Moses, who symbolizes 'the Law') on the right.[30] In the upper part of the icon the two prophets mirror each other, as they engage in a timeless dialogue with a motionless Christ dressed in glistening, intensely white clothes and enclosed in a mandorla of light (the fulfilment of both the Law and Prophets). Byzantines understood the experience of Moses on Sinai as a pre-figuration of the Transfiguration of Christ on Mount Tabor, hence the one mountain is continually superimposed over the other.[31] Old and New Testament narratives converge in space and in time. On the lower part of the icon, time continues to flow. The three apostles below are set against flowering terrain indicative of movement, for each one of them is in a different stage of awakening.[32] On Tabor, the Hebrew word for 'navel' (the centre of the cosmos), past, present and eternity converge in a single moment and the cosmographic and topographic scales converge in a single scene.[33]

---

[25] Lane, *Fierce Landscapes*, 137.   [26] Nes, *Uncreated Light*, 57.

[27] Ibid. 57. Mount Tabor is the object of poetical comparisons on the part of the Psalmist (Psalm 88: 13; Jeremiah 46: 18; Osee 5: 1).

[28] Matt. 17: 2.   [29] Lane, *Fierce Landscapes*, 137.   [30] Matt. 5: 17.

[31] A. M. Talbot, 'Les saintes montagnes à Byzance', in *Le sacré et son inscription dans l'espace à Byzance et en Occident*, M. Kaplan (ed.) (Paris: Byzantina Sorbonensia, 2001), 263–76, at 265. See also A. Andreopoulos, *Metamorphosis: The Transfiguration in Byzantine Theology and Iconography* (Crestwood, NY: St. Vladimir Seminary Press, 2005).

[32] H. Maguire, 'The cycle of images', in L. Safran (ed.), *Heaven on Earth: Art and the Church in Byzantium* (University Park, Pa.: Pennsylvania University Press, 2002), 121–51, at 143.

[33] On the idea of a cosmic axis centred on an *omphalos*, see M. Eliade, 'Centre du monde, temple, maison', in *Le symbolisme cosmique des monuments religieux* (conference proceedings, Rome, 1955), 57–82; Paul Wheatley, *City as Symbol* (University College London, 1967).

Mountains and the desert fathers

A feature characteristic of a number of Byzantine and post-Byzantine icons of the Transfiguration is the two caves in the mountains of the Old Testament. Solrun Nes interpreted them as markers of the apophatic tradition: it was this divine darkness that surrounded Moses and Elijah at the moment of their revelations. Nes also read the two caves as reminders of the spiritual inheritance of the desert fathers, who found their abodes in the mountain fissures of wild desolate landscapes and, like the three disciples of Christ, managed to experience the Uncreated Light of the Transfiguration.[34] The mountain traditions of epiphany were indeed so strong that from the very beginnings of Christianity hermits and holy men sought spiritual refuge and self-purification on desert hilltops and mountains, mimicking the great prophets of the Old Testament.[35] Hermits were drawn to the holy biblical mountains and also settled on other mountains, often chosen for their remoteness, but which in turn drew disciples and pilgrims.[36]

This ascetic tradition originated and first flourished within a 300-kilometre radius from the Mountain of Moses between the third and fourth centuries.[37] One of the first desert fathers and 'literary prototype for all monastic desert experience' was the Egyptian Saint Antony the Great (251–356 CE), who at the age of twenty followed his call to a solitary ascetic life and moved away from society. His abandonment of the world followed different stages and eventually took him to the desolated mountain region of the Upper Thebaid, near the Red Sea.[38] As his biographer Athanasius of Alexandria recorded in the fourth century, having journeyed for three days and three nights, the young Antony came 'to a very lofty mountain, and at the foot of the mountain ran a clear spring, whose waters were sweet and very cold; outside there was a plain and a few uncared-for palm trees'.[39] Antony loved the place and established himself there for the rest of his life.

---

[34] Nes, *Uncreated Light*, 86, 90.

[35] See S. Schama, *Landscape and Memory* (New York: Vintage Books, 1995), 414.

[36] 'Holy mountains', in *The Blackwell Dictionary of Eastern Christianity*, K. Parry, D. Melling, D. Brady, S. Griffith and J. Healey (eds.) (Oxford: Blackwell Publishing, 2001).

[37] Hobbes, *Mount Sinai*, 59.

[38] Antony lived in the interior of the labyrinthine 1,200-metre-high South Galala Plateau in Egypt's northern Eastern Desert; see Hobbes, *Mount Sinai*, 60, and W. Harmless, *Desert Christians: An Introduction to the Literature of Early Monasticism* (Oxford and New York: Oxford University Press, 2004), 62.

[39] P. Schaff and H. Wace (eds.), *A Selected Library of Nicene and post-Nicene Fathers of the Christian Church*, vol. IV: *St. Athanasius: Selected Works and Letters* (Grand Rapids, Mich.: Eerdmans, 1953), 209.

To pilgrims and visitors, however, Inner Mountain (known in the Medieval West as Mount Climax but which Antony simply referred to as 'the Mountain') was less a place of delight than a *locus horridus*, surpassing Mount Sinai in cragginess and inhospitality. Such fame grew through the centuries. A fifteenth-century pilgrim, for example, described the mountain as a most dreadful place occupied by 'women notable for long beards' who spent their time 'most cruelly in hunting, have tigers instead of dogs, and breed leopards and lions'.[40] But a challenging environment was precisely what hermits were after. And so, Antony's words about the delight of solitude 'persuaded many to take up the solitary life. ... From then on, there were monasteries in the mountains, and the desert was made a city by monks, who left their own people and registered themselves for the citizenship in the heavens.'[41] Throughout the fourth century increasing numbers of hermits thus wandered with their disciples through the wilderness of Egypt, the Judean desert and Sinai. For these ascetics mountain and desert in such regions had the same spiritual value.

As part of their vocation, by the mid-third century, hermits in the Sinai region were also endeavouring to find the location of the Mountain of Moses and of other biblical places, an enterprise which was continued by devoted pilgrims in the following century and officially paralleled in the Holy Land by Empress Helena. After putting an end to Christian persecutions and conceding free profession of all religions in the Roman Empire (Edict of Milan, 313 CE), in 326 Constantine put his mother in charge of the identification of the places connected to the life of Christ and marked them with shrines that would forever fix them on the map of the Christian world.[42] As a result of this imperial support of Holy Land pilgrimage, later in the century the mountains that signposted the life of

---

[40] S. Aubrey (trs.), *The Wanderings of Felix Fabri*, vol. 4 (London: Committee of the Palestine Exploration Fund, 1897), 587. The same description is found a century earlier in Gervase of Tilbury's *Otia imperialia*, S. E. Banks and J. W. Binns (tr.) (Oxford: Clarendon Press, 2002), 215.

[41] Athanasius, *Vita Antonii* 14.7, J. M. Bartelink (ed.) [sources Chrétiennes 400] (Paris: Les Éditions du Cerf, 1994), 174.

[42] Helena founded and generously endowed churches of the Nativity in Bethlehem and the Ascension in Jerusalem and was later credited with the finding of the True Cross, in company of which she is normally represented: 'Helena', in *Oxford Dictionary of Byzantium*, Alexander P. Kazhdan (ed.), vol. 2 (Oxford University Press, 1991), 909. Since the fourth century, the most important *locus sanctus* in Jerusalem, however, was the Holy Sepulchre. After its discovery, Constantine directed the building of a basilica, and some years later (*c.* 350 CE) a conical-domed rotunda was carved out of living rock and embellished with columns, a porch and a precious metal sheathing ('Holy Sepulchre', in *Oxford Dictionary of Byzantium*, Kazhdan (ed.), 1870). Constantine also built a basilica on Mount of Olives, where pilgrims collect blessed dust ('Mount of Olives', in *Oxford Dictionary of Byzantium*, Kazhdan (ed.), 1420). J. Herrin, *Byzantium: The Surprising Life of a Medieval Empire* (London: Allen Lane, 2007), 10.

Christ started to become key pilgrimage centres. After its definitive identification by Cyril of Jerusalem in 348 CE, Tabor, for example, was soon surmounted by a basilica and several churches and chapels.[43] Tradition credits Helena with the construction of a tower and a chapel near the site on the Sinai peninsula identified with the burning bush in 330 CE, after she visited the site and responded to the Sinaite monks' pleas for protection from nomads' raids.[44] About thirty years later, a Syrian monk named Julianus Sabus and a party of pious followers from Edessa erected another chapel, this time marking the much disputed summit of Sinai.[45]

The construction of the monastery of Saint Catherine by order of Justinian at the feet of the Mountain of Moses (527–65 CE) for ever sealed the holy status of Sinai, making it one of the primary and certainly most lasting nodes of a network of pilgrimage sites in the Byzantine Empire. Unlike many shrines in the Holy Land, which suffered pillaging and even destruction by the Saracens, thanks to its massive walls and its remoteness, Saint Catherine endured throughout the centuries as an oasis of Christian spirituality in the desert.[46] Tabor remained of course another node of this same network of holy sites, but unlike Sinai, one always more alive as a name (or as an icon) than as an actual spiritual centre, because of historical vicissitudes. By the early ninth century the Mountain of Transfiguration had become the residence of the bishop and 4,340 steps led up to the summit. In the following century, however, the mountain turned into the theatre of a battle for the control of the Holy Land by the Abbasside Caliphate, and during the period of the Crusades it changed hands many times between Muslims and Latin Christians.[47]

Meso-Byzantine mountains

Byzantium's loss of control of the Holy Land and surrounding regions, including Egypt and Sinai in the seventh century, did not mark a decline in ascetic tradition and pilgrimage. It rather encouraged their

---

[43] Church buildings and a monastery are attested on Tabor since the fourth or fifth century, and sixth-century descriptions of the pilgrims speak of three churches on its top. B. Meistermann, *Le Mont Thabor* (Paris: J. Mersch, 1900); see also, 'Tabor', in *Oxford Dictionary of Byzantium*, Kazhdan (ed.), 2004.

[44] Hobbs, *Mount Sinai*, 70.

[45] Ibid. 68. On the controversies on the identification of Mount Sinai, see ibid. 32–53.

[46] The Church of the Holy Sepulchre, for example, was demolished in September 1009 by order of the Fatimid Caliph al-Hakim: R. Greenfield, *The Life of Lazaros of Mt. Galesion: An Eleventh-Century Pillar Saint* (Washington, DC: Dumbarton Oaks, 2000), 8 (a detailed description is given in the *vita* of Saint Lazaros of Galesion: 101–2).

[47] See Meistermann, *Thabor*.

proliferation in other parts of the Empire. Between the ninth and eleventh centuries, when the Arab invasion pushed hermits out of the desert, a number of 'new' (non-biblical) holy peaks started to emerge and previously established ones (dating back to the fifth century) to flourish throughout the Byzantine Empire thanks to charismatic ascetics and imperial donations: Auxentius, Latros and Kyminàs in Bithynia, Olympus in Mysia, Galesion near Ephesus, Ida off the coast of Lesbos, Ganos and Paroria in Thrace, Papikion near Thessalonica, Athos in the Chakidikè, Meteora in Thessaly, the Wondrous Mountain in Syria, and so on.

The geographies of these mountains were as distant from one another as were their locations, and often conditioned their histories.[48] For example, Auxentius (today Kayış Dağı), named after a fifth-century Syrian monk who spent twenty years in a cave near the summit and one of the first mountains of Anatolia to attract monastic communities, was a relatively accessible and by no means striking feature. A barely 428 m hill, Auxentius was one (and not even the highest) of several hills south of Chalcedon (approximately 12 km), not very far from Constantinople. Its diameter did not exceed 5 km, making for a 'small-scale' holy mountain which nevertheless endured into the Palaelogan era.[49] Not-too-distant snow-clad Olympus of Mysia (also inhabited by hermits as early as Auxentius and today a paradise for skiers), by contrast, climbed up to 2,327 m, towering over Brussa. While still relatively close to the capital, Bithynian Olympus was the highest mountain in the northwest part of Asia Minor and, unlike Auxentius, it was a dominant presence in the landscape. By the eighth century the fame of the mountain exceeded its surface and made it the 'Holy Mountain' par excellence until the tenth century, when Athos took pride of place. The term 'Olympus' was loosely employed by the Byzantines to designate not only the monastic communities on the mountain, but also the nearly one-hundred foundations scattered in the entire surrounding region (in the plain of Brussa), gaining the mountain the appellation of 'Olympus of the monks'.[50]

---

[48] See K. Belke, 'Heilige Berge Bithyniens', paper presented at the 21st International Congress of Byzantine Studies, London, 2006 Panel VI.6 'Monastic mountains and deserts'. See also K. Belke (ed.), *Tabula Imperii Byzantini*, vol. 13: *Bithynia and Hellespont* (Vienna: Verlag der Österreichischen Akademie der Wissenschaften, in press).

[49] Talbot, 'Saintes montagnes', 265; see also Belke, 'Heilige Berge'.

[50] Talbot, 'Saintes montagnes', 265; C. Mentzou-Meimaris, 'The contribution of the monasteries on Mount Olympos in Bithinia to the cultural life of Byzantium', in E. Jeffreys and F. K. Hareer (eds.), *Proceedings of the 21st International Congress of Byzantine Studies, London 2006* (Aldershot: Ashgate, 2006), vol. III, 258; 'Holy mountain', in *Oxford Dictionary of Byzantium*, Kazhdan (ed.), 941. See also

By contrast, the 56-km mountain-peninsula of Mount Athos (the easternmost of the three 'fingers' of the Chalkidikē peninsula in northeast Greece) presented well-defined boundaries since its beginnings as a holy mountain in the ninth century. Its coastline acted as a natural boundary around most of its perimeter, whereas its two-kilometre isthmus, which according to Herodotus had once been traversed by a canal cut by Persian King Xerxes, marked the land boundary.[51] Its cragginess, the surrounding stormy Aegean, and distance from major urban centres (Thessalonica, the closest city, lies approximately 100 km away) made Athos the most isolated holy mountain of the Byzantine Empire, and therefore a safe refuge for ninth-century hermits escaping iconoclastic persecution and the Arab invasion of the Egyptian desert, or simply looking for a 'quieter place'.[52]

Rising from sea level, the Athonite peninsula culminates in a 2,033 m peak, one of the most spectacular landmarks in the Aegean.[53] In clear atmospheric conditions the imposing pyramid is visible for over 160 km; in antiquity it thus served as a precious 'natural beacon' for coastal navigation and as one of the most celebrated peaks by classical poets.[54] The stark visibility of the Athonite cone looming on the sea horizon can be compared to the equally dramatic though very different landscape of Meteora, at the north-western edge of the plain of Thessaly in central Greece. Meteora's sandstone rock pinnacles, upon which its six surviving monasteries are nested, rise up several hundred metres over the

---

D. Papachrysanthou, *O Athōnikós monachismós: archés kai orgánōsē* (Athens: Morfōtikó idryma Ethnikēs Trapézēs, 2004), 82–4.

[51] See V. della Dora, 'Geo-strategy and the persistence of antiquity: surveying mythical hydrographies in the eastern Mediterranean, 1798–1869', *Journal of Historical Geography* 33 (2007), 528.

[52] In 859 CE, Saint Euthymios the Younger, one of the earliest documented Christian settlers, moved from Bithynian Olympus to Athos, 'because he had heard of its tranquillity': D. Papachryssánthou, *Actes du Protaton* (Paris: Lethielleux, 1975), 18.

[53] 'From sea level south of the isthmus Megali Vigla rises sharply to a peak of 510 metres within 1.5 kilometres of the frontier, creating a natural barrier against the outside world. This is followed by undulating hills that run down the spine of the peninsula, gradually increasing in altitude until they become a mountain range with heights of between 450 and 900 metres. At the southern tip the range shoots up to a rocky eminence of 2,033 metres before plunging headlong into the sea. The relief is so dramatic that it is perhaps a matter for more surprise that only 20 per cent of the area of the peninsula is above 500 metres': G. Speake, *Mount Athos: Renewal in Paradise* (New Haven and London: Yale University Press, 2002), 29.

[54] E. Livierátos, Evágghelos, 'Athō perimétrou metamorfōseis', in E. Livierátos (ed.), *Orous Athō gēs thalássēs perímetron chartōn metamorfōseis* (Thessalonica: Ethnikē Chartothēkē, 2002), 21–2; E. Churchill Semple, *The Geography of the Mediterranean Region: Its Relation to Ancient History* (London: Constable and Co., 1932), 586–7. On Athos in classical literature, see V. della Dora, *Imagining Mount Athos: Visions of a Holy Mountain from Homer to World War II* (Charlottesville: University of Virginia Press, 2010).

surrounding flat land as cyclopean 'natural pillars' stretching to heaven. As with other Byzantine holy mountains, Meteora's spiritual landscape was carved out of its unique geology through ascetic discourse and practice. Hermits first moved to Meteora around the eleventh century, seeking refuge from Turkish expansion. Where Athonite communities were protected by the sea and high monastic walls, Meteorite monasteries were naturally protected by their height. The only way of access was through a net hauled by a rope. 'Every monastery had its lifting tower built against one edge of its rock, at the top of which was the wooden shed with its platform projecting into space housing the windlass for purposes of communication with the world.' But, as Donald Nicol observed, 'communication with the world was something the Byzantine monk was supposed to regard simply as a regrettable necessity.'[55]

In spite of their very different geographical characteristics and settings, all these holy mountains shared similar developmental patterns and functions. Byzantines considered them as 'places of isolation, away from human society', and often referred to them with the term *erēmos* (desert).[56] Indeed wilderness and isolation were prerequisites for the spiritual *hesychia* (quietness) sought after by hermits and monks. Whether in Asia Minor or in Greece, holy mountains operated as important centres of spiritual resource, pilgrimage and endowment. They were initially attributed an aura of holiness because of the presence of charismatic ascetics and saints (who dwelled in caves, wooden shelters, and even on the top of pillars), and later because of the establishment of organized monastic communities.[57] Unlike Sinai, Tabor and other famous biblical peaks, these holy mountains did not boast any event reported in the Scriptures: they were sanctified through the prayers and miracles of their holy founders, rather than by biblical theophanies. But even so, the fame of some of them ended up surpassing many of their scriptural counterparts – in the tenth century, Mount Athos, for example, became *the* Holy Mountain of Orthodoxy, a title which it continues to retain to our days.[58]

Holy mountains functioned as spiritual beacons and as strongholds of Orthodox faith, in particular as bulwarks against iconoclasm in the eight and ninth centuries (Ida and Olympus) and against unionism in the thirteenth (Athos and Ganos). As Alice Mary Talbot observed, unlike

---

[55] D. Nicol, *Meteora: The Rock Monasteries of Thessaly* (London: Chapman and Hall, 1963), 16.
[56] Talbot, 'Saintes montagnes', 264; see also Nicol, *Meteora*, 25.
[57] The most famous examples are probably those of Saint Athanasius of Mount Athos and Lazaros of Galesion. In the latter case, the holy mountain was created thanks to the perseverance of a single holy man.
[58] Talbot, 'Saintes montagnes', 269.

Athos, the Bithynian mounts were all located near important cities, thus encouraging pilgrimage, indeed often exerting a significant spiritual as well as cultural and even political influence on the world 'below' them.[59] While the 'making' of a holy mountain could often be a contested enterprise, as in the case of Galesion, once established as 'holy', peasants, generals and emperors alike sought the prayers and advice of its elders and monks.[60] Sometimes, as with Bithynian Olympus, the monasteries could also provide a secure shelter for clergy or politically involved laymen persecuted by secular authorities and condemned to a penalty, or to life imprisonment.[61] Holy mountains therefore benefited not only hermits and monks, but Byzantine society at large.

Holy mountains as *loci memoriae*

Memory, Pierre Nora suggested, is rooted in the concrete, in the visible. The most prominent and lasting of all geographical features, mountains can be defined as what the French historian termed 'topographical memory places', sites evoking a sense of continuity with the past, 'compounded of life and death, of the temporal and the eternal ... endless rounds of the collective and the individual, the prosaic and the sacred'.[62] Among their various functions, Byzantine holy mountains had also that of signposting the lives of saints, in the same way New Testament mounts did in the life of Christ. Along with the Holy Land, and other notable holy places such as Rome, or (later) Patmos, since the tenth century, many of these peaks came to represent almost compulsory 'stations' in the life of ascetics aspiring to a perfect life in Christ. In hagiographical accounts, holy mountains often operate as powerful *loci memoriae*, as nodes of spiritual 'memory networks' which could extend spatially for hundreds of kilometres and temporally for several centuries, as the founders of new holy mountains always drew on previous examples of sacred geographies.

[59] For example, Auxentius was located near Constantinople, Olympus near Broussa, Kyminas near Prousias, Latros near Miletus, Galesion near Ephesus: Talbot, 'Saintes montagnes', 266.

[60] Before launching his victorious campaign against the Arabs in Crete in 960, Nicephoros Phokas, for example, asked for the prayers of the monks of Kyminas, Olympus, Athos and 'other famous mountains': Talbot, 'Saintes montagnes', 267; whereas Maria Sklerina, the mistress of Constantine IX Monomachos, financed the foundation of a dependency near Galesion after she heard about Lazaros from her brother, who visited him and remained deeply impressed: Talbot, 'Saintes montagnes', 34. On difficulties in establishing a holy mountain, see, for example, Greenfield, 'Galesion'.

[61] Mentzou-Meimaris, 'Monasteries on Mount Olympos', 258.

[62] P. Nora, *Realms of Memory* (New York: Columbia University Press, 1996), vol. I, 18, 15.

For example, Saint Athanasius, the founder of the first coenobitic monastery on Mount Athos (963), was tonsured on Mount Kyminas by Saint Michael Maleïnos, the uncle of future emperor Nicephorus Phokas. Later on, having left Kyminas with only two small books and his master's hood, he made the rounds of many desolate and solitary places, and guided by God, he came to the extremity of Athos.[63] Similarly, Chrystodoulos, the founder of the monastery of St John the Theologian (1088) on the island of Patmos, had spent his previous monastic life on Mount Latros, a then famous holy mountain 42 km north-east of Miletus which, according to tradition, had been in turn established by a monk from Sinai escaping Arab incursions in the seventh century.[64] After being tonsured in Cyprus, hesychast Gregory the Sinaite left for Mount Sinai; he then moved to Athos, to end up in Paroria, where he founded a monastery (1330) and ended his life (1346?).[65] Or again, the founders of the first monastery on the Meteora, Athanasius Koinovitis and his fellow monks Gregory and Moses, reached the Thessalian 'petrified forest' in 1344, having left the monastery of Iveron on Mount Athos, then besieged by Turkish pirates.[66] And so on.

Intensified by eighth-century iconoclastic persecution and Islamic incursions, mobility remained through the centuries one of the hallmarks of Byzantine (male) monasticism. Byzantine monks were generally allowed to travel from one centre to the other, supported only by charity.[67] A notable example of mobility through holy peaks is recounted by Ioseph Kalothetos, a monk from Esphigmenou on Mount Athos in the fourteenth century and second biographer of Patriarch Athanasios of Constantinople. Kalothetos recounts the Patriarch's 'wanderings' in the first stage of his monastic carrier through a by then well-established network of most famous holy mountains. As a young monk, Athanasius moved between different communities on these mountains, in search of the ideal environment for ascetic life. His venture started from Athos;

---

[63] Athanasius, *Vita A*, in *Vitae duae antiquae Sancti Athanasii Athonitae*, J. Noret (ed.) (Leuven: Leuven University Press, 1982), 18.

[64] 'Patmos', in *Blackwell Dictionary of Orthodox Christianity* and in *Oxford Dictionary of Byzantium*, Kazhdan (ed.), 1596.

[65] 'Gregory the Sinaite', in *Oxford Dictionary of Byzantium*, Kazhdan (ed.), 883.

[66] 'Meteora', in *Oxford Dictionary of Byzantium*, Kazhdan (ed.), 1353. See also G. Dennis, 'Meteora: canonical rule of Athanasios the Meteorite for the monastery of the Transfiguration (Metamorphosis)', in *Byzantine Monastic Foundation Documents: A Complete Translation of the Surviving Founders. Typika and Testaments*, J. Thomas and A. Constantinides Hero (eds.) (Washington, D.C.: Dumbarton Oaks, 2000); N. Nikonános, *Metéōra: ta monastēria kai hē historía tous* (Athens: Ekdotikē Athēnōn, 1987).

[67] In Byzantium the western principle of permanence in a single place was applied more often to nuns than monks. See Herrin, *Byzantium*, 194. On 'useful' monastic travel vs. 'useless' travel, see Papachrysanthou, *Athōnikós monachismós*, 51 n. 111.

then, after a pilgrimage to the Holy Land, he moved on to the Anatolian mountains of Latros, Auxentios and Galesion. After a second visit to Athos, he returned to Galesion, and eventually ended up on Ganos.[68]

The early life of Saint Lazaros (966/67–1053), the founding father of Mount Galesion, is also punctuated by holy mountains. A native of a town near Magnesia, on the western coast of Anatolia, Lazaros received his education at the nearby monasteries of Oroboi and Kalathai. As a young boy, he felt the urge to leave his homeland for the Holy Land. After an adventurous journey through Asia Minor and seven years in a monastery in Attalia (where he was tonsured monk), Lazaros fulfilled his dream and reached Jerusalem, where he remained until the demolition of the church of the Holy Sepulchre by the Fatimid Caliph al-Hakim in 1009.[69] Threatened by the increasing Arab pressure on Christians and guided by divine inspiration, Lazaros headed back to his homeland aged 42.[70] The return journey, which would culminate in his establishment on a pillar on Mount Galesion, was signposted by three other holy mounts. An initial detour to ascend and worship Mount Tabor marked the outset of the journey.[71] After crossing into Byzantine territory at Laodicaea, Lazaros stopped at a second mountain: the Wondrous Mountain (southwest of Antioch). This was a major pilgrimage site hosting a complex built between 541 and 591 around the column of Symeon the Stylite the Younger, a sixth-century saint whose way of life greatly inspired Lazaros and ended up shaping his subsequent career on Galesion.[72] While the Wondrous Mountain can be read as an anticipation of Lazaros' life on Galesion, a third mountain, Argeas, the highest peak of Cappadocia and Asia Minor (3,916 m), was the summit where the saint put himself to test:

When he reached Mt. Argeas, he wanted to climb it but he was stopped by those who lived there because it was winter. Lazaros, however, put his hope in our Lord Jesus Christ and His mother and started to climb. When he was halfway up the mountain, such a dense fog came down around him, as he used to relate, that,

---

[68] Talbot, 'Saintes montagnes', 267. On the early career of Patriarch Athanasius, see A. M. Talbot, 'The Patriarch Athanasius (1289–1293; 1303–1309)', *Dumbarton Oaks Papers* 27 (1973), 16–17. On Mount Ganos, see A. Kuelzer, 'Heilige Berge: das Ganos-Gebirge in Ostthrakien', paper presented at the 21st International Congress of Byzantine Studies, London, 2006, Panel VI.6 'Monastic mountains and deserts'.

[69] See Greenfield, *Life of Lazaros*, 6–9.

[70] 'The Muslims killed a lot of people, monks and laymen; even worse, many people who were afraid of physical death, alas, died spiritually by denying their faith and calling themselves Saracens instead of Christians. So these men [the Christian monks], these lights of the world and imitators of the holy apostles, departed from the Holy Land and were scattered here and there throughout the regions of [the Byzantine world]; and this persecution was the reason for the dispersal of these men over the whole world, just as long ago the murder of St. Stephen was for the apostles': Greenfield, *Life of Lazaros*, 103.

[71] Greenfield, *Life of Lazaros*, 108.   [72] Ibid. 109.

even though he strained his eyes, he could not see the right or left or anywhere else. He did not give up his attempt, however, but bent down and, using his hands to guide him, went up on. While he was climbing like this, he met a bear . . . and neither he nor it sensed the approach of the other until they came so close that they bumped into each other. The only explanation for this was that it was a device of the Evil One intended to frighten him into turning back, or rather God allowing this as a trial of faith and hope. The bear came to halt at their sudden collision and left the path, while Lazaros went on his way unhindered, heartly singing the Davidic psalms. When he had climbed up to the top he found that the door of the chapel had been securely barred. He opened it and went inside; when he had prayed, he came out and closed the door securely, and went down the mountain again.[73]

Holy mountains as ladders to heaven

During his ascent of Mount Argeas, Lazaros (almost prophetically) tasted and successfully overcame some of the temptations he was later to experience (and once again successfully overcome) on Galesion. The imagery of Lazaros' climb bears clear echoes to Moses' ascent of Sinai and Peter, James and John's ascent of Tabor. God manifests His mercy on the saint by miraculously sparing him from the bear in a 'dense fog', beyond vision and rational understanding. Lazaros proceeds with his ascent 'singing the Davidic psalms', and comes into closer contact with the divine on the top of the mountain. The reader though is not allowed to take part in Lazaros' prayer. The saint comes out from the chapel and again locks the door 'securely', before descending.

Lazaros' climb of Argeas can be situated within a much longer tradition of Byzantine mountain ascents inspired by the Scriptures. For the Byzantines, biblical peaks such as Sinai and Tabor were not only actual locations in which epiphanies 'physically' took place. They were first of all metaphorical spaces, 'maps' that guided the ascetic in his spiritual journey (and for this reason, they could also be 'transposed' to other parts of the Empire). In the writings of the Church fathers, ascents were used as metaphors for human deification through ascetic struggle. For example, Gregory of Nyssa (c. 335–after 394 CE) employed the image of Moses' ascent of Sinai to map out the three stages he identified with spiritual life. The saint narrated the path of Salvation as a spatial journey. *Katharsis*, or the purification of the soul from egoistical passions, *Phōtisis*, the enlightenment of the soul by the Holy Spirit, and *Theosis*, or union with God, were compared with the entry into a moonlit desert

---

[73] Ibid. 109–10.

night, followed by a movement to a fog-covered mountain, and finally, into the impenetrable darkness of a thick cloud.[74]

Inspired by the surrounding landscape, Saint John Climacus (525–606 CE), the abbot of Saint Catherine who had lived in a cave on the Mountain of Moses for forty years, took the metaphor further and wrote *The Ladder of Divine Ascent*, a handbook of spiritual life destined to become the most popular text of the Orthodox Church after the Holy Scriptures and service books.[75] The treatise was divided in thirty chapters (or rungs of the ladder), which corresponded to the thirty virtues with their opposite vices, as classified by Saint John – thirty as the age of Christ before baptism. The ascetics who successfully managed to resist the demons of temptations as they ascended the ladder were to be taken 'from terrestrial things to things divine'.[76] By contrast, those who fell victim to their passions would plunge into the abyss of sin (or the mouth of hell, as illustrated on manuscript illuminations and, since the eleventh century, on icons) (Fig 12.3).[77]

Tabor, the transfigured mountain, was an equally recurrent image in patristic writing. As Saint Gregory Palamas (1296–1359) wrote:

The blessed Moses, by virtue of the glory of the Holy Spirit which shone on his face [when he descended from Sinai] and which no man could bear to gaze upon, showed by this sign how the bodies of the saints would be glorified after the resurrection of the righteous ... Glorified by the divine light, the saints will always be with the Lord (Macarius of Egypt). According to the great Denys, that was the same light which illumined the chosen apostles on the Mountain [Tabor]: 'When we become incorruptible and immortal', he says, 'and attain the blessed state of conformity with Christ, we will ever be with the Lord as Scripture says' (I Thess. 4:17), gaining fulfilment in the purest contemplations of His visible theophany, just as it illuminated the disciples of the most divine Transfiguration. ... As Gregory the Theologian remarked, 'in my view, He will come as He appeared or was manifested to the disciples on the Mountain, the divine triumphing over the corporeal'.[78]

---

[74] Lane, *Fierce Landscapes*, 101–2.

[75] Archbishop Damianos, 'The icon as a ladder of divine ascent in form and color', in H. Evans, *Byzantium: Faith and Power, 1261–1557* (New York: Yale University Press, 2004), 335.

[76] Archimandrite Ignatios, *Agiou Iwannou tou Sinaitou Klimax* (Ōropos Attikēs: I. M. Paraklētou, 1998), 28.

[77] On visual renderings of the Klimax, see K. Weitzmann, 'Sinai peninsula: icon painting from the sixth to the twelfth century', in K. Weitzmann, M. Chatzidakis, K. Miatev and S. Radojčić (eds.), *Icons from South Eastern Europe and Sinai* (London: Thames and Hudson, 1968), xiii; E. Draghici-Vasilescu, 'Spiritual ascent in a Sinaite monastery: the icon of the Heavenly Ladder', *Series Byzantina* 7 (2011), 101–14.

[78] Gregory Palamas, *The Triads* (New York: Paulist Press, 1983), III.i.10, p. 72.

Figure 12.3 'The Ladder of Divine Ascent', icon from the monastery of Saint Catherine in Sinai, c. 1150.
Source: Reproduced by kind permission of the Holy Monastery of Saint Catherine, Sinai.

Byzantines superimposed biblical mountain imageries on non-biblical holy mountains and on the lives of their inhabitants. Saint Iōannikios' climb of Bithynian Olympus after the battle of Markellai (792), for example, was likened by his biographer to Moses' ascent of Sinai,[79] whereas Steven the Younger (*c.* 713–64 CE) compared Auxentius and the Monacheion hill to Mount Horeb, Carmel, Sinai, Tabor and Lebanon.[80] Athos, by contrast, was not only likened to biblical mountains, but also physically imprinted with their geographies. The subsidiary peak of Athos, a sort of 'Athos in miniature', having the same shape of the main peak reduced to almost half its size, was named 'New Carmel', and its summit 'peak of the Prophet Elijah'.[81] Here Saint Gerasimos (1506–79), like the biblical prophet, was said to have conducted 'a heroic existence, battling constantly against nature's elements – wind, thunder and lightening, rain and snow – and against the full guile of demons'.[82] The summit of Athos was instead traditionally associated to Tabor (and today it is topped by a small chapel dedicated to the Transfiguration).[83] Like the Mount of Transfiguration, the summit of Athos was commonly described by the Byzantines as a place for divine revelation. Here Saint Euthymius the Georgian (955–1028 CE) served a liturgy in which 'everyone who was there was covered with a heavenly light which made them fall down, unable to endure the brightness', whereas St Maximos Kausokalyviotēs (fourteenth century CE) was fed by the Mother of God with heavenly bread, after he endured demons and thunderstorms in a three-day solitary vigil. Similarly, after a long night of prayer, martyr monk James (sixteenth century), whose soul 'had already climbed Mount Tabor and seen the spiritual Jerusalem', had a vision in which he was told to go to Aetolia, where he was to die as a martyr.[84]

---

[79] Talbot, 'Saintes montagnes', 270. Saint Ioannikios is often portrayed on miniatures with a mountain. In two manuscripts of the *mēnologion* of Symeon Metaphrastes, the mountain is accompanied by the female personification of Mount Olymp: 'Ioannikios', in *Oxford Dictionary of Byzantium*, Kazhdan (ed.), 1006.

[80] Talbot, 'Saintes montagnes', 269.

[81] R. Dawkins, *The Monks of Athos* (London: G. Allen and Unwin Ltd., 1936), 259.

[82] Archimandrite Ioannikios, *An Athonite Gerontikon: Sayings of the Holy Fathers of Mount Athos* (Thessalonica: Publications of the Holy Monastery of Saint Gregory Palamas, 1997), 251. Saint Athanasius is also compared to Elijah: 'You inhabit Mount Athos as Elijah [inhabited] Mount Carmel' (*Menologion*, 5 July, 5th ode).

[83] Great Meteoron (*c.* 1350), the first and highest monastery in the Meteora complex (400 m above Kalambaka), is also dedicated to the Transfiguration. On the history of Meteora, see Dennis, 'Meteora' and Nikos Nikonanos, *Metéōra: ta monastēria kai hē historia tous* (Athens: Ekdotikē Athēnōn, 1992).

[84] See F. Halkin, 'Deux vies de S. Maxime le Kausokalybe, ermite au Mont Athos (XIVe s.)', *Analecta Bollandiana*, 54 (1936), 77–9. On James' vision, see *Orthodox Life Magazine* 29.6 (1979), 1–7. The liturgy celebrated by Saint Euthymius the Georgian is narrated in Ioannikios' *Athonite Gerontikón*, 282–3.

Ascents of holy mountain peaks were often used to illustrate saints' powers, or spiritual progress. Saint Euthymios the Younger (823–98 CE), for example, had attempted to dissuade his brethren from ascending Athos' peak, for which he could see no justification, but they stubbornly insisted despite the wintry weather. The monks ended up being caught in a snowstorm and would have died of the cold, if Euthymios (who apparently had decided to accompany them) had not managed to kindle a fire by merely blowing on a pile of firewood.[85] Saint Lazaros' ascent of Galesion, by contrast, is narrated as a lifetime enterprise and as a 'physical' manifestation of the saint's spiritual ascent. The mountain, Lazaros' biographer recounted, 'happened to be not only impassable and craggy and very rugged, but it was in addition waterless, and for these reasons able to offer much tranquillity to the person who went up there'.[86] Having thus found his ideal environment, Lazaros first settled himself in a cave once occupied by another holy man. After a few months, he moved on to a pillar he ordered a brother monk to build nearby, 'in the middle of a dry stream bed, in the open air as he wished'.[87] As a number of disciples started to gather around the saint, the first monastic community on Galesion (the Saviour) grew up around the pillar. Twelve years later, Lazaros left his pillar for a new one, which he had ordered to be built higher up the mountain, after a disagreement with some monks of the community over the continuous visits of a nun from Ephesus to the saint. Another community (of the Theotokos) gradually developed around the second pillar.[88] After an unspecified number of years, Lazaros (now probably in his seventies) moved further up the mountain on a third pillar, where the same pattern was repeated and a new community (the Resurrection, the third and larger on the mountain) was founded.[89]

Like Saint John Climacus and Gregory of Nyssa, Lazaros equated altitude with beatitude. For him, the way that leads to eternal life was as hard and rough as Mount Galesion.[90] Ascetics who went to see him in search of spiritual advice were told by the saint not to go down from the mountain 'into the world' for any reasons:

'If, as you say, you want to do this because you have been discouraged by being alone on the mountain', said the father to [a hermit], 'then go to the monastery of the Saviour and live there with the brothers as long as you like. But just don't go down from the mountain lest you meet with some unexpected evil.'[91]

---

[85] L. Petit, 'Vie et office de St. Euthyme le Jeune', *Revue de l'Orient chrétien*, 8 (1903), 201–2.
[86] Greenfield, *Life of Lazaros*, 123.   [87] Ibid. 141.   [88] Ibid. 12.   [89] Ibid. 13.
[90] Ibid. 208.   [91] Ibid. 148.

The monk, however, was pulled down from the mountain by the evil of his own will, just like the stumbling ascetics on the icon of Saint John's Ladder. As he was walking in the fields below the mountain, he fell into temptation and ate some cooked meat he had come across there.[92]

Visiting the saint for confession (or for spiritual advice) was itself both a spiritual and a physical exercise, whose 'geographies' reiterated those described by Gregory of Nyssa and Climacus. Having found repentance and contrition of their souls, visitors struggled to ascend the rugged mountain in order to reach Lazaros' pillar. On the spot, they also had to ascend a ladder in order to see the saint, as this was the only way to reach his cell on the top of the pillar. As with Saint John's *Klimax*, Lazaros' ladder served as a bridge between earth and the saint's 'heavenly dwelling', as revealed to a monk in a vision:

[Monk Photios] saw a building complex between heaven and earth, which boasted wonderful constructions ... Up above this complex, near heaven, he seemed to see another complex that was much more glorious and splendid than the former; indeed the human tongue is unable to describe its beauty and splendour. Between both complexes was a ladder set up that flashed more brightly than the sun's rays. The holy father [Lazaros] was sometimes to be seen lingering in the lower complex and apparently delivering words of instruction to some people, ... at other times ascending the ladder to the habitation near heaven and hastening to enter heaven.[93]

Holy mountains as Gardens of Eden

Holy mountains were not only ladders to be ascended. They were also fierce landscapes to be tamed, wildernesses to be transformed into gardens – once again, both physically and metaphorically. Most of the early desert fathers used to tend small garden plots where they lived, to have a food supply for their own use and for their visitors. The extreme arid environment of the Near East shaped the understanding of the garden as something that must be carved out of the wilderness and guarded from its depredations. Saint Antony the Great himself raised his vegetables in the desolate Inner Mountain.[94] Egeria, a Spanish nun visiting Sinai in the late fourth century, described 'a very pleasant garden in front of the church [in what will later become the monastic complex of Saint Catherine], containing excellent and

---

[92] Ibid. 148.   [93] Ibid. 175–6.

[94] A. M. Talbot, 'Byzantine monastic horticulture: the textual evidence', in *Byzantine Garden Culture*, A. Littlewood and H. Maguire (eds.) (Washington, D.C.: Dumbarton Oaks Research Library and Collection, 2002), 47.

abundant water' and recounted that 'although the mountain is rocky throughout, so that it has not even a shrub on it, yet down below [near its feet . . .], there is a little plot of ground where the holy monks diligently plant little trees and orchards, and set up oratories with cells near to them, so that they may gather fruits which they have evidently cultivated with their own hands from the soil of the very mountain itself'.[95] Gardens became an essential part of all Byzantine monasteries, and by the twelfth century, Archbishop Eustathios of Thessalonica criticized hermits who 'withdrew to mountains and, like the Cyclopes, did not plow or plant anything'.[96]

Gardening was not simply a necessary source of living for the monks: it meant mapping a spiritual landscape on the physical one. As with mountain climbing, gardening was a metaphor for inner transformation, one that (once again) found its origins in the Scriptures. The Old Testament, especially Isaiah's prophecies, abounds with imagery of wilderness transformed by God into fertile land.[97] As Claudia Rapp observed, the monks who made the Egyptian desert their home were perceived as a testament to the transformative power God continued to exercise down to their day: the transformation of the physical land was a reflection of the monks' beautifully transformed nature.[98] Indeed, as Gregory of Nyssa wrote in his homily XV on Canticles, the sinner was like a wasted field, but through repentance he could once again turn into 'a garden cultivated by Christ'.[99]

Carving a cultivated plot out of a wild landscape was understood by the fathers as an act of domestication not only of nature, but also of passions. Saint Athanasius' foundation of the Great Lavra, the first coenobitic monastery on Mount Athos, on one of the most desolate and least accessible locations on the peninsula, is certainly the most notorious and perhaps dramatic example of 'domestication' of a holy mountain. When he set to cast the foundations of the monastery, the first thing

---

[95] M. McClure and C. Feltoe, *The Pilgrimage of Etheria* (New York: MacMillan, 1920), 8, 5. See also J. Wilkinson, *Egeria's Travels* (Oxford: Aris and Phillips, 2006). It is far from certain that Egeria was either a nun or from Spain.

[96] Eustáthios quoted in Talbot, 'Byzantine monastic horticulture', 46.

[97] Is. 51: 3; Is. 35: 1–2; Is. 41: 19–20.   [98] Rapp, 'Desert, city and countryside'.

[99] 'The [true husbandman] is he who at the beginning in Paradise cultivated human nature, which the heavenly Father planted. But the wild boar [Psalm 80:13, from the forest] has ravaged our garden and spoiled the planting of God. That is why [the husbandman] has descended a second time to transform the desert into a garden, ornamenting it by planting virtues and making it flourish with the pure and divine stream of solicitous instruction by means of the word': Gregory of Nyssa, quoted in G. Williams, *Wilderness and Paradise in Christian Thought: The Biblical Experience of the Desert in the History of Christianity and the Paradise Theme in the Theological Idea of the University* (New York: Harper and Brothers Publishers, 1962), 40.

Athanasius did was to 'cut down trees in the thick forest and to make level areas in the rough ground'. Water for irrigation was conducted from a lofty site through a system of pipes and trenches running along the mountain's steep slopes for about 12 km.[100] Athanasius envisaged the construction of his Lavra as a titanic struggle to tame wilderness – one that paralleled his ascetic inner struggle to suppress his passions. In the *typikon* (foundational charter) of the Lavra (963 CE), the saint recorded:

> But how much hard work, the afflictions I suffered, the trials and hardships I endured, the expenditures I put out for quarrying of stone, excavating, heaping up earth, transporting stones, the rooting up, the cutting down, the removal of branches, bushes, and trees, in order to build the holy church of the most holy Mother of God, and setting the entire Lavra in place, to discuss all this in detail would take longer than the time at my disposal. It is enough that the Lord alone knows exactly what I mean.[101]

If so much struggle granted Athanasius and his fellow monks the kingdom of heaven, the physical result 'in this world' was the re-creation of the new 'miniature terrestrial paradise' that was meant to be the Byzantine monastery.[102] The Edenic garden represented one of the moral spaces most central to the Judaeo-Christian tradition. The garden is defined by the boundary the planter sets between the cultivated land and wasteland. Figuratively, the primary act of gardening is that of fixing a line between known and unknown, rational and irrational, good and evil. Used by Septuagint translators to render Eden in the Old Testament, the ancient Greek word *paradeisos* originally designated an 'enclosed park'. In ancient Greek literature, the historian Xenophon (431–355 BCE) for the first time employed the term to indicate 'a large well-watered field containing trees, flowers and animals, and surrounded by a wall' – like the monastery, a well-defined 'insular' space separated from the surrounding world.[103]

'Islands on the land', mountains themselves were also narrated by the Church fathers and later by Byzantine literati as 'natural' gardens of Eden, or *loci amoeni*. While the desert was generally perceived as an open and uniform, if not monotonous, space without the delineation of natural boundaries, mountains' varied environment made for unique

[100] Talbot, 'Byzantine monastic horticulture'.

[101] Athanasius, '*Typikón* of Athanasios the Athonite for the Lavra Monastery', G. Dennis (tr.), in Thomas and Constantinides Hero, *Byzantine Monastic Foundation Documents*, 252.

[102] 'The enclosure wall was the primary necessity of the coenobium and its important structural and functional features were also confirmed in monastic *typika*': S. Popovic, 'Dividing the indivisible: the monastery space – secular and sacred', *Recueil des travaux de l'Institut d'études Byzantines* 44 (2007), 52.

[103] On the etymology of Eden, see Scafi, *Mapping Paradise*, 34–5.

self-contained microcosms. In such spaces the early fathers found a perfect setting for the poetic contemplation of the handcrafts of the Creator. As Alexander von Humboldt noted, 'the fathers of the Church in their rhetorically correct and often poetically imaginative language ... taught that the Creator showed Himself great in inanimate no less than animate nature, and in the wild strife of elements no less than in the still activity of the organic development.'[104] 'If the aspect of columnades of sumptuous buildings would lead thy spirit astray, look upward to the vault of heaven, and around thee in the open fields', Saint John Chrysostom wrote.[105] Indeed, for Gregory of Nyssa, 'he who contemplates [nature] with the eye of the soul, feels the littleness of man amid the greatness of the universe.'[106]

Mountains allowed the hermit to access both the variety and the majestic grandeur of the cosmos while remaining safely anchored within a self-enclosed space (like the Garden of Eden). For example, Basil of Caesarea (330–79 CE) wrote to Gregory of Nazianzus:

I believe I may at last flatter myself with having found the end of my wanderings. ... A high mountain, clothed with thick woods, is watered to the north by fresh and ever-flowing streams. At its foot lies an extended plain, rendered fruitful by the vapors with which it is moistened. The surrounding forest, crowded with trees of different kinds, encloses me as in a strong fortress. The wilderness is bounded by two deep ravines; on the one side, the river, rushing in foam down the mountain, forms an almost impassable barrier, while on the other all access is impeded by a broad mountain ridge. My hut is situated on the summit of the mountain that I can overlook the whole plain, and follow throughout its course the Iris ... Shall I describe to thee the fructifying vapors that rise from the moist earth, or the cool breezes wafted over the rippled face of the waters? Shall I speak of the sweet song of the birds, or of the rich luxuriance of the flowering plants? What charms me beyond all else is the calm repose of this spot. It is only visited occasionally by huntsmen; for my wilderness nourishes herds of deer and wild goats, but not bears and wolves. What other spot could I exchange for this? Alcmaeon, when he had found the Echynades, would not wander further.[107]

The mountain described by Saint Basil encompasses all the elements of terrestrial paradise: trees, water streams, animals (wild but not dangerous to man). At the same time, as with any 'high mountain', its peak is also close to the vault of heaven and its stars, what the saint calls in his homilies on the *Hexaemeron* 'those everlasting blossoms of heaven' that

---

[104] A. von Humboldt, *Cosmos: A Sketch of the Physical Description of the Universe*, E. Otté (tr.) (Baltimore and London: The Johns Hopkins University Press, 1997 [1858]), vol. II, 39.
[105] Quoted ibid. 42.   [106] Quoted ibid.   [107] Quoted ibid. 40–1.

elevate the soul from the visible to the invisible – a reminder, according to von Humboldt of 'the mildness of the constantly clear nights of Asia Minor'.[108] Basil's hut, located on the top of the mountain, functions as an observatory of the beauty of Creation. From his lonely mountain hut, the saint's eye wanders over the humid leafy roof of the forest below.[109] From above, it dominates a self-contained microcosm and makes the holy man feel one with the Creator of the cosmos. Basil's mountain thus becomes a sort of 'anti-Mount of Temptation' (the highest mountain where Christ was tempted by the Evil with 'all the kingdoms of the Earth and the glory of them').[110] Unlike the tempted Christ, Basil is offered an edifying God's-eye view, as he contemplates the works of God, and not those of men. The landscape traversed by the saint's gaze is that of the transfigured cosmos announced by Apostle Paul, in which 'God shall be all in all'.[111]

Basil's mountain is not only a biblical Garden of Eden. As von Humboldt once again observed, the poetical and mythical allusion at the close of the letter falls as 'an echo from another and earlier world'; a world the saint was intimately familiar with from his early studies in rhetoric in Athens.[112] The imagery of biblical Eden overlaps with that of the classical *locus amoenus*, populated by melodious birds and deer, and permanently kissed by the soft breeze of eternal spring.[113] In a sense, it anticipates a Byzantine ekphrastic tradition of non-biblical (yet 'formal') holy mountains as *loci amoeni* that became popular between the eleventh and fourteenth centuries. This tradition is strongly connected to the imagery as well as popularity of pleasure gardens in middle (and late) Byzantine culture. Byzantines had a long-standing fascination with pleasure gardens. Byzantine emperors and noblemen had their own personal 'paradises' built at home in Constantinople, according with the latest fashions and their personal tastes. Byzantine gardens ranged from small plots to extensive parks embellished with fountains, canals and exotic animals. Sometimes they could also be terraced with the upper level still providing space for groves of trees that appear to be 'suspended in the air', as the eleventh-century polymath Michael Psellos wrote.[114]

---

[108] Quoted ibid. 41.   [109] Ibid.   [110] Matt. 4: 8.   [111] I Cor. 15: 28.

[112] Von Humboldt, *Cosmos*, 41; see also H. Beyer, 'Der Heilige Berg in der byzantinischen Literatur', *Jahrbuch der Oesterreichischen Byzantinistik* 30 (1981), 177; S. Hildebrand, *The Trinitarian Theology of Basil of Caesarea* (Washington, D.C.: Catholic University of America Press, 2007), 19.

[113] See 'locus amoenus' in *Oxford Dictionary of Byzantium*, Kazhdan (ed.), 880.

[114] Maguire, 'Paradise withrawn', 23; H. Maguire, 'Gardens and parks in Constantinople', *Dumbarton Oaks Papers* 54 (2000), 251-64, at 261.

Psellos himself was the initiator of the aforementioned meso-Byzantine ekphrastic tradition of holy mountains as *loci amoeni*. Byzantine *ekphraseis* of *loci amoeni* were literary rhetorical descriptions which found their roots in classical literature (the first *ekphrasis* being Homer's description of King Alkinoos' wonderful garden and Calypso's grotto). They served as useful fictions for self-edification, whose primary goal was to make present not the actual picture, but rather the spiritual reality behind it.[115] In his funeral oration to Patriarch John Xiphilinos, Psellos describes Bithynian Olympus as 'a second paradise' and 'second heaven', full (like Basil's mountain) of woods, water streams and singing birds:

This mountain looks like the one the Greeks call heavenly flowery place, where only the best souls dwell; where, as it is believed, are clouds and the ether which is decorated with highest stars similar to blossoms. . . . Wherever you turn your gaze you are filled with delight because [the mountain] is [equally] divided between slopes and plains and neither competes to get the winning place. And if one walks on the plain area, he gives a negative vote to the mountains. If he gets to the hilly area, he scorns the horse-trodden fields. Everything is covered by vegetation on the mountains, even though the waters that spring from the earth flow elsewhere and pour into the rivers, and elsewhere they sprout and water plants. Plane-trees, cypresses, and every other highest tree stand up on line, as if under command; others weave their branches in their thick tufts, making their shadow look like a true paradise . . . If the wild cedars, the myrtles and the *schoinaria* receive someone under their shadow, with what breaths of fresh air do they restore him! And with what songs do birds delight his ears when at high noon they sweetly cheep amidst the leaves! As to the long shade of the osier, only Plato's tongue could praise it.[116]

Similar descriptions can be found three centuries later, this time referring to Mount Athos. In a chrysobull issued in 1312 Emperor Andronicus Palaeologus II defined the mountain-peninsula 'a second paradise, a starry sky, and the abode of all virtues'. Few decades later, Nicephorus Gregoras, Andronicus' chronicler and *chartophylax* (secretary), portrayed Athos as a self-sufficient agricultural institution modelled on Plato's ideal polis. In his description, nature and the virtues of monastic life fused in a harmony of colours, scents and sounds, 'as if flowing out of a treasure house'.

---

[115] L. James and R. Webb, 'To understand ultimate things and enter secret places: ekphrasis and art in Byzantium', *Art History* 14 (1991), 1–17; Maguire, 'Paradise withrawn'; J. Heffernan, *The Museum of Words: The Poetics of Ekphrasis from Homer to Ashbery* (University of Chicago Press, 1993).

[116] M. Psellos, 'Epitaphios eis ton makariōtaton patriarchēn kyr Iōannēn ton Xiphilinon', K. Sathas (ed.), *Mesaiōnikē Bibliothēkē*, 4 (1972 [1874]), 442–3. See also Beyer, 'Der Heilige Berg', 180 and Talbot, 'Saintes montagnes', 275. Psellos stayed with Xiphilinos, his old friend and colleague at the university, on Olympus in 1055. The previous year he had entered the monastery of Theotokos tēs Ōraias Pēgēs as a monk, but he was soon recalled to court by Empress Theodora.

An amiable mantle of varied woods and cultivated lands covered the peninsula during all the seasons: like Eden and the *loci amoeni* chanted by ancient authors, Athos was blessed by eternal spring. Among its woods, in the early morning, one could hear the music of the nightingale, 'as if chanting and praising the Lord together with the monks'. Little water streams flowed quietly, in harmony with the monks' silent way of life, because, the author explained, 'Mount Athos *per se* offers many occasions for inner quietness (*hēsychia*) to those who desire to live a celestial life on earth'.[117]

Conclusion

Psellos' Edenic description of Bithynian Olympus presents a stark contrast with fierce and inhospitable Galesion (as described in the life of Lazaros), in spite of being written at about the same time. The former mount is portrayed as an idyllic place for rest and contemplation, whose plants reflect the botanical totality of Eden and whose verdant plains 'win' over cragginess; the latter as a ladder ascended by monks, pilgrims, and all sort of laypeople seeking repentance. Similarly, Gregoras' depiction of Athos as a *jardin des delices* stands in marked contrast with Athanasius' account, in spite of describing the same place. Once again, Gregoras' Athos is a green Edenic utopia in which there is no place for the rocky mountain summit, whereas for Athanasius it is a secluded wild spot to be tamed. Of course, as Alice Mary Talbot observed, such different tropes can be justified by the each author's aims and educational background. Broadly speaking, descriptions of holy mountains as *loci amoeni* were the work of literati (or saints, like Basil, educated in the Classics), whereas holy mountains as arenas for ascetic struggle were mostly employed in hagiographic literature to stress the virtues of the mountain's founding fathers and successive occupants.[118] Earlier descriptions by the fathers of the Church, such as Basil the Great's, seem to blur this binary, offering poetic edification through sceneries that are at once beautiful *and* sublime.

Whether portrayed as ladders to heaven or as gardens of Eden, Byzantine holy mountains were symbolic spaces that allowed the pious reader to map out his or her own ascetic path, for, as John of Damascus wrote in the eighth century, it is 'through bodily sight [that] we reach spiritual contemplation' – and geography delights us because of its way of helping us localize and grasp the abstract by way of

---

[117] The original text by Gregoras is quoted in Z. Papantōnios, *Aghion Oros* (Athens: n.p., 1934), 83.

[118] Talbot, 'Saintes montagnes', 275.

the visible.[119] Holy mountains were nevertheless at the same time 'real' places that owed much of their meaning and effectiveness as ascetic dwellings (and/or as settings for poetic contemplation) to their different locations and physical geographies. So strong was their association with monastic life that in middle-Byzantine Greek the term 'mountaineer' was sometimes used as a synonym for 'monk'.[120] Holy mountains played a crucial role in Byzantine spiritual culture, as well as in the lives of holy men and ordinary laypeople. Located within the boundaries of an Empire destined to vanish, Byzantine holy mountains were of course not immune from the turns of history. Many of these peaks disappeared from the map as holy mountains – Bithynian Olympus, Galesion, Latros and other holy peaks in Asia Minor all fell to the Turks in the fourteenth century. Others never gained a 'top ranking' in the holy mountains list (as in the case of Ida, for example).[121] But others endured. Meteora, for instance, flourished beyond the sixteenth century, a period when other monastic republics declined as a result of insecurity or Turkish pressure. Athos too knew periods of revival after the Fall of Constantinople, and so did Sinai.[122] No longer competing with the great Anatolian centres, nor with other famous holy peaks, Sinai and Athos (and to a lesser extent Meteora) became, together with the Holy Land, the great geographical foci of Orthodox Christianity: not only ladders to heaven and gardens of Eden, but true spiritual islands and beacons to which the faithful in the Ottoman-dominated lands increasingly turned their eyes and souls.[123]

---

[119] John of Damascus quoted in Herrin, *Byzantium*, 105.

[120] Nicol, *Meteora*, 25, n.5.

[121] Ida was a refuge for monks during iconoclastic persecutions (earliest testimonies date back to the 730s), but it never reached the fame of Olympus and other holy mountains. See Papachrysanthou, *Athōnikós monachismós*, 84.

[122] See 'Meteora', in *Oxford Dictionary of Byzantium*, Kazhdan (ed.), and 'Holy mountains', *Blackwell Dictionary of Eastern Christianity*. After the Fall of Constantinople (1453) and of Sinai to the Turks (1517), Saint Catherine and the monastic centres of Athos and Meteora managed to retain their autonomy in exchange for the payment of an annual tribute to the Porte.

[123] E. Contescu, 'L'image du Mont-Athos dans l'exonarthex de Polovraci', *Balkan Studies* 14 (1973), 311. I would like to thank Claudia Rapp for her helpful suggestions on the first draft of this chapter and the Fathers of Docheiariou for their precious inputs and help with translations from Byzantine Greek.

13 Journeying to the world's end?

Imagining the Anglo-Irish frontier in Ramon de Perellós's *Pilgrimage to St Patrick's Purgatory*

Sara V. Torres

In September 1397, the Aragonese courtier and diplomat Ramon, first viscount of Perellós and second viscount of Roda, made a pilgrimage to St Patrick's Purgatory. In his narrative account of the journey, he describes travelling from Avignon to Paris, across the English Channel, from Dover westwards through England to Wales, across the Irish Sea, then finally from Dublin northwards to Lough Derg.[1] While this account draws heavily on Henry of Saltrey's *Tractatus de Purgatorio Sancti Patricii* (*c.* 1180–4), it also contains original material, including descriptions of Perellós's encounters with Anglo-Irish and Irish lords.[2] Perellós's text displays features common in medieval travel writing, from the interest in topography and history of conquest, to the ontological appraisal of exotic wonders, to a chivalric emphasis on courage.[3] To these we can add a diplomat's attention to the delineation of geopolitical spaces and to the construction of cultural and historiographical legacy.

---

[1] For the text of Ramon de Perellós's *Vitage al Purgatori*, see Dorothy M. Carpenter, 'The Pilgrim from Catalonia/Aragon: Ramon de Perellós, 1397', in *The Medieval Pilgrimage to St Patrick's Purgatory: Lough Derg and the European Tradition,* Michael Haren and Yolande de Pontfarcy (eds.) (Enniskillen: Clogher Historical Society, 1999), 99–119; hereafter cited parenthetically in the text. Carpenter's English-language composite translation of the non-derivative portion of Perellós's text is from the earliest extant manuscript version, which is in Occitan: Auch, Archives Départmentales du Gers MS. I-4066; Carpenter supplements a missing manuscript folio (xxix) with Ramon Miquel y Planas's edition of the earliest Catalan text in *Llegendes de l'altre vida: viatges del Cavaller Owein y de Ramón de Perellós al Purgatori de Sant Patrici* (Barcelona: F. Giró, 1914).

[2] For the text of the *Tractatus de Purgatorio Sancti Patricii,* see Henry of Saltrey, *Saint Patrick's Purgatory: A Twelfth Century Tale of a Journey to the Other World,* Jean-Michel Picard (tr.) and Yolande de Pontfarcy (ed.) (Dublin: Four Courts Press, 1985); hereafter cited parenthetically in the text. For the textual history of the *Tractatus,* see Jacques Le Goff, *The Birth of Purgatory,* Arthur Goldhammer (tr.) (University of Chicago Press, 1984 [1981]), 397–8, n. 20.

[3] On medieval travel writing, see Mary B. Campbell, *The Witness and the Other World: Exotic European Travel Writing, 400–1600* (Ithaca: Cornell University Press, 1988). For pilgrims' narratives, see Diana Webb, *Pilgrims and Pilgrimage in the Medieval West* (London: I. B. Tauris, 1999) and Dee Dyas, *Pilgrimage in Medieval English Literature: 700–1500* (Rochester, NY: Brewer, 2001).

Perellós's journey has two parts, the first over land and sea, and the second subterranean. Together, these two movements demonstrate the implicit 'political investment' that marks a geographical endeavour.[4] Within the cartographic imagination of his itinerary, Ireland exists as a geographical extremity, an alternative terrain on the periphery of English linguistic and juridical cultures. Perellós's narrative attends to the delineation of national and imperial frontiers, exposing the many potential topographical, political and cultural dimensions of boundaries – and the problematic porousness that compromises their policing. Perellós anachronistically draws on Henry of Saltrey's hagiographical memory of St Patrick's conversion of the Irish in order to describe the resistance of the Gaelic Irish to Plantagenet rule. The central strategy of his pilgrimage travelogue is achieved through a sustained ethnological analogy between the Berbers of North Africa and the highly exoticized, 'nomadic' Gaelic Irish. In claiming that the Gaelic Irish are comparable to 'Saracens', Perellós reproduces hostile Anglo-Irish rhetoric about their 'Irish enemies', whose encroachment upon lands held by the Anglo-Irish colonial aristocracy during the fourteenth century made direct Plantagenet intervention necessary (first in the person of Lionel of Clarence, and then by Richard II himself).

The second part of Perellós's narrative describes his descent into the cave at Lough Derg. St Patrick's Purgatory serves as the aperture between the earthly world fixed in historical time and the otherworldly space of Purgatory, with its markedly different relationship with historical time. According to tradition, the entrance to Purgatory at Lough Derg was created through divine intervention to assist Patrick in his work of evangelization and conversion. Patrick represents both a Rome-based ecclesiastical authority and a British and Gaulish cultural heritage straining to extend its influence. His mission to make the Irish Christian is at once spiritual and cultural; he must subdue them from pagan 'barbarity' to culture as he incorporates them geographically into western Christendom. Long after the fifth-century mission of Patrick, this language of 'subduing', as crystallized in conversion legends, was appropriated by the Anglo-Norman kings (especially Henry II) eager to exert their political influence over Ireland, and continued to be used by Plantagent rulers seeking to culturally and linguistically 'subdue' the politically resistant Irish. The place of St Patrick's Purgatory witnesses the historical moment of Irish conversion and thus engages with multiple

---

[4] See Sylvia Tomasch, 'Introduction: medieval geographical desire', in *Text and Territory: Geographical Imagination in the European Middle Ages*, Sylvia Tomasch and Sealy Gilles (eds.) (Philadelphia: University of Pennsylvania Press, 1998), 1–12, at 5.

temporalities that both record and relive the spiritual and cultural conquest of the Roman church over Ireland. Perellós's narrative strategies evoke the Anglo-Norman historiographical agenda of works such as Gerald of Wales's *Expugnatio Hibernica* which seek to redeploy the antiquated discourses of barbarism to serve Henry's project of political incorporation.

The geographical placement of St Patrick's Purgatory embodies a fundamental spatial paradox; it occupies a middling space in the hierarchy of heaven, earth and hell, yet it is consigned to an earthly cartographic periphery. The cognitive tension produced by these convergent spatial models of political extremity and spiritual centrality puts pressure on the utility of purgatory as a space of historiographical reflection in which central matters of polity receive dialectical evaluation. Like the twelfth-century underground caves discussed by Monika Otter in *Inventiones*, the Purgatory at Lough Derg serves as a site of historiographical self-reflection; the preservation of the past is achieved not through the material 'preservation of objects from the "other world" – a world that, when it has a name, is often historicized as Roman antiquity', but rather through a textual transmission that privileges the topographical qualities of a landscape divinely marked by conversion and conquest.[5] Perellós's text identifies Ireland as both a political reality and a spiritual model, its two spatial resonances signalling both actual topographies and the spiritual journeys of penance.

The postmortem displacement of the Aragonese court to the Purgatory – Perellós discovers not only the soul of his deceased king Juan I there, but also several personages familiar from his court – speaks to the political utility of this liminal space at the edge of the European map. The revelatory potential of Purgatory, heightened by the ontological 'otherness' of its peripheral placement, enables Perellós to develop a rhetorical language of concealment and privileged access. Perellós performs the journey of the *Tractatus*'s protagonist, the penitent knight Owein, in order to discern the spiritual state of his own deceased king. His pilgrimage narrative, which itself arises from a documentary impulse to gloss history, functions as an authenticating text for Catalo-Aragonese theological and political authority at the same time that it exculpates his counsel in the king's misgovernment.

Perellos's geographical and spiritual topographies model advantageous reshapings of the late fourteenth-century political terrain. His pilgrimage narrative performs ideological work by creating analogies between the

---

[5] Monika Otter, *Inventiones: Fiction and Referentiality in Twelfth-Century English Historical Writing* (Chapel Hill: University of North Carolina Press, 1996), 129.

political geographies of England and the Crown of Aragon that minimize both the domestic turmoil at the courts of each kingdom and the implications of schismatic disaffection between them. The Crown of Aragon and England – sometime allies, sometime enemies in the intermittent conflicts of the Hundred Years War – engaged in a diplomatically ambivalent relationship heavily dictated by the shared boundaries of each with France, and the consequent religious-ecclesiastical and cultural repercussions of policies dictating French alliance or enmity.[6] Perellós would have been aware of the significance of traversing schismatic, national and physical borders as he travelled to and through Ireland. He variously elides and calls attention to such perceptions, the ultimate effect of which is to correlate the political boundaries of England with those of the Crown of Aragon. As a political agent of the Crown of Aragon, Perellós's diplomatic sensibilities were attuned to the multifaceted role of central administration in establishing and maintaining political hegemony over such frontierlands, as well as to the long-term investments necessary to and inherent in territorial expansion. Perellós's attention to issues touching upon the consolidation of royal power in a satellite territory can be seen as contingent to his own domestic preoccupations; Aragon, like England, had expended resources maintaining a colonial or military presence on territorial islands (Sardinia, the Kingdom of Majorca, and, eventually, Sicily). Through ethnological analogies, Perellós creates an imaginative religious frontier along the regional and hazily defined Anglo-Irish borders in Ireland, drawing a cartographic parallel between the British Isles and Iberia, and the Atlantic and Mediterranean worlds, which subtly underscores the failure of both to secure their boundaries from hostile military encroachments.

Underlying the spatial terms of Perellós's voyage are its spiritual dimensions. His stated purpose, to find the soul of his deceased lord in Purgatory, transcends the traditional disciplinary boundaries of geography, yet is intimately dependent upon the interdisciplinarity – the shared epistemic boundaries – of cartography and theology.[7] The doctrinally

---

[6] In the case of England, these borders with France included not only those along the Channel but also the land borders of the English duchy of Aquitaine. For Anglo-Aragonese relations, see P. E. Russell, *The English Intervention in Spain and Portugal in the Time of Edward III and Richard II* (Oxford: Clarendon, 1955), 5, and Nigel Saul, *Richard II* (New Haven and London: Yale University Press, 1997), 225–6. See also T. N. Bisson, *The Medieval Crown of Aragon: A Short History* (Oxford: Clarendon, 1986); Teofilo F. Ruiz, *Spain's Centuries of Crisis: 1300–1474* (Malden, Mass.: Blackwell, 2007) and Joseph F. O'Callaghan, *A History of Medieval Spain* (Ithaca: Cornell University Press, 1975).

[7] See Daniel K. Connolly, 'Imagined pilgrimage in the itinerary maps of Matthew Paris', *The Art Bulletin* 81 (1999), 598–622, at 606.

sanctioned existence of purgatory as a physical place, and the hagio-graphic identification of this place as Lough Derg, gives Perellós access to knowledge beyond the bounds of worldly human scholastic pursuit. Moreover, the perception of Ireland (especially to a Mediterranean visitor) as cartographically peripheral, bolstered by the emphatic assertions of Anglo-Norman writers regarding its exotic 'barbarity', facilitates the placement of this liminal 'underworld' at the threshold of civilization.[8]

Perellós's desire to find his lord the King and determine whether or not he was in Purgatory indicates the extent to which his journey is also an intellectual exploration of just rule in narrative form. Beyond the broadly political implications of territorial expansion and domination, Perellós's 'soul-searching' for and on behalf of his monarch is indicative of an interest in the ethical and moral constraints upon a monarch. Like Dante before him, Perellós uses the geography of Purgatory to evaluate a temporal ruler. Through his narrated performance of loyalty and discretion, Perellós displays an investment in postmortem 'life' of the king in a way that is immediately relevant to Perellós's own personal interests. Perellós casts Juan I in the role of individual penitent, endowing him with an interiority that is at once public and private – for the king suffers in St Patrick's Purgatory along with his courtiers. Finally, Perellós identifies his own place within the world he describes – as a traverser of boundaries and a privileged observer of and commentator on political power, both at home and abroad.

Pilgrimage to Lough Derg: Ireland and geographical extremity

The depiction of the Irish as uncultured and resistant to reform – whether religious or social – is ingrained in Henry of Saltrey's *Tractatus*, as well as in earlier hagiographic texts such as St Bernard's *Vita sancti Malachiae* (1152).[9] The *Tractatus*, in both Henry of Saltrey's original Latin and its many vernacular recensions and adaptations, was the main form of textual transmission of the legend of St Patrick's Purgatory

---

[8] On the portrayal of the Gaelic Irish as 'other', see: Keith D. Lilley, 'Imagined geographies of the "Celtic Fringe" and the cultural construction of the "Other" in medieval Wales and Ireland', in *Celtic Geographies: Old culture, New times*, David C. Harvey, Rhys Jones, Neil McInroy and Christine Milligan (eds.) (London: Routledge, 2002), 21–36; Kathleen Biddick, 'The cut of genealogy: pedagogy in the blood', *Journal of Medieval and Early Modern Studies* 30 (2000), 449–62, at 454–5; Anngret Simms, 'Core and periphery in medieval Europe: the Irish experience in a wider context', in *Common Ground: Essays on the Historical Geography of Ireland: Presented to T. Jones Hughes*, William J. Smyth and Kevin Whelan (eds.) (Cork: Cork University Press, 1988), 22–40.

[9] See Haren and de Pontfarcy (eds.), *Saint Patrick's Purgatory*, 28–9.

before Jacobus da Voragine's *Legenda aurea*. According to the *Tractatus*, God miraculously created the purgatory at Lough Derg in order to aid St Patrick in his evangelization of the pagan Irish. Patrick 'worked earnestly at dissuading from evil the savage souls of the men of this land, frightening them by the torments of hell and strengthening them in good by promising them the joys of paradise' (45–6). Neither miracles nor preaching, the Irish claimed, would be sufficient to compel them to convert, 'unless one of them could witness both the torments of the wicked and the joys of the just ... they would have more confidence in things seen than in promises' (47). The burden of proof fell upon Patrick to provide eyewitness evidence of postmortem retribution and reward, which he was able to do through miraculous intervention (47–8). The *Tractatus*'s emphasis on the truculence of the Irish is retained in the *Legenda aurea*: 'He preached throughout Ireland but with very meager results, so he besought the Lord to show some sign that would terrify the people and move them to repentance'.[10]

The *Tractatus*, with its emphasis on the theology of absolution, unites the devotional context of penitential pilgrimage with a strong interest in chivalric endeavour.[11] Its protagonist Owein is armed with 'the weapons of spiritual chivalry' (56), which include the invocation of Jesus' name to protect him from the demonic torment of the purgatory. Owein's encounters with the demons are described in military terms; we recognize the trope of the hero's arming familiar from the many Virgilian echoes in the medieval literary tradition:

Thus instructed for a new kind of chivalry, the knight, who in the past had bravely fought men, is now ready to give battle bravely to demons. Shielded by Christ's arms he waits to see which of the demons will first challenge him to a fight. He is arrayed with the breastplate of justice; as a head bears a helmet, his spirit is crowned with the hope of victory and eternal salvation and he is protected by the shield of faith. Also, he holds the sword of the spirit which is the word of God: he piously invokes the Lord Jesus Christ to protect him with his royal battlements and prevent him from being overcome by his relentless enemies. (54–5)

The knight Owein's violent past is opposed to his new asceticism as a 'soldier of Christ' (63), facilitating the transition from a purely monastic, Cistercian devotional focus to one that encompasses the secular subject. This thematic concordance of pilgrimage and chivalry is also evident in Owein's pilgrimage to the Holy Land after emerging from St Patrick's

---

[10] Jacobus de Voragine, *The Golden Legend: Readings on the Saints*, 2 vols., William Granger Ryan (tr.) (Princeton University Press, 1993), vol. I, 194.
[11] See Richard W. Kaeuper, *Holy Warriors: The Religious Ideology of Chivalry* (Philadelphia: University of Pennsylvania Press, 2009).

Purgatory, despite having already been purged of his sins. Vernacular recensions of the *Tractatus* likewise demonstrate a generic indebtedness to romance, as is the case with the English version of St Patrick's Purgatory in the Auchinleck Manuscript (National Library of Scotland, Adv MS 19.2.1).[12] Historical pilgrimage accounts also bear witness to this chivalric emphasis, such as those of the Italians, Malatesta 'Ungaro' of Rimini and his companion Niccolò Beccaria of Ferrera (1358), and the Hungarian nobleman Laurence of Páztho (1411).[13] In some cases, the manuscript contexts of such pilgrimage accounts offer insights into the generic valences of the texts. *The Vision of William of Stranton*, from the first decade of the fifteenth century, survives in a manuscript also containing *Mandeville's Travels*, the Middle English *Vision of Tundale*, and *Sir Gowther*.[14] Travel narratives relating to Lough Derg continued to exert a distinctly chivalric appeal even after the Middle Ages.[15]

The *Vitage* of Ramon de Perellós stands out among the surviving pilgrims' narratives in its generous treatment of the journey to Lough Derg. By contrast, both Froissart's sceptical recording of William de Lisle's visit to St Patrick's Purgatory and William Stranton's vision discuss only the purgatory and its immediate environs; other pilgrims, such as Ghillebert de Lannoy, describe the journey to Lough Derg without any hint of exoticizing embellishment.[16] Perellós's narrative is therefore exceptional in its close textual integration of Saltrey's text (via a French recension) with his own original material, which includes ethnographic descriptions of the Gaelic Irish.[17] In particular, his heightened perception of the cultural alterity of the Gaelic Irish is central to Perellós's foregrounding of geopolitical issues. Perellós's claim that the Gaelic Irish are analogous to Saracens draws on the theme of

---

[12] Robert Easting makes this point in 'The English tradition', in *The Medieval Pilgrimage to St Patrick's Purgatory*, 58-82, at 64.

[13] See Diana Webb, *Pilgrims and Pilgrimage in the Medieval West* (London: I. B. Tauris, 1999), 226–7.

[14] See Easting, 'The English tradition', 79. The textual history of *The Vision of William of Stranton* is given in Easting, 'The English tradition', 67–8; the *Vision* survives in two British Library manuscripts, MS Royal 17 B.xliii and MS Additional 34193. For the text, see Robert Easting (ed.) *St. Patrick's Purgatory: Two Versions of Owayne Miles and the Vision of William of Stranton together with the long text of the Tractatus de purgatorio Sancti Patricii* (Oxford: Published for the EETS by Oxford University Press, 1991).

[15] See de Pontfarcy (ed.), *Medieval Pilgrimage to St Patrick's Purgatory*, 83.

[16] For the text of Froissart, see *The Chronicles of Froissart translated by John Bourchier, Lord Berners*, The Globe Edition, G. C. Macaulay (ed.) (London: Macmillan, 1913), 425, cited in Easting, 'The English tradition', 66–7. For de Lannoy, see *Oeuvres de Ghillebert de Lannoy*, C. Potvin (ed.) (Louvain: Lefever, 1878), cited in de Pontfarcy, *Medieval Pilgrimage to St Patrick's Purgatory*, 93.

[17] For Perellós's use of Henry of Saltrey, see de Pontfarcy, *Medieval Pilgrimage to St Patrick's Purgatory*, 19, 88–9.

conversion as codified in the origin of the St Patrick's Purgatory as a form of religious proof to the pre-Christian Irish. Perellós emphasizes and reinforces the cultural and religious differences between the Gaelic Irish and the Anglo-Irish by participating in a discourse which culturally marginalizes the Gaelic Irish by designating them as 'barbarous'.

In his description of his travels in Ireland, Perellós participates in a geographical tradition that accentuates the cartographic extremity of the British Isles, especially that of Ireland. As Kathy Lavezzo has argued, Norman English writings exoticize Ireland and its people in order to emphasize England's proximity to Continental lands.[18] This tradition receives its most famous and influential expression in Gerald of Wales's *Topography of Ireland*, in which Gerald famously asks 'what new things, and what secret things not in accordance with her usual course, had nature hidden away in the farthest western lands?'[19] Gerald consistently structures an analogy between 'the marvels of the West which, so far, have remained hidden away and almost unknown' (57, cap. 33) and those of the East.[20] Perellós, like Gerald of Wales before him, comments upon the wondrous qualities of the Gaelic Irish, claiming that: 'They are among the most beautiful men and women I have ever seen anywhere in the world' (110). Elsewhere, he describes the uncivilized nature of both the 'common people' and the royalty, claiming that 'both the women and the men show their shameful parts without any shame' (111). The corporeal exposure of the Gaelic Irish – both the beautiful and the abject, the royal and the lowly – to the foreigner's gaze allows Perellós confidently to characterize them as uncivilized. His mixed wonder at the attractiveness of the Irish and disgust at their practices recalls Gerald's own dualisms:

Moreover, I have never seen among any other people so many blind by birth, so many lame, so many maimed in body, and so many suffering from some natural defect. Just as those that are well formed are magnificent and second to none,

[18] Kathy Lavezzo, *Angels on the Edge of the World: Geography, Literature, and English Community, 1000–1534* (Ithaca: Cornell University Press, 2006); see also James S. Romm, *The Edges of the Earth in Ancient Thought: Geography, Exploration, and Fiction* (Princeton University Press, 1992), 140–2.

[19] Gerald of Wales [Giraldus Cambrensis], *The History and Topography of Ireland*, John J. O'Meara (ed. and tr.) (Harmondsworth: Penguin, 1982 [1951]), 31; hereafter cited parenthetically in the text. For texts, see also James F. Dimock (ed.), *Giraldi Cambrensis Opera*, 5 (London: Longman, Green, Reader and Dyer, 1867); A. B. Scott and F. X. Martin (eds. and trs.), *Expugnatio Hibernica: The Conquest of Ireland by Giraldus Cambrensis* (Dublin: Royal Irish Academy, 1978). See also Birkholz, ch. 10, this vol., and Robert Bartlett, *Gerald of Wales: 1146–1223* (Oxford University Press, 1982).

[20] See also Gerald of Wales, *History and Topography of Ireland*, 54 (cap. 27, 28) and 56 (cap. 32).

so those that are badly formed have not their like anywhere. And just as those who are kindly fashioned by nature turn out fine, so those that are without nature's blessing turn out in a horrible way. (117–18, cap. 109)

The exotic texture of Perellós's descriptions of the Irish as beautiful, naked, and savage is heightened by the grammar of cultural analogues he constructs in which the Gaelic Irish are equivalent to the Saracens. His rhetorical gesture is equivalent to the trope in medieval travel literature described by Mary B. Campbell: 'Medieval travelers tended to take another course, employing "similitudes" to frequently grotesque effect ... the truth thus rendered is an *exotic truth*, bearing witness to an alienated experience. But the medieval writer is more deeply shocked, or at least his shock is more rhetorically blatant.'[21] Perellós's sexualization of the exotic, which in turn exposes anxiety about – and desire for – the consolidation of political territory, appears to us now as an *avant la lettre* enactment of an Orientalist worldview. It is certainly representative of what Sylvia Tomasch has termed 'geographical desire'.[22]

Perellós draws attention to the fact that the incorporation of Ireland has never been fully realized by the English, despite their efforts and internecine warfare.[23] He describes Ireland as 'the land of the savage Irish where King O'Neill reigned supreme', and writes of O'Neill's retainers:

They are armed with coats of mail and round iron helmets like the Moors and Saracens. Some of them are like the Bernese. They have swords and very long knives and long lances, like those of that ancient country which were two fathoms in length. Their swords are like those of the Saracens, the kind we call Genoese ... They are very courageous. They are still at war with the English and have been for a long time. The king of England is unable to put an end to it, for they have had many great battles. Their way of fighting is like that of the Saracens who shout in the same manner ... Their dwellings are communal and most of them are set up near the oxen, for that is where they make their homes in the space of a single day and they move on through the pastures, like the swarms of Barbary in the land of the Sultan. (110–11)

Perellós's itinerary thus functions as a spatial enactment of his conception of Christian territory and privileges Avignon as the cultural and ecclesiastical capital of the Church. Perellós's journey along a north–south route

---

[21] Campbell, *The Witness and the Other World*, 3.

[22] Tomasch, 'Introduction', 2. See also Mary Baine Campbell, '"Nel mezzo del cammin di nostra vita": the palpability of *Purgatorio*', in *Text and Territory*, Tomasch and Gilles (eds.), 15–28, at 15.

[23] Rees Davies, 'Frontier arrangements in fragmented societies: Ireland and Wales', in *Medieval Frontier Societies*, Robert Bartlett and Angus MacKay (eds.) (Oxford: Clarendon Press, 1989), 77–100, at 77–8.

shifts the paradigm of international cultural movement and exchange from a conventional east–west trajectory (long familiar from the political *translatio* theories of the latter Middle Ages) perpendicularly; instead of travelling from Rome to Santiago (or vice versa), he journeys from Aragon to Ireland, via Avignon. Moreover, Perellós's explicit references to Saracens reorient the north–south and east–west geographic paradigms that define the cartographic relationships between pagans and Christians. His descriptions bear out what Kathryn L. Lynch describes as 'a richer and more dialogic, rather than sharply oppositional, relationship between the cultures we think of as "West" and "East" throughout much of the Middle Ages'.[24] As Suzanne Akbari writes,

> It would be a mistake ... to conflate a binary overtly based on religious difference with the binary of Orient and Occident. The East is where 'they' are; it is, as Mary Campbell puts it, 'essentially Elsewhere.' It does not follow, however, that West is where 'we' are ... A whole, homogenous East is posited, which only gradually gives rise to its mirror image: a cold (because northerly) European West.[25]

Yet even in this 'northerly European West', the cultural capital implicit in the stereotyped Saracen may be deployed, exposing the cultural referent of 'barbarity' as more broad than its strict lexical value might suggest.

Perellós's insistent characterization of the Gaelic Irish as culturally analogous to the Saracens aligns his own experiences in the Mediterranean and along Iberian political and religious frontiers with his pilgrimage to Lough Derg. He frames Anglo-Irish conflict in a manner that also evokes internecine Iberian crusades, foregrounding the chivalric motifs of the pilgrimage and endowing the journey with the generic aura of crusade romance.[26] In Perellós's *Vitage*, diplomacy is presented as a chivalric enterprise with deep moral and feudal valences. By modelling loyalty, even beyond the grave, Perellós identifies the political agent as an ideal, valorous knight. Perellós's allusions to the Arthurian 'spaces' of England heighten the *Vitage*'s generic echoes of romance, while his depiction of Ireland further endows the narrative with an affinity to

---

[24] Kathryn L. Lynch (ed.), *Chaucer's Cultural Geography* (New York: Routledge, 2002), 5.

[25] Suzanne Conklin Akbari, 'From due east to true north: Orientalism and Orientation', in *The Postcolonial Middle Ages*, Jeffrey Jerome Cohen (ed.) (New York: Palgrave, 2000), 19–34, at 20.

[26] See Joseph F. O'Callaghan, *Reconquest and Crusade in Medieval Spain* (Philadelphia: University of Pennsylvania Press, 2003); Norman Housley, 'Frontier societies and crusading in the late Middle Ages', in *Mediterranean Historical Review* 10 (1995), 104–19, rpt. in *Crusading and Warfare in Medieval and Renaissance Europe* (Aldershot: Variorum, 2001), and Manuel González Jiménez, 'Frontier and settlement in the kingdom of Castile (1085–1350)', in *Medieval Frontier Societies*, Robert Bartlett and Angus MacKay (eds.) (Oxford: Clarendon Press, 1989), 49–74.

crusade literature.[27] Perellós's references to Saracens had a personal immediacy. Over two decades before his journey to Lough Derg, in 1374, he had been taken prisoner in the Moorish kingdom of Granada and was finally ransomed by Peter the Ceremonious.[28] Yet his cultural analogy or 'similitude' is not as facile as it might appear. The history of political alliances between the Iberian kingdoms in the late Middle Ages involves a complicated and ever-shifting network of alliances and animosities. Indeed, in the so-called 'War of the Two Pedros', Pedro IV of Aragon counted the King of Granada, Abū-Saʿīd, among his allies against Castile.[29] Perellós's characterization of the Saracens as savage disregards the Emirate of Granada's long history of cultural production.[30] Whether his condemnation is fuelled by anxieties about Aragonese boundaries or by a personal hostility based upon his experience of captivity we cannot know. It is likely that his allusions to Saracens might be related implicitly to heightened Aragonese interest in crusade at the end of the fourteenth century, a manifestation of the cyclical resurgence of Reconquista ideologies (often fuelled by immediate and self-serving political goals) throughout the later Middle Ages.[31]

Frontier in late medieval Ireland

Perellós's itinerary brought him from the administrative and ecclesiastical centres of England to its periphery at Plantagenet domains in Ireland. As a foreign observer, he bore witness to the shifting political geography of Ricardian Britain in the late 1390s, including the royal preference shown for the West Country, the preoccupation with the status of the Lordship of Ireland, and strengthening diplomatic relations with France.[32] His journey traverses the cultural and political 'frontier zones' of Ireland and Wales, lands which, as Rees Davies writes, 'stood at one of the peripheries of the area of feudal imperialism associated with the Norman conquest and colonization and indeed seemed to slow down

---

[27] Carpenter, 'The Pilgrim from Catalonia/Aragon', 107.   [28] Ibid. 110.

[29] Bisson, *The Medieval Crown of Aragon*, 113.

[30] See Leonard Patrick Harvey, *Islamic Spain 1250 to 1500* (University of Chicago Press, 1990) and Antonio Luis Cortés Peña, *Historia de Granada*, 4 vols (Granada: Editorial Don Quijote, 1983–7). For religious conflict in the Crown of Aragon see Bisson, *The Medieval Crown of Aragon*; Robin J. E. Vose, *Dominicans, Muslims, and Jews in the Medieval Crown of Aragon* (Cambridge University Press, 2009); Jarbel Rodriguez, *Captives and their Saviors in the Medieval Crown of Aragon* (Washington, D.C.: Catholic University of America Press, 2007); and Elka Klein, *Jews, Christian Society, and Royal Power in Medieval Barcelona* (Ann Arbor: University of Michigan Press, 2006).

[31] See María Mercedes Rodríguez Temperley, 'Narrar, Informar, Conquistar: Los *Viajes de Juan de Mandevilla* en Aragon', *Studia Neophilologica* 73 (2001), 184–96.

[32] Michael J. Bennett, 'Richard II and the Wider Realm', in *Richard II: The Art of Kingship* (Oxford: Clarendon Press, 1999), 187–204, at 188–9.

and even to frustrate its apparently remorseless advance'.[33] Despite the efforts of the English, the conquest of Ireland remained incomplete in the late Middle Ages, and the frontier between the English and Gaelic Irish lands receded over the course of the fourteenth century in favour of the latter.[34] This territorial loss was accompanied by a Gaelic cultural resurgence that was perceived as increasingly threatening to the 'English of Ireland', or Anglo-Irish.[35] The transformation of loyal English subjects into so-called 'English rebels', via intermarriage with Gaelic Irish or the adoption of Gaelic Irish cultural practices and institutions, also prompted legislative intervention by the English government.

The Statute of Kilkenny (1366) contrasts the political conditions in earlier post-conquest Ireland, in which the king's subjects '[lived] in [due] subjection', with the current state of Ireland, 'whereby the said land, and the liege people thereof, the English language, the allegiance due to our lord the King, and the English laws there, are put in subjection and decayed, and the Irish enemies exalted and raised up, contrary to reason'.[36] The Statute of Kilkenny sought to halt the process by which 'loyal' English subjects become 'degenerate Irish' through a reinforcement of racial divisions, which officials hoped would solidify linguistic, cultural and political ties to England: 'Also, it is ordained, and established, that no alliance by marriage, gossipred, fostering of childred, concubinage or by amour, not in any other manner, be henceforth made between the English and the Irish of one part, or of the other part'.[37] This legislation suppresses the historical role played by marriages and other alliances between the Irish and Anglo-Normans in land acquisition during the earlier colonization and raises questions about the political and cultural identity of the 'loyal English' in Ireland.[38] As Robin Frame

---

[33] Davies, 'Frontier arrangements', 96. See also P. J. Duffy, 'The nature of the medieval frontier in Ireland', *Studia Hibernica*, 22–3 (1982–3), 21–38; Brendan Smith, *Colonisation and Conquest in Medieval Ireland: The English in Louth, 1170–1330* (Cambridge University Press, 1999).

[34] Art Cosgrove, *Late Medieval Ireland, 1370–1541* (Dublin: Helicon Limited, 1981), 1; Davies 'Frontier arrangements', 77–8.

[35] Cosgrove, *Late Medieval Ireland*, 1–2.

[36] Crowley, *The Politics of Language in Ireland, 1366–1922*, 14. For the Statute of Kilkenny see Crowley, *The Politics of Language in Ireland, 1366–1922*, 12–16; Henry Berry, *Statutes and Ordinances and Acts of the Parliament of Ireland* (Dublin: His Majesty's Stationery Office, 1907), 430–69. See also Robin Frame, 'English officials and Irish chiefs in the fourteenth century', *English Historical Review* 90 (1970), 748–77; Robin Frame, *Political Development of the British Isles, 1100–1400* (Oxford: Clarendon Press, 1990).

[37] Crowley, *The Politics of Language in Ireland, 1366–1922*, 15.

[38] Robin Frame, '"Les Engleys nées en Irlande": the English political identity in medieval Ireland', *Transactions of the Royal Historical Society*, 6th ser., 3 (1993), 83–103, at 84.

writes, 'The public identity of the English of Ireland had acquired a historical aspect, as they sought to explain to others, and perhaps to themselves, who they were, and how they had come to be where they were'.[39] The encroachment of the Gaelic Irish upon historically colonial territories prompted some 'English of Ireland' to forge stronger political ties amongst themselves, resulting in 'the strengthening of a feeling of embattled Englishness among those who lived in the encircled heartlands of the lordship'.[40] The Statute of Kilkenny, for example, encourages the Anglo-Irish to live together peacefully and to appeal to English common law for the resolution of disputes.[41] The perception and articulation of an 'Anglo-Irish identity' remained a complex issue, however, and one which was only complicated by the rash of legislation in England regarding Irish repatriation.

Perellós's narrative registers the political and military tensions of late 1397, by which time Richard II's hoped-for peace following the submission of the Irish chieftains seemed increasingly improbable.[42] The defensive, wary tone of the Statute of Kilkenny, which 'present[ed] an image of an English world under cultural as well as military siege',[43] still accurately described the hostilities present over twenty years later when the Catalan travelled to Ireland, whose 'commons ... are in the different marches at war' (16).[44] Perellós points out the failure of England to incorporate 'the land of the savage Irish where King O'Neill reigned supreme' (109). His descriptions of the Gaelic Irish are decidedly military, calling attention to their weaponry and describing their living conditions as similar to encampments; as warriors, the Irish are threateningly mobile. They resist colonial domination or cartographic containment, they disrupt boundaries, and they refuse to be incorporated into the body politic of England, despite the political rhetoric of its king. Perellós disparages their resistance, which he characterizes as even more contemptible due to a false sense of cultural superiority: 'they consider their own customs to be better than ours [i.e. the French, Aragonese and Castilians] and more advantageous than any others in the whole world' (111).

---

[39] Frame, 'Les Engleys nées en Irlande', 84. A comparison of Anglo-Irish and Welsh Marcher culture and law is outside the scope of this chapter; on Welsh Marches, see, for example, Birkholz, ch. 10, this vol.

[40] Frame, 'Les Engleys nées en Irlande', 94.

[41] Crowley, *The Politics of Language in Ireland, 1366–1922*, 15–16.

[42] For the submissions of the Irish chiefs, see Edmund Curtis, *Richard II in Ireland, 1394–95, and Submissions of the Irish Chiefs* (Oxford: Clarendon Press, 1927). See also Nigel Saul, 'Richard II and the vocabulary of kingship', *English Historical Review* 110 (1995), 854–77; Dorothy Johnson, 'Richard II and the submissions of Gaelic Ireland', *Irish Historical Studies* 32 (1980): 1–20; Cosgrove, *Late Medieval Ireland*, 28.

[43] Frame, 'Les Engleys nées en Irlande', 100.

[44] Crowley, *The Politics of Language in Ireland, 1366–1922*, 16.

The depiction of the Gaelic Irish as degenerate in their religious practices is a central element in English colonial discourses on Ireland. Henry II's twelfth-century invasion took place 'against a background of disquiet about Irish morals and Church order that had long been voiced, with the encouragement of the Irish reforming clergy, by Canterbury and Rome'. Henry succeeded in gaining recognition from the Irish bishops;[45] later, legislation was put into effect which made it increasingly difficult for Irishmen to enter into high ecclesiastical service.[46] As Kathleen Biddick writes, the Statute of Kilkenny 'prohibited marriage *between* various Christians and denied both domestic and spiritual miscegenation and in so doing fabricate[d] blood as a juridical substance ... The statutes juridically constituted Englishness, even at the expense of "Christianness"'.[47] Perellós echoes the condemnation of Gaelic Irish religious deviance when he claims that the Irish revere their own archbishop 'as if he were a Pope' (108). He makes no mention, however, of religious differences between England and Aragon, which positioned themselves on opposite sides of the schismatic controversy; his elision of this delicate issue might reflect the Anglo-French diplomatic efforts to bring an end to ecclesiastical division.

The closely intertwined spheres of papal and royal authority can be seen over twenty years later, when the Irish parliament in 1421 issued a memorandum to the king requesting him to authorize a crusade against the Gaelic Irish on the grounds of their broken oaths of submission of 1395:

several great chieftains of Irish families ... McMorogho, O'Neel, O'Brien of Thomond, O'Conoghor of [Connaught], and diverse other Irish, humbly submitted of their own free will, and became liege men to [Richard II] and his heirs, kings of England ... And since that time, as formerly, the said persons have become disloyal and rebellious ... therefore ... your said lieges pray you to inform and complain to our said most holy father the Pope ... with a view to having a crusade on that account against the said Irish enemies ...[48]

The crusade proposed against the Irish, though it did not materialize in any royal action, emphasizes the extent to which the perceived 'savageness' of the Irish is implicated in a broader discourse of religious alterity and crusade; at the same time, it signals shifting attitudes regarding the political valences of crusade.[49] The crusade proposal also

---

[45] Frame, 'Les Engleys nées en Irlande', 86. [46] Ibid. 89.

[47] Biddick, 'The cut of genealogy', 453.

[48] *Statute rolls of the Parliament of Ireland. Reign of King Henry the Sixth*, H. F. Berry (ed.) (Dublin, 1910), 564–7. See Art Cosgrove, 'Ireland', in *The New Cambridge Medieval History: Vol. 7*, C. T. Allmand (ed.) (Cambridge University Press, 2008), 496–513, at 509.

[49] Elizabeth Matthew, 'Henry V and the proposal for an Irish crusade', in *Ireland and the English World in the Late Middle Ages: Essays in Honour of Robin Frame*, Brendan Smith (ed.) (Houndmills: Palgrave Macmillan, 2009), 161–75.

serves as a counterpoint to early modern political writings that emphasize the colonization of Ireland and the war with Catholic Spain as mutually contingent, both because of the ambiguous status of papal donations in a Protestant state and because of the perceived threat of a Spanish naval invasion that would capitalize on England's vulnerable maritime boundary with Ireland.[50] As Barbara Fuchs has shown, Spenser's discussion of Irish ethnic origin in *A View of the Present State of Ireland* draws on the historiographical tradition of early Spanish colonization of Ireland, dating back to Nennius and Geoffrey of Monmouth: 'The Irish genealogy that Irenius introduces is hardly innocent: the Scythian connection, often reiterated, will account for Irish savagery and nomadism; the Spanish connection, meanwhile, must be qualified and exposed as fraudulent in order to neutralize dangerous Irish pleas for Spanish help as England fights both of them simultaneously.'[51] Perellós's descriptions of the Irish as Saracens, informed as it was by both the Aragonese diplomatic relationship with England and his own experience of Iberian religious frontiers, offers a late medieval antithesis of early modern ideas that Ireland served as a boundary between the hostile Iberian and English kingdoms. In Perellós's account, Ireland is emblematic of the boundaries of both England, and, by analogue, the Crown of Aragon, rather than serving as a shared boundary between them. While Perellós makes Ireland remote to both Aragon and England, he pushes the geographical extremity to the edges of the earth itself by making it the entrance of Purgatory.

In Perellós's *Vitage*, the political utility of geographic peripheries is emphasized through his scrutiny of the tenuous Plantagenet lordship over Ireland. The priority of centres in core-periphery models is troubled by Perellós's exposure of frontiers and political courtly and urban centres as mutually contingent. At the time of Perellós's journey, even England, whose strong Plantagenet administrative and juridical legacy was based in London, was experiencing significant shifts in its status as a 'centre' due to the city's perennial resistance against Richard II's authority. It seems likely that Perellós's political sensibilities were informed by the royal prerogatives of his former king, Pedro IV, whose 'vision of his ancestor's acquisitions not as peripheral but as vital to the survival of the Crown of Aragon itself was reflected in the policies of his successors over the following century'.[52] As Pedro IV wrote in 1380 to his heir Juan

---

[50] Barbara Fuchs, 'Spanish lessons: Spenser and the Irish Moriscos', *Studies in English Literature, 1500–1900* 42 (2002), 43–62, at 44–5.

[51] Ibid. 52.

[52] Suzanne F. Cawsey, *Kingship and Propaganda: Royal Eloquence and the Crown of Aragon c. 1200–1450* (Oxford University Press, 2002), 14–15.

(the king Perellós later sought in St Patrick's Purgatory), 'If Sardinia is lost, Majorca, without its food supply from Sicily and Sardinia, will be depopulated and will be lost, and Barcelona will also be depopulated, for Barcelona could not live without Sicily and Sardinia, nor could its merchants trade if the isles were lost'.[53] The Crown of Aragon lacked a definite, stable political centre, owing to its typically itinerant court, multiple urban centres, and the accretive territorial history resulting from *Reconquista* ideology, in which boundaries are always theoretically in flux; the economic and political contingencies uniting the traditional domains of the Crown of Aragon with its thirteenth- and fourteenth-century annexations were thus a factor in royal policy and administration during the reigns of Pedro IV and Juan I.

Finding centre: political legacy in Dante's *Purgatorio* and Perellós's *Vitage*

The emergence of Purgatory as an earthly *place*, and therefore within the disciplinary domains of cartographic and geographical knowledge, allows for the creation of what Jacques Le Goff has called a 'dual geography . . . a geography of this world coupled with a geography of the next'.[54] Henry of Saltrey's *Tractatus* played an important role in the developing theology of purgatory.[55] The wide dissemination of the *Tractatus* and its many vernacular recensions was augmented by the incorporation of material from the *Tractatus* into Jacobus da Voragine's *Legenda aurea*[56] and into historical writing in the late Middle Ages, including Roger of Wendover's *Flores historiarum*, Matthew Paris's *Chronica majora*, James of Vitry's *Historia orientalis*, Vincent of Beauvais's *Speculum historiale*, Stephen of Bourbon's *Tractatus de diversis materiis praedicabilibus*, Humbert of Romans's *De dono timoris*, and Gossouin of Metz's *L'image du monde*.

The entrance to either a general or specific purgatory at Lough Derg, positioned in the far northwest of Europe, appears on the opposite side of the *mappa mundi* as another physical location important to salvation history – that of earthly paradise, traditionally represented in the Far East. The distance between the two 'intermediate locations', both part of this world yet also fundamentally 'other' to it, stretches across the known world in opposite cardinal directions.[57] Both locations are separate and

[53] Pedro IV/III of Aragon, *Pere III of Catalonia (Pedro IV of Aragon): Chronicle*, M. and J. N. Hillgarth (eds. and trs.), 2 vols., continuously paginated (Medieval Sources in Translation, 3–4; Toronto, 1980), 35, cited in Cawsey, *Kingship and Propaganda*, 14.

[54] Le Goff, *Birth of Purgatory*, 177.   [55] Ibid. 177–208.

[56] Ibid. 199–200; *Medieval Pilgrimage to St Patrick's Purgatory*, de Pontfarcy (ed.), 86.

[57] Alessandro Scafi, 'Mapping Eden: cartographies of the earthly paradise', in *Mappings*, Denis Cosgrove (ed.) (London: Reaktion Books, 1999), 50–70, at 52; see Alessandro Scafi, *Mapping Paradise: A History of Heaven on Earth* (London: British Library, 2006).

defined by boundaries.[58] At times, Eden is represented on *mappaemundi* as an island, and therefore corresponds with Ireland, which, according to the writers mentioned above, contains a place of purgation. At other times, Eden is portrayed as a walled garden, enclosed from the rest of the world.[59] Unlike terrestrial paradise, St Patrick's Purgatory was not a ubiquitous place name on medieval world maps; rather, its location is implicit, based upon the textual geography described in theological and devotional writing, and, later, in pilgrims' accounts. Its subterranean location, an integral part of the doctrine's evolution, makes its representation on the terrain of earth, like that of inaccessible Eden, paradoxical. Together, they are integral to the spiritual geography of the life of man, in the genealogical as well as in the personal sense. Yet while binding the extremities of the earthly and the spiritual worlds, Ireland – as an entrance to Purgatory – becomes soteriologically central and thus occupies a middle position in spiritual topographies.

Up to this point, I have emphasized the geographical position of Ireland as one of periphery – whether on the edge of northern Europe on world maps or as a textually exoticized frontier zone in Norman French and English writing – at the very least, far from the administrative 'centre' of south-eastern England. On the salvific terrain of the allegorically understood spiritual 'pilgrimage' of man's soul, however, purgatory occupies a role that, though liminal, is central rather than peripheral – a transitional phase between earthly life and heavenly afterlife. This can be seen, for example, in the visual representation on a parchment fragment of a *mappa mundi* found in the records of the Duchy of Cornwall and described by Harley and Woodward:

Forming a border along the bottom edge of the fragment is a series of finely executed line drawings of figures apparently depicting the stages of life; each figure delivers a cautionary message. They include a woman at vespers, an old man bent with age, a figure in purgatory holding a bowl of fire, and an angel...[60]

Likewise, in the tripartite division of world into a vertical hierarchy of Heaven, Purgatory, and Hell, purgatory, closest to earth, occupies the middle position.

The tradition of St Patrick's Purgatory is both a source and context for Dante's *Purgatorio*.[61] Dante conceives of Purgatory as an intermediate

---

[58] Scafi, 'Mapping Eden', 57ff.

[59] For a discussion of terrestrial paradise in the Byzantine geographical imagination, see Della Dora, ch. 12, this vol.

[60] J. Brian Harley and David Woodward (eds.), *The History of Cartography Volume 1: Cartography in Prehistoric, Ancient, and Medieval Europe and the Mediterranean* (University of Chicago Press, 1987), 307.

[61] Campbell, 'Nel mezzo del cammin di nostra vita', 17; Le Goff, *Birth of Purgatory*, 200.

zone, a juxtaposition of geographical centrality and extremity.[62] Unlike Henry of Saltrey's, though, Dante's Purgatory is not situated underground. Rather, it is a mountain, the inverse concave of Lucifer's fall to earth, the antipodean mirror of Jerusalem, at whose summit lies the Earthly Paradise.[63] Mary B. Campbell has written about the formal and narrative emphasis on the *Purgatorio* as the middle canticle of the *Divine Comedy*:

> Dante isolates his middle too by the firmly formal division of the poem into three equal canticles. As I hope to show, he puts his weight there, in that segment of narrative which is the narrative territory of travel literature, in the middle book concerning the middle (and material) realm of the Other World, to which he traveled 'in the middle of the journey of our life'.[64]

The generic proximity of the *Purgatorio* to travel literature suggests for the poet-narrator Dante an allegory for poetic composition. Dante's sea-voyage at the beginning of the *Purgatorio* is at once a geographical journey from the solitary island to the antechamber of Purgatory, a spiritual ascent through the levels of Purgatory towards Paradise, and a metaphor for poetic exertion.[65]

Dante's *Divine Comedy* shares an investment in the evaluation of political figures with the tradition of visionary literature from the Middle Ages, in which we can include many other voyages to otherworlds and underworlds both literal and dreamt. From the Irish kings in the *Vision of Tundale* and the Carolingian sovereigns in the *Vision of Charles the Fat* to Book Four of Gregory the Great's *Dialogues*, we can see how Dante, in his treatment of the tyrants of the *Inferno* and the princes of the *Purgatorio*, is drawing on a larger tradition that deploys the political potential of purgatory narratives to censure monarchs.[66] Dante uses the poet and political satirist Sordello to identify princes in Canto VII of the *Purgatorio*. Sordello emphasizes both the territorial and ethical failures of rulers as part of a larger treatment of issues surrounding political unity in the *Divine Comedy*.[67]

---

[62] Le Goff, *Birth of Purgatory*, 337.   [63] Ibid. 335.

[64] Campbell, 'Nel mezzo del cammin di nostra vita', 16.

[65] A point made by Allen Mandelbaum (ed.), *The Divine Comedy of Dante Alighieri* (Berkeley: University of California Press, 1982), viii–ix.

[66] Le Goff, *Birth of Purgatory*, 95, 191, 207.

[67] Teodolinda Barolini, 'Bertran de Born and Sordello: the poetry of politics in Dante's *Comedy*', *Publications of the Modern Language Association* 94 (1979), 395–405, at 395–6. See also Claire E. Honess, '*Salus, venus, virtus*: poetry, politics, and ethics from the *De vulgari eloquentia* to the *Commedia*', *Italianist* 27 (2007), 185–205; Caron Cioffi, 'Fame, prayer, and politics: Virgil's Palinurus in Purgatorio V and VI', *Dante Studies* 110 (1992), 179–200.

Dante's geography of Purgatory, like that of the visions mentioned above, is one of otherworldly displacement. Here we may return to Perellós's *Vitage*, which, like the *Purgatorio* and the *Vision of Tundale*, depicts the afterlife of a deceased ruler's soul. Perellós's final discovery of his former king and deceased courtiers in the caves of St Patrick's Purgatory amounts to a veritable displacement of the Aragonese court, far from the political centre of his realm and the origin of his journey, to Lough Derg. Here, issues of court politics and jurisdiction prevail, and Perellós is no longer at a periphery, so to speak, but rather at the place where theological and governmental answers crucial to his own political career might be found.

The court Perellós leaves at the beginning of his journey was in an uproar, owing to the recent death of Juan I of Aragon and the subsequent prosecution of various court advisers.[68] Perellós tells us in his account that 'Because of his [Juan I's] death I was, contrary to God's will, as sorrowful and as sad as any servant could be at the death of his lord. I set my heart at that moment on going to St Patrick's Purgatory in order to find out, if it were possible, whether my lord was in purgatory and the torments he was suffering' (106). In the narrative model of the pilgrimage based on Henry of Saltrey's *Tractatus*, the pilgrim to St Patrick's Purgatory undergoes a series of sufferings and temptations for the purposes of spiritual purgation. However, the emphasis in the *Vitage* is not on Perellós himself as a confessional subject; rather, Perellós marks the king as the penitential subject and himself as an observer and interlocutor, even as a questing knight. His comment, 'my great desire to know in what state my lord the king was, and also to purge my sins', stresses the fate of the king, which has been foregrounded throughout the narrative, rather than his own penitence. This is significant because Perellós's own status as a royal adviser in the Crown of Aragon was under scrutiny. At the time that Perellós left for St Patrick's Purgatory, he was under investigation for treason by court reformers in Aragon.[69] Though charges against him never formally materialized, the elision of any hint of guilt or defamation on his own part in the *Vitage* is central to understanding the historical context of Perellós's voyage, which serves as a performance of exemplary loyalty as it simultaneously purges him of any suspected political or moral taint.

Perellós's account resists full disclosure of the spiritual fate of the king. While he demonstrates no qualms about reporting the punishments undergone by others he recognizes in Purgatory, such as a lecherous

---

[68] Carpenter, 'The Pilgrim from Catalonia/Aragon', 99, 101.   [69] Ibid. 101.

Franciscan named Brother Francis del Puech and his own niece, Na Aldosa de Quaralt, exposed for her vanity, he does not take such liberties with regard to the king's fate. He describes his reunion with his deceased sovereign thus:

There I spoke at great length to my lord the king who, by God's grace, was on the road to salvation. I do not wish to say on any account the reason he was suffering those torments. But I will tell you that the great kings and princes in this world ought, above all else, to avoid committing injustice in order to give pleasure of favour to anyone, male or female, no matter how close [they may be] to them, [even if they be] male or female relatives of [their own family]. For I saw several women and men there of whom I do not care to speak, but I thanked God that they were on the way to salvation. (115–16)

Perellós will not identify the actual transgressions of the king, nor will he describe the nature of his torments or the degree of his suffering. His rhetorical *praeteritio*, however, gives much away. Perellós clearly indicates that the king's sins, those unspoken 'injustices', resulted from his misguided susceptibility to courtly influences. His emphasis on both sexes and on family relations as objects of favouritism would be suggestive to an audience familiar with courtly life in Aragon. Perellós's placement of Juan in Purgatory (as opposed to Hell) vindicates him from grievous moral or political failure by witnessing that he was, 'by God's grace ... on the road to salvation'. If he is to be grouped with the misguided and weak princes of the *Purgatorio*, then he is at least not to be counted among the tyrants of the *Inferno*.

Perellós's journey to St Patrick's Purgatory 'in order to find out, if it were possible, whether my lord was in purgatory and the torments he was suffering' (106), emerges from a historiographical, even documentary, impulse to ascertain questions vital to the legacy of both temporal and papal monarchies. These questions include the degree to which King Juan I's misgovernment was the fault of severe ethical failings and whether the ecclesiastical authority of the Avignon Pope Benedict XIII (Pedro de Luna of Aragon) (1394–1423) was legitimate. Perellós's strong support for the Aragonese Pope is evident in the *Vitage*, and the theological implications of schismatic heresy are at stake in Perellós's efforts to vindicate Juan I.[70] Perellós passed through Avignon on this journey during a historical moment when the Anglo-French pressure to end the schism had culminated, only a year in advance of the subtraction of French obedience from the Avignon Papacy of 1398–1403.[71]

---

[70] Ibid. 101 n. 8.

[71] Joëlle Rollo-Koster, 'The politics of body parts: contested topographies in late-medieval Avignon', *Speculum* 78 (2003), 66–98; Noël Valois, *La France et le Grand Schisme d'Occident*, 4 vols. (Paris, 1896–1902), vol. III; Howard Kaminsky, 'The politics of

Perellós's text can be seen as a form of historical glossing, relevant to the tradition of royal chronicles and historical writing in the Crown of Aragon. As in Dante's *Purgatorio*, spirituality's translation into political centrality becomes a crucial strategy. Perellós does not feature himself as a universal pilgrim in the manner of Guillaume de Deguileville. His narrative resists allegory in favour of historical contingency and specificity. This is a text that names (at least some) names. The historical specificity required of a diplomat's records is transferred onto the devotional content of Perellós's pilgrimage narrative. The complicity of diplomats and historiographers in textual constructing and memorializing political boundaries is shown as integral to the claims of royal authority. This dynamic is demonstrated in correspondence between Pedro IV and his chronicler Bernat Descoll. Despite numerous military setbacks, the king anxiously anticipated the recording of his successful annexation of Sardinia, and wrote to Descoll in August 1375: 'We hope, with God's aid, shortly to conquer the whole island ... And so leave enough space, so that the conquest we will make of the island can be continued here.'[72] Perellós's *Vitage* wrests control over the textual production and dissemination of historical legacy by appealing to the subjectivity of the king. Perellós endows him with an interiority that is both public and private, creating a porous boundary between the individual penitent and exterior observer. This gesture calls into question the relationship between the king's interiority and national spaces, giving the philosophical dilemma of the king's 'two bodies' a geographical dimension. The king's ethical conscience is intimately tied to the military defence of political 'interiors' in a way that recalls Foucault's investment in the idea of territory as the discursive foundation of political power structures.[73] One of the hallmarks of Juan I's reign is that, partially in order to finance his courtly extravagance, he compromised the borders of his composite realm; in other words, he committed the same errors of Sordello's 'weak' princes in Dante's *Purgatorio* VII.[74] The implicit movement in Dante from the ceding of boundaries to the ethics of kingship has a correlation with the political sensibility of the *Vitage*.

The royal advisers who have privileged access to courtly spaces are also implicated in the historiographical legacy of a reign – as the prosecutors

France's subtraction of obedience from Pope Benedict XIII, 27 July, 1398', *Proceedings of the American Philosophical Society* 115 (1971), 366–97.

[72] Pedro IV/III of Aragon, Pere III of Catalonia (Pedro IV of Aragon): Chronicle, 608, cited in Cawsey, *Kingship and Propaganda*, 14.

[73] Michael Foucault, *Power/Knowledge: Selected Interviews and Other Writings, 1972–1977*, Colin Gordon (ed. and tr.) (London: Harvester, 1980).

[74] Carpenter, 'The Pilgrim from Catalonia/Aragon', 99.

of Juan's courtiers insisted. Literary texts bear witness to their authors' innocence in the face of intrigue and function as a courtly equivalent to legal documents. Such is the case with Bernat Metge, a notary in the Royal Chancellery of the Crown of Aragon under Juan I, who was tried for corruption and treason and ultimately absolved. Like Perellós, he used a literary, textual medium to stage a dialogue between himself and the deceased King Juan calculated to demonstrate his own guiltlessness. Metge's text, though, was not a travelogue like Perellós's *Vitage*, but rather a dream-vision, *Lo somni* (1399).[75] Bernat Metge's articulation of his interior psychic spaces contrasts with the public nature and international scope of Perellós's literal and textual venture. In both texts, however, the dynamics of courtly life are shown to have an impact on the policies of the kingdom, and the ethics of dominion have spiritual as well as political or dynastic repercussions.

Textual geographies and geographies of texts

And there I was, incredible to myself,
among people far too eager to believe me
and my story, even if it happened to be true.[76]

Perellós's *Vitage* creates a textual itinerary, based on the twelfth-century Cistercian *Tractatus* and the fruits of lived experience, which responds to the political situations of the late fourteenth-century European world. The mental mappings and imagined geographies created by Perellós's itinerary guide readers through geographical spaces marked by historical record and textual precedent. Perellós's England is at once the England of Arthur, of Beckett, and of Richard II. The ways in which place is conceived of in Perellós's itinerary are indebted to the trends of courtly literacy at Continental courts, specifically those of the Valois French and of the count-kings of the Crown of Aragon; Pedro IV and his sons, '[l]ike many other contemporary European rulers ... tapped into the international interlibrary loan system of the later medieval world'.[77] The conceived spaces of travel narratives are largely determined by a geographical imagination born of reading habits, an imagination at whose

---

[75] Bernat Metge, *The Dream of Bernat Metge*, Richard Vernier (tr.) (Aldershot, England: Ashgate, 2002); *Lo Somni*, Josep Maria de Casacuberta (ed.) (Barcelona: Editorial Barcino, 1980). W. H. Hutton discusses the possibility that Metge was familiar with Dante's *Divine Comedy* when he composed *Lo Somni*, 'The influence of Dante in Spanish literature', *The Modern Language Review* 3 (1908), 105–25, at 109.

[76] Seamus Heaney, 'Sweeney Redivivus', in Seamus Heaney, *Station Island* (New York: Farrar, Straus, Giroux, 1985), 98, ll. 12–14.

[77] Cawsey, *Kingship and Propaganda*, 33.

'centre' is a library. Moreover, the circulation of people and of manuscripts inscribes itself upon a textual tradition by affecting the material conditions of reading.[78] We can draw a comparison between the *Vitage*'s relationship with historical pilgrimage and the Hereford Cathedral Map's use of written itineraries as textual sources. The Hereford Map, which drew on pilgrims' texts, became 'a repository of contemporary geographical information of use for planning pilgrimages and stimulating the intended traveler'; the *Vitage* contributed to the chivalric resonances of pilgrimage accounts to Lough Derg and to the popularity of the pilgrimage itself.[79] Ultimately, the *Vitage* offers an example of how textual geographies may determine historical travel and suggests the importance of the relationship between literary court culture and diplomacy.[80] Perellós's wide diplomatic travels facilitated the introduction of manuscripts into the royal court in Aragon, presided over by the highly literate King Juan I.[81] The practice of diplomacy, like the experience of pilgrimage, is shown to be informed by texts. The diplomat, like the *peregrinus* (etymologically, a foreigner in an exotic terrain), achieves a privileged status based on his geographical displacement.

Matthew Paris's maps (*c.* 1250s), which visually represent a pilgrimage itinerary from England to Jerusalem, offer a useful comparison with the imagined geography of Perellós's *Vitage*. As Daniel K. Connolly writes,

The Benedictine brother who perused these pages understood this map primarily through its performative possibilities, as a dynamic setting, the operation of whose pages, texts, images, and appendages aided him in effecting an imagined pilgrimage that led through Europe to the Crusader city of Acre and eventually to a complex representation of Jerusalem.[82]

Yet as both Connolly and Katharine Breen point out, the sacred geography which positions Jerusalem as the '*umbilicus mundi*, the navel of the world and the geometric centre of most medieval world maps' contrasts with its geographical unavailability as the culmination of a Christian pilgrimage in historical time:[83]

[78] E.g. see Birkholz, 'Hereford maps, Hereford lives'.
[79] Harley and Woodward (eds.), *History of Cartography*, vol. I, 288; see also Gerald R. Crone, 'New light on the Hereford Map', *Geographical Journal* 131 (1965), 447–62.
[80] Perellós himself was raised at French court and served as a page of Charles V, whose vast library was renowned. See Carpenter, 'The Pilgrim from Catalonia/Aragon', 100; see also Léopold V. Delisle, *Recherches sur la librairie de Charles V*, 2 vols. (Paris: H. Champion, 1907).
[81] Carpenter, 'The Pilgrim from Catalonia/Aragon', 100–1; see also Bisson, *The Medieval Crown of Aragon*, 121.
[82] Connolly, 'Imagined pilgrimage', 598.
[83] Katharine Breen, 'Returning home from Jerusalem: Matthew Paris's first map of Britain in its manuscript context', *Representations* 89 (2005), 59–93, at 81ff.; Connolly,

The most prominent of the texts abutting the city explains in red letters that 'tutes cestes parties ... ore sunt en la subiecciun des sarrazins [all these parts ... are now under the subjection of the Saracens],' leaving the land 'tute corrumpeue e pasture au diable [entirely corrupted and a pasture for the devil].' The blocks of text that surround Jerusalem occupy space that Christians explicitly cannot, and the city itself remains undefiled only in the minds of the faithful.[84]

Because Jerusalem cannot be inhabited by Christians, it becomes, in Matthew's maps, 'the center of a collocation of texts'.[85] In the spiritual geography of Matthew's imagined Jerusalem pilgrimage, the delineation of centres and peripheries is challenged by the unavailability of its goal, which must either be rendered into pure text or, as Breen argues, subjected to a shift in directionality so that the map of England (significantly, oriented northward), rather than the earthly or heavenly Jerusalem, becomes the end point of the journey.

Perellós's *Vitage* likewise participates in a conceptualization of space that simultaneously engages with spiritual and political geographies by reimagining the contested delineation of centres and peripheries. Jerusalem, beyond the bounds of Christendom, and Rome, whose status as an ecclesiastical centre was displaced by that of Avignon, are both 'unavailable' as spiritual destinations for Perellós. His solution, as I have suggested above, is to employ a new orientation along a north–south axis that would privilege Avignon as both a centre of ecclesiastical authority and a point of origin (and ultimate return), and, crucially in distinction to real or imagined Jerusalem pilgrimages, identify a destination on the periphery of the world rather than at its geographical centre or *umbilicus mundi*. Perellós's intervention, like the celebrated Catalan Atlas (*c.* 1375)[86] housed in the library of Charles V, which depicts the South at its top edge, offers alternative ways of envisioning geographical space and cartographic orientation. Moreover, Perellós situates the cultural authority of Aragon – itself once a Marcher domain on the periphery of the Carolingian empire – as embedded both at the core of Christendom (through the Aragonese Avignon Pope) and at its periphery in Ireland (Juan I's soul in St Patrick's Purgatory). Perellós's negotiation of spiritual and political geographies echoes those of Matthew Paris, who uses the Jerusalem pilgrimage to reimagine the political centrality of England, or, as Breen argues more aggressively, 'to sacralize the geography of Britain',

'Imagined pilgrimage', 598: 'This image of Jerusalem was seen as both the unavailable center of earthly pilgrimage and as a goal of spiritual contemplations, which focused on it as a figure of the Heavenly Jerusalem.'

[84] Breen, 'Returning home from Jerusalem', 83.   [85] Ibid. 82.

[86] L'Atlas Catalan (*BNF*, esp. 30), http://www.expositions.bnf.fr/ciel/catalan/index.htm, accessed 7 March 2011.

by making Aragon the implicit focus of a voyage to the peripheries of English domains and ecclesiastical authority.[87] The *Vitage* offers a way of imagining space that privileges national affiliation, and in so doing, identifies the politically-threatening as 'other' through discourses of barbarousness. Perellós's ethical quest leads him from the hostile edges of the kingdom – and beyond it to the peripheries of the known world – to the spiritual delineation of the interior spaces of the king's own soul. The spiritual geographies created and traversed in his narrative offer an epistemological terrain which locates a Purgatory on the edge of the world while preserving its contingency to central issues of polity.[88]

[87] Breen, 'Returning home from Jerusalem', 61.
[88] I am grateful to Christopher Baswell, Christine Chism, Henry Ansgar Kelly, Keith Lilley and Joseph Falaky Nagy for comments on earlier drafts of this chapter.

Select bibliography

Allen, Rosamund (ed.), *Eastward Bound: Travel and Travellers, 1050–1550.* Manchester University Press, 2004.

Connolly, Daniel. K., *The Maps of Matthew Paris: Medieval Journeys through Space, Time and Liturgy.* Woodbridge: Boydell Press, 2009.

Cosgrove, Denis, *Apollo's Eye: A Cartographic Genealogy of the Earth in the Western Imagination.* Baltimore: Johns Hopkins University Press, 2001.

Edson, Evelyn, *Mapping Time and Space: How Medieval Mapmakers Viewed Their World.* London: The British Library, 1997.

Gautier Dalché, Patrick, *Géographie et culture: La représentation de l'espace du VIe au XIIe siècle.* Aldershot: Ashgate, 1997.

'Maps in words: The descriptive logic of medieval geography, from the eighth to the twelfth-century', in Paul D. A. Harvey (ed.), *The Hereford World Map: Medieval World Maps and Their Context.* London: The British Library, 2006: 223–42.

'The reception of Ptolemy's Geography (end of the fourteenth to beginning of the sixteenth century)', in David Woodward (ed.), *The History of Cartography, Volume 3. Cartography in the European Renaissance, Part 1.* University of Chicago Press, 2007: 285–364.

Glacken, Clarance, *Traces on the Rhodian Shore: Nature and Culture in Western Thought from Ancient Times to the End of the Eighteenth Century.* Berkeley: University of California Press, 1967.

Hanawalt, Barbara A. and Kobialka, Michal (eds.), *Medieval Practices of Space.* Minneapolis: University of Minnesota Press, 2000.

Harley, J. Brian and Woodward, David (eds.), *The History of Cartography Volume One: Cartography in Prehistoric, Ancient, and Medieval Europe and the Mediterranean.* University of Chicago Press, 1985.

Hiatt, Alfred, *Terra Incognita: Mapping the Antipodes before 1600.* Chicago University Press, 2008.

Howe, Nicholas, *Writing the Map of Anglo-Saxon England: Essays in Cultural Geography.* New Haven: Yale University Press, 2008.

Kimble, George H. T., *Geography in the Middle Ages.* London: Methuen, 1938.

Kline, Naomi Reed, *Maps of Medieval Thought: The Hereford Paradigm.* Woodbridge: Boydell Press, 2001.

Lavezzo, Kathryn, *Angels on the Edge of the World: Geography, Literature, and English Community, 1000–1534.* Ithaca: Cornell University Press, 2006.

Lilley, Keith D., 'Geography's medieval history: a neglected enterprise?', *Dialogues in Human Geography* 1 (2011), 147–62.

Lozovsky, Natalia, *'The Earth is our Book': Geographical Knowledge in the Latin West ca. 400–1000*. Ann Arbor: University of Michigan Press, 2000.

'Roman geography and ethnography in the Carolingian empire', *Speculum*, 81 (2006), 325–64.

Merrills, Andrew, *History and Geography in Late Antiquity*. Cambridge University Press, 2005.

Smith, Donald K., *The Cartographic Imagination in Early Modern England: Re-writing the World in Marlowe, Spenser, Raleigh and Marvell*. Aldershot: Ashgate, 2008.

Raaflaub, Kurt A. and Talbert, Richard J. A., *Geography and Ethnography: Perceptions of the World in Pre-Modern Societies*. Oxford: Wiley-Blackwell, 2010.

Scafi, Alessandro, *Mapping Paradise: A History of Heaven on Earth*. London: British Library, 2006.

Talbert, Richard J. A. and Unger, Richard W. (eds.), *Cartography in Antiquity and the Middle Ages: Fresh Perspectives, New Methods*. Leiden: Brill, 2008.

Tomasch, Sylvia and Gilles, Sealy (eds.), *Text and Territory: Geographical Imagination in the European Middle Ages*. Philadelphia: University of Pennsylvania Press, 1998.

Wallace, David, *Premodern Places: Calais to Surinam, Chaucer to Aphra Behn*. Oxford: Blackwell, 2004.

Wey Gómez, Nicolás, *The Tropics of Empire: Why Columbus Sailed South to the Indies*. Cambridge, Mass.: MIT Press, 2008.

Wright, John K., *The Geographical Lore of the Time of the Crusades: A Study in the History of Medieval Science and Tradition in Western Europe*. New York: American Geographical Society, 1925.

Index

16015425R00197

Printed in Poland
by Amazon Fulfillment
Poland Sp. z o.o., Wrocław